Advances in Pattern Recognition

For other titles published in this series, go to
http://www.springer.com/series/4205

Stan Z. Li

Markov Random Field Modeling in Image Analysis

Third Edition

 Springer

Stan Z. Li
Center for Biometrics and Security Research &
  National Laboratory of Pattern Recognition
Institute of Automation
Chinese Academy of Science
Beijing 100190, China
Stan.ZQ.Li@gmail.com

Series editor
Professor Sameer Singh, PhD
Research School of Informatics, Loughborough University, Loughborough, UK

ISBN: 978-1-84996-767-9   e-ISBN: 978-1-84800-279-1
DOI: 10.1007/978-1-84800-279-1

Advances in Pattern Recognition Series ISSN 1617-7916

British Library Cataloguing in Publication Data
A catalogue record for this book is available from the British Library

Printed on acid-free paper

Springer Science+Business Media
springer.com

In Memory of My Mother

"An excellent book — very thorough and very clearly written."

— Stuart Geman

"I have found the book to be a very valuable reference. I am very impressed by both the breadth and depth of the coverage. This must have been a truly monumental undertaking."

— Charles A. Bouman

Foreword by Anil K. Jain

The objective of mathematical modeling in image processing and computer vision is to capture the intrinsic character of the image in a few parameters so as to understand the nature of the phenomena generating the image. Models are also useful to specify natural constraints and general assumptions about the physical world; such constraints and assumptions are necessary to solve the "inverse" problem of three-dimensional scene interpretation given two-dimensional image(s) of the scene. The introduction of stochastic or random field models has led to the development of algorithms for image restoration, segmentation, texture modeling, classification, and sensor fusion. In particular, Gibbs and Markov random fields for modeling spatial context and stochastic interaction among observable quantities have been quite useful in many practical problems, including medical image analysis and interpretation of remotely sensed images. As a result, Markov random field models have generated a substantial amount of excitement in image processing, computer vision, applied statistics, and neural network research communities.

This monograph presents an exposition of Markov random fields (MRF's) that is likely to be extensively used by researchers in many scientific disciplines. In particular, those investigating the applicability of MRF's to process their data or images are bound to find its contents very useful. The main focus of the monograph, however, is on the application of Markov random fields to computer vision problems such as image restoration and edge detection in the low-level domain, and object matching and recognition in the high-level domain. Using a variety of examples, the author illustrates how to convert a specific vision problem involving uncertainties and constraints into essentially an optimization problem under the MRF setting. In doing so, the author introduces the reader to the various special classes of MRF's, including MRF's on the regular lattice (e.g., auto-models and multilevel logistic models) that are used for low-level modeling and MRF's on relational graphs that are used for high-level modeling.

The author devotes considerable attention to the problems of parameter estimation and function optimization, both of which are crucial in the MRF paradigm. Specific attention is given to the estimation of MRF parameters in the context of object recognition, and to the issue of algorithm selection for MRF-based function optimization. Another contribution of the book is a study on discontinuities, an important issue in the application of MRF's to image analysis. The extensive list of references, high-level descriptions of algorithms, and computational issues associated with various optimization algorithms are some of the attractive features of this book.

On the whole, the contents of this monograph nicely complement the material in Kindermann and Snell's book *Markov Random Fields and Their Applications* and Chellappa and Jain's edited volume entitled *Markov Random Fields: Theory and Applications*. In my opinion, the main contribution of this book is the manner in which significant MRF-related concepts are lucidly illustrated via examples from computer vision.

Anil K. Jain

East Lansing, Michigan

June 8, 1995

Foreword by Rama Chellappa

Noncausal spatial interaction models were first introduced by Peter Whittle in his classic 1954 paper. Whittle applied 2D noncausal autoregressive models for representing spatial data and pointed out the inconsistency of ordinary least-squares methods for parameter estimation. He then proceeded to derive maximum likelihood estimation techniques and hypothesis testing procedures. In a lesser known paper, Rosanov in 1967, discussed the representation of 2D discrete Gaussian Markov random fields (MRF's). The 2D discrete Gaussian MRF models were introduced to the engineering literature in 1972 by John Woods, who used them for 2D spectral analysis. A big impetus to theoretical and practical considerations of 2D spatial interaction models, of which MRF's form a subclass, was given by the seminal works of Julian Besag. Since the early 1980s, MRF's have dominated the fields of image processing, image analysis and computer vision.

Before the first edition of this book appeared, there were only two books on the topic, one by Kindermann and Snell and the other an edited volume by Chellappa and Jain. The former focused on the mathematical representation and the latter was mostly tailored to the needs of advanced researchers. The first edition of this book filled a great void, in that one could readily use it in a classroom. I have done so for more than a decade while teaching image processing and computer vision courses for graduate students. The author greatly succeeded in finding the delicate balance between theory and applications of MRF's and presented all the major topics in a reader-friendly manner.

Due to the overwhelmingly positive response to the first edition the second edition of the book appeared in due course with several additions. The new topics covered include a formal approach to textures analysis using MRF's, Monte Carlo Markov Chain (MCMC)-based algorithms and their variations.

In this third edition, the author has included detailed discussions on graphical models and associated inference techniques and pointed out their relationships to MRF-based approaches. A second example of such an outreach effort is the inclusion of discussions on conditional random field models, which are increasingly being used in the graphics and vision community. Another illustrative example is the discussion related to graph flow algorithms, which have become very popular in many image analysis and computer vision problems. When links to such diverse applications are made, one often loses the focus; the author should be congratulated for maintaining a cohesive picture of all the topics related to MRF's, new and old, and their inter-relationships. Although one cannot be sure if another enlightened author such as Stan Li may not come along and best this book, I am confident that for many years to come this book will be the one that will be read and reread for anything and everything on MRF's. The author has done a tremendous service to the students and researchers who are interested in learning about MRF's.

Rama Chellappa

College Park, Maryland

June 2, 2008

Preface to the Third Edition

Most important advances in MRF modeling made in the past decade or so are included in this edition. Mathematical MRF models are presented in a newly added chapter. The following are the added contents.

- Mathematical MRF Models:
    conditional random field, discriminative random fields,
    strong MRF, $\mathcal{K}$-MRF and Nakagami-MRF,
    MRF's and Bayesian networks (graphical models)
- Low-Level Models:
    stereo vision, spatio-temporal models
- High-Level Models:
    face detection and recognition
- Discontinuities in MRF's:
    total variation (TV) models
- MRF Model with Robust Statistics:
    half-quadratic minimization
- Minimization – Local Methods:
    belief propagation, convex relaxation
- Minimization – Global Methods:
    graph cuts

I would like to thank the following colleagues for assisting me in writing this edition. Linjing Li made a great effort in putting materials together. It would have been very difficult to deliver the manuscript to the publisher without a long delay without Linjing's tremendous help. Xiaotong Yuan and Junyan Wang provided their comments on the additions.

Finally, I would like to thank Rama Chellappa for giving his positive and encouraging feedback on his use of the book in teaching image processing and computer vision courses and for his kindness in writing the foreword to the third edition.

Preface to the Second Edition

Progress has been made since the first edition of this book was published five years ago. The second edition has included the most important progress in MRF modeling in image analysis in recent years, such as Markov modeling of images with "macro" patterns (e.g., the FRAME model), Markov chain Monte Carlo (MCMC) methods, and reversible jump MCMC. Work done by the author in this area after publication of the first edition is also included. The author would like to thank Song Chun Zhu for valuable discussions and suggestions.

Preface to the First Edition

Since its beginning, image analysis research has been evolving from heuristic design of algorithms to systematic investigation of approaches. Researchers have realized: (1) The solution to a vision problem should be sought based on *optimization* principles, either explicitly or implicitly, and (2) *contextual constraints* are ultimately necessary for the understanding of visual information in images. Two questions follow: how to define an optimality criterion under contextual constraints and how to find its optimal solution.

Markov random field (MRF), a branch of probability theory, provides a foundation for the characterization of contextual constraints and the derivation of the probability distribution of interacting features. In conjunction with methods from decision and estimation theory, MRF theory provides a systematic approach for deriving optimality criteria such as those based on the *maximum a posteriori* (MAP) concept. This MAP-MRF framework enables us to systematically develop algorithms for a variety of vision problems using rational principles rather than ad hoc heuristics. For these reasons, there has been increasing interest in modeling computer vision problems using MRF's in recent years.

This book provides a coherent reference to theories, methodologies, and recent developments in solving computer vision problems based on MRF's, statistics, and optimization. It treats various problems in low- and high-level computational vision in a systematic and unified way within the MAP-MRF framework. The main issues of concern are how to use MRF's to encode contextual constraints that are indispensable to image understanding; how to derive the objective function, typically the posterior distribution, for the optimal solution to a problem; and how to design computational algorithms for finding the optimal solution.

As the first thorough reference on the subject, the book has four essential parts for solving image and vision analysis problems using MRF's: (1) introduction to fundamental theories, (2) formulations of various image models in the MAP-MRF framework, (3) parameter estimation, and (4) optimization methods.

Chapter 1 introduces the notion of visual labeling and describes important results in MRF theory for image modeling. A problem is formulated in terms of Bayes labeling of an MRF. Its optimal solution is then defined as the MAP configuration of the MRF. The role of optimization is discussed. These form the basis on which MAP-MRF models are formulated.

Chapter 2 formulates MRF models for low-level vision problems, such as image restoration, reconstruction, edge detection, texture, and optical flow. The systematic MAP-MRF approach for deriving the posterior distribution is illustrated step by step.

Chapter 3 addresses the issue of discontinuities in low-level vision. An important necessary condition is derived for any MRF prior potential function to be adaptive to discontinuities to avoid oversmoothing. This gives rise to the definition of a class of *adaptive interaction functions* and thereby a class of MRF models capable of dealing with discontinuities.

Chapter 4 provides a comparative study on discontinuity adaptive MRF priors and robust M-estimators based on the results obtained in Chapter 3. To tackle the problems associated with M-estimators, a method is presented to stabilize M-estimators w.r.t. the initialization and convergence.

Chapter 5 presents high-level MRF models for object recognition and pose determination. Relational measurements are incorporated into the energy function as high-level constraints. The concept of line process is extended for the separation of overlapping objects and the elimination of outlier features.

Chapter 6 describes various methods for both supervised and unsupervised parameter estimation, including the coding method, pseudo-likelihood, least squares method, and expectation maximization. A simultaneous image labeling and parameter estimation paradigm is also presented that enhances the low-level models in Chapter 2.

Chapter 7 presents a theory of parameter estimation for optimization-based object recognition. Two levels of criteria are proposed for the estimation: correctness and optimality. Optimal parameters are learned from examples using supervised learning methods. The theory is applied to parameter learning for the MRF recognition.

Chapters 8 and 9 present local and global methods, respectively, for energy optimization in finding MAP-MRF solutions. These include various algorithms for continuous, discrete, unconstrained, and constrained minimization as well as strategies for approximating global solutions.

The final version of this manuscript benefited from comments on earlier versions by a number of people. I am very grateful to Anil K. Jain and Kanti V. Mardia for their valuable suggestions. I would like to thank Kap Luk Chan, Lihui Chen, Yi-Ping Hung, Eric Sung, Han Wang, Ming Xie, and Dekun Yang. Their corrections have had a very positive effect on the book. I am particularly indebted to Yunjun Zhang, Weiyun Yau, and Yihong Huang for their proofreading of the whole manuscript. Finally, I owe a deep debt of gratitude to my wife for her understanding, patience, and support.

Contents

Chapter 1

Introduction

Modeling problems in this book are addressed mainly from the computational viewpoint. The primary concerns are how to define an objective function for the optimal solution to a image analysis or computer vision problem and how to find the optimal solution. The solution is defined in an *optimization* sense because the perfect solution is difficult to find due to various uncertainties in the process, so we usually look for an optimal one that optimizes an objective in which constraints are encoded.

Contextual constraints are ultimately necessary in the interpretation of visual information. A scene is understood in the spatial and visual contexts; the objects are recognized in the context of object features in a lower-level representation; the object features are identified based on the context of primitives at an even lower-level; and the primitives are extracted in the context of image pixels at the lowest level of abstraction. The use of contextual constraints is indispensable for a capable vision system.

Markov random field (MRF) theory provides a convenient and consistent way of modeling context-dependent entities such as image pixels and correlated features. This is achieved through characterizing mutual influences among such entities using conditional MRF distributions. The practical use of MRF models is largely ascribed to a theorem stating the equivalence between MRF's and Gibbs distributions that was established by Hammersley and Clifford (1971) and further developed by Besag (1974). This is because the joint distribution is required in most applications but deriving the joint distribution from conditional distributions turns out to be very difficult for MRF's. The MRF-Gibbs equivalence theorem points out that the joint distribution of an MRF is a Gibbs distribution, the latter taking a simple form. This gives us not only a mathematically sound but also mathematically tractable means for statistical image analysis (Grenander 1983; Geman and Geman 1984). From the computational perspective, the local property of MRF's leads to algorithms that can be implemented in a local and massively parallel manner.

S.Z. Li, *Markov Random Field Modeling in Image Analysis,*
Advances in Pattern Recognition, DOI: 10.1007/978-1-84800-279-1_1,
© Springer-Verlag London Limited 2009

Furthermore, MRF theory provides a foundation for multi-resolution computation (Gidas 1989).

For the reasons above, MRF's have been widely employed to solve vision problems at all levels. Most of the MRF models are for low-Level processing. These include image restoration and segmentation, surface reconstruction, edge detection, texture analysis, optical flow, shape from X, active contours, deformable templates, data fusion and visual integration, and perceptual organization. The use of MRF's in high-level vision, such as for object matching and recognition, has also emerged. A unified framework for solving image and vision analysis problems from low level to high level is proposed by Li (Li 1991; Li 1994b). The interest in MRF modeling in image and vision analysis and synthesis is increasing, as reflected by books as well as journal and conference papers.[1]

MRF theory tells us how to model the a priori probability of context-dependent patterns, such as textures and object features. A particular MRF model favors the its own class of patterns by associating them with larger probabilities than other pattern classes. MRF theory is often used in conjunction with statistical decision and estimation theories so as to formulate objective functions in terms of established optimality principles. *Maximum a posteriori* (MAP) probability is one of the most popular statistical criteria for optimality and in fact has been the most popular choice in MRF vision modeling. MRF's and the MAP criterion together give rise to the MAP-MRF framework adopted in this book as well as in most other MRF works. This framework, advocated by Geman and Geman (1984) and others, enables us to develop algorithms for a variety of vision problems systematically using rational principles rather than relying on ad hoc heuristics. See also introductory statements in (Mardia 1989; Chellappa and Jain 1993; Mardia and Kanji 1994).

An objective function is completely specified by its *form* i.e., the parametric family and the *parameters* involved. In the MAP-MRF framework, the objective is the joint posterior probability of the MRF labels. Its form and parameters are determined in turn, according to the Bayes formula, by those of the joint prior distribution of the labels and the conditional probability of the observed data. In the previous paragraph, "a particular MRF model" means a particular probability function (of patterns) specified by the functional form and the parameters. Two major parts of MAP-MRF modeling are to derive the form of the posterior distribution and to determine the parameters in it so as to completely define the posterior probability. Another important part is to design optimization algorithms for finding the maximum of the posterior distribution.

This book is organized in four parts in accordance with the motivations and issues brought out above. The first part (Chapters 1 and 2) introduces

---

[1]There are numerous recent publications in this area. They are not cited here to keep the introductory statements neat. They will be given subsequently.

basic notions, mathematical background, and useful MRF models. The second part (Chapters 3–6) formulates various MRF models in low and high-level vision in the MAP-MRF framework and studies the related issues. The third part (Chapters 7–8) addresses the problem of MRF parameter estimation. Part four (Chapters 9–10) presents search algorithms for computing optimal solutions and strategies for global optimization.

In the rest of this chapter, basic definitions, notations, and important theoretical results for the MAP-MRF modeling will be introduced. These background materials will be used throughout the book.

1.1 Labeling for Image Analysis

Many image analysis and interpretation problems can be posed as labeling problems in which the solution to a problem is a set of labels assigned to image pixels or features. Labeling is also a natural representation for the study of MRF's (Besag 1974).

1.1.1 Sites and Labels

A *labeling problem* is specified in terms of a set of *sites* and a set of *labels*. Let $\mathcal{S}$ index a discrete set of m sites

$$\mathcal{S} = \{1, \ldots, m\} \tag{1.1}$$

in which $1, \ldots, m$ are indices. A site often represents a point or a region in the Euclidean space such as an image pixel or an image feature such as a corner point, a line segment, or a surface patch. A set of sites may be categorized in terms of their "regularity." Sites on a lattice are considered spatially *regular*. A rectangular lattice for a 2D image of size $n \times n$ can be denoted by

$$\mathcal{S} = \{(i, j) \mid 1 \le i, j \le n\} \tag{1.2}$$

Its elements correspond to the locations at which an image is sampled. Sites that do not present spatial regularity are considered *irregular*. This is the usual case for features extracted from images at a more abstract level, such as detected corners and lines.

We normally treat the sites in MRF models as unordered. For an $n \times n$ image, pixel (i, j) can be conveniently reindexed by a single number k, where k takes on values in $\{1, 2, \ldots, m\}$ with $m = n \times n$. This notation for a single-number site index will be used in this book even for images unless an elaboration is necessary. The interrelationship between sites is maintained by a so-called *neighborhood system* (to be introduced later).

A label is an event that may happen to a site. Let $\mathcal{L}$ be a set of *labels*. A label set may be categorized as being continuous or discrete. In the continuous case, a label set may correspond to the real line $\mathbb{R}$ or a compact interval of it

$$\mathcal{L}_c = [X_l, X_h] \subset \mathbb{R} \tag{1.3}$$

An example is the dynamic range for an analog pixel intensity. It is also possible for a continuous label to take a vector or matrix value, for example $\mathcal{L}_c = \mathbb{R}^{a \times b}$ where a and b are dimensions.

In the discrete case, a label assumes a discrete value in a set of M labels

$$\mathcal{L}_d = \{\ell_1, \cdots, \ell_M\} \tag{1.4}$$

or simply

$$\mathcal{L}_d = \{1, \cdots, M\} \tag{1.5}$$

In edge detection, for example, the label set is $\mathcal{L} = \{$edge,nonedge$\}$.

Besides the continuity, another essential property of a label set is the ordering of the labels. For example, elements in the continuous label set $\mathbb{R}$ (the real space) can be ordered by the relation "smaller than". When a discrete set, say $\{0, \dots, 255\}$, represents the quantized values of intensities, it is an ordered set because for intensity values we have $0 < 1 < 2 < \dots < 255$. When it denotes 256 different symbols such as texture types, it is considered to be unordered unless an artificial ordering is imposed.

For an ordered label set, a numerical (quantitative) measure of similarity between any two labels can usually be defined. For an unordered label set, a similarity measure is symbolic (qualitative), typically taking a value of "equal" or "nonequal". Label ordering and similarity not only categorize labeling problems but more importantly affect our choices of labeling algorithms and hence the computational complexity.

1.1.2   The Labeling Problem

The labeling problem is to assign a label from the label set $\mathcal{L}$ to each of the sites in $\mathcal{S}$. Edge detection in an image, for example, is to assign a label f_i from the set $\mathcal{L} = \{$edge,nonedge$\}$ to site $i \in \mathcal{S}$, where elements in $\mathcal{S}$ index the image pixels. The set

$$f = \{f_1, \dots, f_m\} \tag{1.6}$$

is called a *labeling* of the sites in $\mathcal{S}$ in terms of the labels in $\mathcal{L}$. When each site is assigned a unique label, $f_i = f(i)$ can be regarded as a function with domain $\mathcal{S}$ and image $\mathcal{L}$. Because the support of the function is the whole domain $\mathcal{S}$, it is a *mapping* from $\mathcal{S}$ to $\mathcal{L}$, that is,

$$f : \mathcal{S} \longrightarrow \mathcal{L} \tag{1.7}$$

A labeling is also called a *coloring* in mathematical programming. Figure 1.1 illustrates mappings with continuous and discrete label sets.

In the terminology of random fields (Section 2.1.2), a labeling is called a *configuration*. In vision, a configuration or labeling can correspond to an image, an edge map, an interpretation of image features in terms of object features, or a pose transformation, and so on.

When all the sites have the same label set $\mathcal{L}$, the set of all possible labelings (that is, the configuration space) is the Cartesian product

$$\mathbb{F} = \underbrace{\mathcal{L} \times \mathcal{L} \cdots \times \mathcal{L}}_{m \text{ times}} = \mathcal{L}^m \qquad (1.8)$$

where m is the size of $\mathcal{S}$. In image restoration, for example, $\mathcal{L}$ contains admissible pixel values that are common to all pixel sites in $\mathcal{S}$, and $\mathbb{F}$ defines all admissible images. When $\mathcal{L} = \mathbb{R}$ is the real line, $\mathbb{F} = \mathbb{R}^m$ is the m-dimensional real space. When $\mathcal{L}$ is a discrete set, the size of $\mathbb{F}$ is combinatorial. For a problem with m sites and M labels, for example, there exist a total number of M^m possible configurations in $\mathbb{F}$.

In certain circumstances, admissible labels may not be common to all the sites. Consider, for example, feature-based object matching. Supposing there are three types of features – points, lines, and regions, then a constraint is that a certain type of image feature can be labeled or interpreted in terms of the same type of model feature. Therefore, the admissible label for any site is restricted to one of the three types. In an extreme case, every site i may have its own admissible set $\mathcal{L}_i$ of labels, and this gives the configuration space

$$\mathbb{F} = \mathcal{L}_1 \times \mathcal{L}_2 \cdots \times \mathcal{L}_m \qquad (1.9)$$

This imposes constraints on the search for the configurations wanted.

1.1.3   Labeling Problems in Image Analysis

In terms of the regularity and the continuity, we may classify a vision labeling problem into one of the following four categories:

LP1: Regular sites with continuous labels.

LP2: Regular sites with discrete labels.

LP3: Irregular sites with discrete labels.

LP4: Irregular sites with continuous labels.

The first two categories characterize low-Level processing performed on observed images and the two others do high-level processing on extracted token features. The following describes some vision problems in terms of the four categories.

Restoration or smoothing of images having continuous pixel values is an LP1. The set $\mathcal{S}$ of sites corresponds to image pixels, and the set $\mathcal{L}$ of labels is a real interval. The restoration is to estimate the true image signal from a degraded or noise-corrupted image.

Restoration of binary or multilevel images is an LP2. Similar to the continuous restoration, the aim is also to estimate the true image signal from

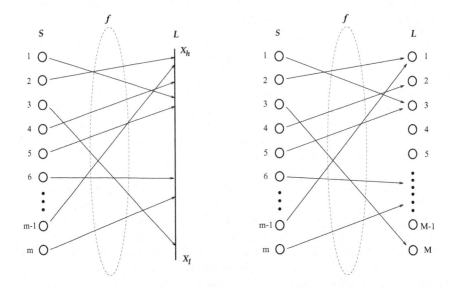

Figure 1.1: A labeling of sites can be considered as a mapping from the set of sites $\mathcal{S}$ to the set of labels $\mathcal{L}$. The Figure shows mappings wa continuous label set (left) and discrete label set (right).

the input image. The difference is that each pixel in the resulting image here assumes a discrete value and thus $\mathcal{L}$ in this case is a set of discrete labels.

Region segmentation is an LP2. It partitions an observation image into mutually exclusive regions, each of which has some uniform and homogeneous properties whose values are significantly different from those of the neighboring regions. The property can be, for example, gray tone, color, or texture. Pixels within each region are assigned a unique label.

The prior assumption in the problems above is that the signal is smooth or piecewise smooth. This is complementary to the assumption of abrupt changes made for edge detection.

Edge detection is also an LP2. Each edge site, located between two neighboring pixels, is assigned a label in {edge, nonedge} if there is a significant difference between the two pixels. Continuous restoration with discontinuities can be viewed as a combination of LP1 and LP2.

Perceptual grouping is an LP3. The sites usually correspond to initially segmented features (points, lines and regions) that are irregularly arranged. The fragmentary features are to be organized into perceptually more significant features. A label in {connected,disconnected} is assigned between each pair of the features, indicating whether the two features should be linked.

Feature-based object matching and recognition is an LP3. Each site indexes an image feature such as a point, a line segment, or a region. Labels are discrete in nature, and each of them indexes a model feature. The resulting configuration is a mapping from the image features to those of a model object.

Pose estimation from a set of point correspondences might be formulated as an LP4. A site is a given correspondence. A label represents an admissible (orthogonal, affine, or perspective) transformation. A prior (unary) constraint is that the label of transformation itself must be orthogonal, affine or perspective. A mutual constraint is that the labels $f_1, \cdots, f_m$ should be close to each other to form a consistent transformation.

For a discrete labeling problem of m sites and M labels, there exist a total number of M^m possible labelings. For a continuous labeling problem, there are an infinite number of them. However, among all labelings, there are only a small number of them that are good solutions, and maybe just a few are optimal in terms of a criterion. How to define the optimal solution for a problem and how to find it are two important topics in the optimization approach to visual labeling.

1.1.4 Labeling with Contextual Constraints

The use of contextual information is ultimately indispensable in image understanding (Pavlidis 1986). The use of contextual information in image analysis and pattern recognition dates back to Chow (1962) and Abend, Harley, and Kanal (1965). In Chow (1962), character recognition is considered as a statistical decision problem. A nearest neighborhood dependence of pixels on an image lattice is obtained by going beyond the assumption of statistical independence. Information on the nearest neighborhood is used to calculate conditional probabilities. That system also includes parameter estimation from sample characters; recognition is done by using the estimated parameters. The work by Abend, Harley, and Kanal (1965) is probably the earliest work using the Markov assumption for pattern recognition. There, a Markov mesh model is used to reduce the number of parameters required for processing using contextual constraints. Fu and Yu (1980) used MRF's defined on an image lattice to develop a class of pattern classifiers for remote-sensing image classification. Another development of context-based models is relaxation labeling (RL) (Rosenfeld et al. 1976). RL is a class of iterative procedures that use contextual constraints to reduce ambiguities in image analysis. A theory is given in Haralick (1983) to explain RL from a Bayes point of view.

In probability terms, contextual constraints may be expressed locally in terms of conditional probabilities $P(f_i \mid \{f_{i'}\})$, where $\{f_{i'}\}$ denotes the set of labels at the other sites $i' \neq i$ or globally as the joint probability $P(f)$. Because local information is more directly observed, it is normal for a global inference to be made based on local properties.

In situations where labels are independent of one another (no context), the joint probability is the product of the local ones

$$P(f) = \prod_{i \in \mathcal{S}} P(f_i) \qquad\qquad (1.10)$$

The above implies conditional independence

$$P(f_i \mid \{f_{i'}\}) = P(f_i) \quad i' \neq i \qquad\qquad (1.11)$$

Therefore, a global labeling f can be computed by considering each label f_i locally. This is advantageous for problem solving.

In the presence of context, labels are mutually dependent. The simple relationships expressed in (1.10) and (1.11) do not hold any more. How to make a global inference using local information becomes a nontrivial task. Markov random field (MRF) theory, to be introduced in Chapter 2, provides a mathematical foundation for solving this problem.

1.2   Optimization-Based Approach

Optimization has been playing an essential and important role in image analysis. There, a problem is formulated as optimizing some criterion, explicitly or implicitly. The extensive use of optimization principles is due to various uncertainties in the imaging and vision processes, such as noise and occlusion in the sensed image and ambiguities in visual interpretation. Exact or perfect solutions hardly exist. Inexact but optimal (in some sense) solutions are usually sought instead.

In the pioneer vision system of Roberts (1965), object identification and pose estimation are performed using the simplest least squares (LS) fitting. Nowadays, optimization is pervasive in all aspects of vision, including image restoration and reconstruction (Grimson 1981; Terzopoulos 1983a; Geman and Geman 1984; Leclerc 1989; Hung et al. 1991), shape from shading (Ikeuchi and Horn 1981), stereo, motion and optical flow (Ullman 1979; Horn and Schunck 1981; Hildreth 1984; Murray and Buxton 1987; Barnard 1987), texture (Hassner and Slansky 1980; Kashyap et al. 1982; Cross and Jain 1983), edge detection (Torre and Poggio 1986; Tan et al. 1992), image segmentation (Silverman and Cooper 1988; Li 1990a), perceptual grouping (Lowe 1985; Mohan and Nevatia 1989; Herault and Horaud 1993), interpretation of line drawings (Leclerc and Fischler 1992), object matching and recognition (Fischler and Elschlager 1973; Davis 1979; Shapiro and Haralick 1981; Bhanu and Faugeras 1984; Ben-Arie and Meiri 1987; Modestino and Zhang 1989; Nasrabadi et al. 1990; Wells 1991; Friedland and Rosenfeld 1992; Li 1992c; Li 1994a), and pose estimation (Haralick et al. 1989).

In all of the above examples cited, the solution is explicitly defined as an optimum of an objective function by which the goodness, or otherwise cost, of the solution is measured. Optimization may also be performed implicitly: The solution may optimize an objective function but in an implicit way that may or may not be realized. Hough transform (Hough 1962; Duda and Hart 1972; Ballard 1981; Illingworth and Kittler 1988) is a well-known technique for detecting lines and curves by looking at peaks of an accumulation function. It was later found to be equivalent to template matching (Stockman and Agrawala 1977) and can be reformulated as a maximizer of some probabilities such as the likelihood (Haralick and Shapiro 1992). Edge detection was performed using some simple operators like derivatives of Gaussians (Rosenfeld and Kak 1976). The operators can be derived by using regularization principles in which an energy function is explicitly minimized (Poggio et al. 85b).

The main reason for the extensive use of optimization is the existence of uncertainties in every vision process. Noise and other degradation factors, such as those caused by disturbances and quantization in sensing and signal processing, are sources of uncertainties. Different appearances and poses of objects, their mutual-occlusion and self-occlusion, and possible shape deformation also cause ambiguities in visual interpretation. Under such circumstances, we can hardly obtain exact or perfect solutions and have to resort to inexact yet optimal solutions.

Because of the role of optimization, it is importance to study image analysis problems from the viewpoint of optimization and to develop methodologies for optimization-based modeling. Section 1.2.1 presents discussions on optimization-based vision.

1.2.1 Research Issues

There are three basic issues in optimization-based vision: problem representation, objective function, and optimization algorithms. Concerning the first issue, there are two aspects of a representation: descriptive and computational. The former concerns how to represent image features and object shapes, which relates to photometry and geometry (Koenderink 1990; Mundy and Zisserman 1992; Kanatani 1993) and is not an emphasis of this book. The latter concerns how to represent the solution, which relates to the choice of sites and label set for a labeling problem. For example, in image segmentation, we may use a chain of boundary locations to represent the solution; we may alternatively use a region map to do the same job. Comparatively speaking, however, the region map is a more natural representation for MRF's.

The second issue is how to formulate an objective function for the optimization. The objective function maps a solution to a real number measuring the quality of the solution in terms of some goodness or cost. The formulation determines how various constraints, which may be pixel properties such as intensity and color and/or context such as relations between pixels or object features, are encoded into the function. The formulation defines the optimal solution.

The third issue is how to optimize the objective, i.e., how to search for the optimal solution in the admissible space. Two major concerns are (1) the problem of local minima existing in nonconvex functions and (2) the efficiency of algorithms in space and time. They are somewhat contradictory, and currently there are no algorithms that guarantee the global solution with good efficiency.

These three issues are related to one another. In the first place, the scheme of representation influences the formulation of the objective function and the design of the search algorithm. On the other hand, the formulation of an objective function affects the search. For example, suppose two objective functions have the same point as the unique global optimum but one of them is convex, whereas the other is not; obviously the convex one is much more desired because it provides convenience for the search.

In the following presentation, we will be mainly dealing with minimization problems. An objective function is in the form of an energy function and is to be minimized.

1.2.2   Role of Energy Functions

The role of an energy function in minimization-based vision is twofold: (1) as the quantitative measure of the global quality of the solution and (2) as a guide to the search for a minimal solution. As the quantitative cost measure, an energy function defines the minimal solution as its minimum, usually a global one. In this regard, it is important to formulate an energy function so that the "correct solution" is embedded as the minimum. We call this the correctness of the formulation.

To understand an optimization approach, one should not mix problems in formulation and those in search. Differentiating the two different kinds of problems helps debug the modeling. For example, if the output of an optimization procedure (assuming the implementation is correct) is not what is expected, there are two possible reasons: (1) The formulation of the objective function is not a correct one for modeling the reality and (2) the output is a low-quality local minimum. Which one is the problem should be identified before the modeling can be improved.

The role of an energy function as a guide to the search may or may not be complete. In real minimization, for example, when the energy function is smooth and convex w.r.t. its variables, global minimization is equivalent to local minimization and the gradient of the energy function provides sufficient information about where to search for the global solution. In this case, the role of guiding the search can be complete. However, when the problem is nonconvex, there is no general method that can efficiently utilize the energy function to guide the search. In this case, the role as the search-guide is limited.

In certain cases, it may be advantageous to consider the formulation of an energy function and the search simultaneously. This is to formulate the function appropriately to facilitate the search. The work on graduated non-convexity (GNC) (Blake and Zisserman 1987) is an example in this regard. There, the energy function is deformed gradually from a convex form to its target form in the process of approximating the global solution using a gradient-based algorithm.

Local minimization in real spaces is the most mature area in optimization and many formal approaches exist for solving it. This is not so for combinatorial and global minimization. In the latter case, heuristics become an important and perhaps necessary element in practice. In the heuristic treatment of global minimization, rather restrictive assumptions are made. An example is the bounded model (Baird 1985; Breuel 1992). It assumes that a measurement error is upper-bounded by a certain threshold (within the threshold, the error may be assumed to be evenly distributed). Whether the assumption is valid depends on the threshold. It is absolutely true when the threshold is infinitely large. But, in practice, the threshold is almost always set to a value that is less than that required to entirely validate the bounded error assumption. The lower the value, the higher the efficiency is but the less general the algorithm becomes.

In the hypothesis-verification approach, efficient algorithms are used to generate hypothetical solutions, such as the Hough transform (Hough 1962; Duda and Hart 1972), interpretation tree search (Grimson and Lozano-Prez 1987) and geometric hashing (Lamdan and Wolfson 1988). The efficiency comes from the fast elimination of infeasible solutions, or pruning of the solution space, by taking advantage of heuristics. In this way, a relatively small number of solution candidates are picked up relatively quickly and are then verified or evaluated thoroughly, for example, by using an energy function. In this strategy, the energy function is used for the evaluation only, not as a guide to the search.

Note that the advantage of formal approaches is in the evaluation, and the advantage of heuristic approaches is in the search. A good strategy for the overall design of a specialized system may be to use a heuristic algorithm to quickly find a small number of solution candidates and then evaluate the candidates found using an energy function derived formally to give the best solution.

1.2.3 Formulation of Objective Functions

In pattern recognition, there are two basic approaches to formulating an energy function: parametric and nonparametric. In the parametric approach, the types of underlying distributions are known and the distributions are parameterized by a few parameters. Therefore, the functional form of the energy can be obtained and the energy function is completely defined when the parameters are specified.

In the nonparametric approach, sometimes called the distribution-free approach, no assumptions about the distributions are made. There, a distribution is either estimated from the data or approximated by a pre-specified basis functions with several unknown parameters to be estimated. In the latter case the prespecified basis function will determine the functional form of the energy.

Despite the terms parametric and nonparametric, both approaches are somewhat parametric in nature. This is because in any case, there are always parameters that must be determined to define the energy function.

The two most important aspects of an energy function are its form and the parameters involved. The form and parameters together define the energy function, which in turn defines the minimal solution. The form depends on assumptions about the solution f and the observed data d. We express this using the notation $E(f \mid d)$. Denote the set of parameters involved by θ. With θ, the energy is expressed further as $E(f \mid d, \theta)$. In general, given the functional form for E, a different d or θ defines a different energy function, $E(f \mid d, \theta)$, w.r.t. f and hence a (possibly) different minimal solution f^*.

Since the parameters are part of the definition of the energy function $E(f \mid d, \theta)$, the minimal solution $f^* = \arg\min_f E(f \mid d)$ is not completely defined if the parameters are not specified, even if the functional form is known. These parameters must be specified or estimated by some means. This is an important area of study in MRF modeling.

1.2.4   Optimality Criteria

In formal models, as opposed to heuristic ones, an energy function is formulated based on established criteria. Because of inevitable uncertainties in imaging and vision processes, principles from statistics, probability and information theory are often used as the formal basis. When the knowledge about the data distribution is available but the prior information is not, the *maximum likelihood* (ML) criterion may be used, $f^* = \arg\max P(d \mid f)$. On the other hand, if only the prior information is available, the *maximum entropy* criterion may be chosen, $f^* = \arg\max - \sum_{i=1}^{m} P(f_i) \ln P(f_i)$. The maximum entropy criterion simply takes this fact into account that configurations with higher entropy are more likely because nature can generate them in more ways (Jaynes 1982).

When both the prior and likelihood distributions are known, the best result is achieved by maximizeing a Bayes criterion according to Bayes statistics (Therrien 1989). Bayes statistics is a theory of fundamental importance in estimation and decision-making. Although there have been philosophical and scientific controversies about their appropriateness in inference and decision-making (see (Clark and Yuille 1990) for a short review), Bayes criteria (the MAP principle in particular) are the most popular ones in computer vision and in fact, MAP is the most popular criterion in optimization-based MRF modeling. The theorem of equivalence between Markov random fields and the Gibbs distribution established in Section 2.1.4 provides a convenient way to specify the joint prior probability, solving a difficult issue in MAP-MRF labeling.

In the principle of *minimum description length* (MDL) (Rissanen 1978; Rissanen 1983), the optimal solution to a problem is that which needs the smallest set of vocabulary in a given language to explain the input data. The MDL has a close relationship to the statistical methods such as the ML and MAP (Rissanen 1983). For example, if $P(f)$ is related to the description length and $P(d \mid f)$ related to the description error, then the MDL is equivalent to the MAP. However, it is more natural and intuitive when prior probabilities are not well defined. The MDL has been used for vision problems at different levels such as segmentation (Leclerc 1989; Pentland 1990; Darrell et al. 1990; Dengler 1991; Keeler 1991) and object recognition (Breuel 1993).

1.3 The MAP-MRF Framework

Bayes statistics is a fundamental theory in estimation and decision-making. According to this theory, when both the prior distribution and the likelihood function of a pattern are known, the best that can be estimated from these sources of knowledge is the Bayes labeling. The maximum a posterior (MAP) solution, as a special case in the Bayes framework, is sought in many applications.

The MAP-MRF framework is advocated by Geman and Geman (1984) and others (Geman and McClure 1985; Derin and Elliott 1987; Geman and Graffigne 1987; Dubes and Jain 1989; Besag 1989; Szeliski 1989; Geman and Gidas 1991). Since the paper of Geman and Geman (1984), numerous statistical image analysis problems have been formulated in this framework. This section reviews related concepts and derives probabilistic distributions and energies involved in MAP-MRF labeling. For more detailed materials on Bayes theory, the reader is referred to books such as (Therrien 1989).

1.3.1 Bayes Estimation

In Bayes estimation, a risk is minimized to obtain the optimal estimate. The Bayes risk of estimate f^* is defined as

$$R(f^*) = \int_{f \in \mathbb{F}} C(f^*, f) P(f \mid d) df \tag{1.12}$$

where d is the observation, $C(f^*, f)$ is a cost function, and $P(f \mid d)$ is the posterior distribution. First of all, we need to compute the posterior distribution from the prior and the likelihood. According to the Bayes rule, the posterior probability can be computed by using the formulation

$$P(f \mid d) = \frac{p(d \mid f) P(f)}{p(d)} \tag{1.13}$$

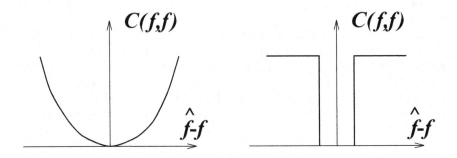

Figure 1.2: Two choices of cost functions.

where $P(f)$ is the prior probability of labelings f, $p(d \mid f)$ is the conditional p.d.f. of the observations d, also called the likelihood function of f for d fixed, and $p(d)$ is the density of d, which is a constant when d is given.

The cost function $C(f^*, f)$ determines the cost of estimate f when the truth is f^*. It is defined according to our preference. Two popular choices are the quadratic cost function

$$C(f^*, f) = \|f^* - f\|^2 \tag{1.14}$$

where $\|a - b\|$ is a distance between a and b, and the δ (0-1) cost function

$$C(f^*, f) = \begin{cases} 0 & \text{if } \|f^* - f\| \leq \delta \\ 1 & \text{otherwise} \end{cases} \tag{1.15}$$

where $\delta > 0$ is any small constant. A plot of the two cost functions is shown in Fig. 1.2.

The Bayes risk under the quadratic cost function measures the variance of the estimate

$$R(f^*) = \int_{f \in \mathbb{F}} \|f^* - f\|^2 P(f \mid d) df \tag{1.16}$$

Letting $\frac{\partial R(f^*)}{\partial f^*} = 0$, we obtain the minimal variance estimate

$$f^* = \int_{f \in \mathbb{F}} f P(f \mid d) df \tag{1.17}$$

which is the mean of the posterior probability.

For the δ cost function, the Bayes risk is

$$R(f^*) = \int_{f:\|f^* - f\| > \delta} P(f \mid d) df = 1 - \int_{f:\|f^* - f\| \leq \delta} P(f \mid d) df \tag{1.18}$$

When $\delta \to 0$,(1.18) is approximated by

$$R(f^*) = 1 - \kappa P(f \mid d) \tag{1.19}$$

where κ is the volume of the space containing all points f for which $\|f^* - f\| \le \delta$. Minimizing (1.19) is equivalent to maximizing the posterior probability. Therefore, the minimal risk estimate is

$$f^* = \arg\max_{f \in \mathbb{F}} P(f \mid d) \tag{1.20}$$

which is known as the MAP estimate. Because $p(d)$ in (1.13) is a constant for a fixed d, $P(f \mid d)$ is proportional to the joint distribution

$$P(f \mid d) \propto P(f, d) = p(d \mid f)P(f) \tag{1.21}$$

Then the MAP estimate is equivalently found by

$$f^* = \arg\max_{f \in \mathbb{F}} \{p(d \mid f)P(f)\} \tag{1.22}$$

Obviously, when the prior distribution, $P(f)$, is flat, the MAP is equivalent to the maximum likelihood.

1.3.2 MAP-MRF Labeling

In the MAP-MRF labeling, $P(f \mid d)$ is the posterior distribution of an MRF. An important step in Bayes labeling of MRF's is to derive this distribution. Here we use a simple example to illustrate the formulation of a MAP-MRF labeling problem. The problem is to restore images from noisy data. Assuming that the image surfaces are flat, then the joint prior distribution of f is

$$P(f) = \frac{1}{Z} e^{-U(f)} \tag{1.23}$$

where $U(f) = \sum_i \sum_{i' \in \{i-1, i+1\}} (f_i - f_{i'})^2$ is the *prior energy* for the type of surface. Assuming that the observation is the true surface height plus the independent Gaussian noise, $d_i = f_i + e_i$, where $e_i \sim N(\mu, \sigma^2)$, then the likelihood distribution is

$$p(d \mid f) = \frac{1}{\Pi_{i=1}^m \sqrt{2\pi\sigma^2}} e^{-U(d \mid f)} \tag{1.24}$$

where

$$U(d \mid f) = \sum_{i=1}^{m} (f_i - d_i)^2 / 2\sigma^2 \tag{1.25}$$

is the *likelihood energy*. Now the posterior probability is

$$P(f \mid d) \propto e^{-U(f \mid d)} \tag{1.26}$$

where

$$U(f \mid d) \;=\; U(d \mid f) + U(f) \tag{1.27}$$

$$\;=\; \sum_{i=1}^{m} (f_i - d_i)^2 / 2\sigma_i^2 + \sum_{i=1}^{m} (f_i - f_{i-1})^2$$

is the *posterior energy*. The MAP estimate is equivalently found by minimizing the posterior energy function

$$f^* = \arg\min_{f} U(f \mid d) \tag{1.28}$$

There is only one parameter in this simple example, σ_i. When it is determined, $U(f \mid d)$ is fully specified and the MAP-MRF solution is completely defined.

1.3.3 Regularization

The MAP labeling with a prior potential of the form $[f^{(n)}]^2$ (to be introduced in Section 2.4) is equivalent to the regularization of an ill-posed problem. A problem is mathematically an *ill-posed problem* (Tikhonov and Arsenin 1977) in the Hadamard sense if its solution (1) does not exist, (2) is not unique, or (3) does not depend continuously on the initial data. An example is surface reconstruction or interpolation. A problem therein is that there are infinitely many ways to determine the interpolated surface values if only the constraint from the data is used. Additional constraints are needed to guarantee the uniqueness of the solution to make the problem well-posed. An important such constraint is *smoothness*. By imposing the smoothness constraint, the analytic regularization method converts an ill-posed problem into a well-posed one. This has been used in solving low-Level problems such as surface reconstruction from stereo (Grimson 1981), optical flow (Horn and Schunck 1981), shape-from-shading (Ikeuchi and Horn 1981), and motion analysis (Hildreth 1984). A review of earlier work on regularization in vision is given by Poggio, Torre, and Koch (85a). Relationships between regularization and smoothing splines for vision processing have been investigated by Terzopoulos (1986b) and Lee and Pavlidis (1987).

A regularization solution is obtained by minimizing an energy of the form

$$E(f) = \sum_{i \in \mathcal{A}} [f(x_i) - d(x_i)]^2 + \lambda \int_a^b [f^{(n)}(x)]^2 \mathrm{d}x \tag{1.29}$$

where $\mathcal{A}$ is a set of indices to the sample data points, x_i's are the locations of the data points, $\lambda \geq 0$ is a weighting factor and $n \geq 1$ is an integer number. The first term on the right-hand side, called the closeness term, imposes the constraint from the data d. The second term, called the smoothness term or the regularizer, imposes the a priori smoothness constraint on the solution.

It is desired to minimize both, but they may not each be minimized simultaneously. The two constraints are balanced by λ. Any f minimizing (1.29) is a smooth solution in the so-called Sobolev space (Tikhonov and Arsenin 1977) $W_2^{n}{}^2$. Different types of regularizers impose different types of smoothness constraints. Section 2.4 has described smoothness priors for some classes of surfaces. A study of the quadratic smoothness constraints is given in (Snyder 1991).

Under certain conditions, the MAP labeling of MRF's is equivalent to a regularization solution (Marroquin et al. 1987). These conditions are (1) that the likelihood energy is due to additive white Gaussian noise and (2) that the prior assumption is the smoothness. Letting $n = 1$ and discretizing x, (1.29) becomes the posterior energy (1.27) for the restoration of flat surfaces (in 1D). When the regularizer takes the form of the rod (2.51), it encodes the smoothness of planar surfaces (in 1D). The 2D counterparts are the membrane (2.50), (2.55) and the plate (2.56).

The MRF theory is more general than regularization in that it can encode not only the smoothness prior but also priors of other constraints. MRF-based models have to be chosen when the priors are due to something other than the smoothness, e.g., texture modeling and analysis. However, regularization can deal with spatially continuous fields, whereas MRF's, which in this book are spatially discrete, cannot. When the problem under consideration involves concepts such as discontinuities, the analytical regularization is more suitable for the analysis. This is why we will study the issue of discontinuities in Chapter 5, from regularization viewpoint.

1.3.4 Summary of the MAP-MRF Approach

The procedure of the MAP-MRF approach for solving computer vision problems is summarized in the following:

1. Pose a vision problem as one of labeling in categories LP1–LP4, and choose an appropriate MRF representation f.

2. Derive the posterior energy to define the MAP solution to a problem.

3. Find the MAP solution.

The process of deriving the posterior energy is summarized as the following four steps:

1. Define a neighborhood system $\mathcal{N}$ on $\mathcal{S}$ and the set of cliques $\mathcal{C}$ for $\mathcal{N}$.

2. Define the prior clique potentials $V_c(f)$ to give $U(f)$.

3. Derive the likelihood energy $U(d \mid f)$.

---

[2]In the Sobolev space, every point f is a function whose $n-1$ derivative $f^{(n-1)}$ is absolutely continuous and whose nth derivative $f^{(n)}$ is square integrable.

    4. Add $U(f)$ and $U(d \mid f)$ to yield the posterior energy $U(f \mid d)$.

The prior model depends on the type of the scene (e.g., the type of surfaces) we expect. In image analysis, it is often one of the Gibbs models (to be introduced in Chapter 2). The likelihood model depends on physical considerations such as the sensor process (transformations, noise, etc.). It is often assumed to be Gaussian. The parameters in both models need to be specified for the definitions of the models to be complete. The specification can be something of an art when done manually, and it is desirable that it be done automatically.

    In the subsequent chapters, we are concerned with the following issues:

1. Choosing an appropriate representation for the MRF labeling.

2. Deriving the a posteriori distribution of the MRF as the criterion function of the labeling solution. It mainly concerns the specification of the forms of the prior and the likelihood distributions. The involved parameters may or may not be specified at this stage.

3. Estimating the parameters involved in the prior and the likelihood distributions. The estimation is also based on some criterion, very often maximum likelihood. In the unsupervised case, it is performed together with MAP labeling.

4. Searching for the MRF configuration to maximize the target (e.g., posterior) distribution. This is mainly algorithmic. The main issues are the quality (globalness) of the solution and the efficiency.

1.4   Validation of Modeling

Having formulated an energy function based on probability and statistics, with both its form and parameters given, one would ask the following question: Is the formulation correct? In other words, does the energy minimum f^* correspond to the correct solution? Before answering this, let us make some remarks on the questions. First, "the energy minimum" f^* is meant to be a global one. Second, "the correct solution", denoted $f_{correct}$, may be referred to in a subjective sense of our own judgment. For example, in edge detection, we run a program and get an edge map as a solution; but we find that some edges are missing and some false edges appear. This is a case where the solution differs from our subjectively correct solution.

    In the ideal case, the global minimal solution f^* should correspond to the correct solution $f_{correct}$

$$f_{correct} = f^* = \min_f E(f \mid d, \theta) \qquad (1.30)$$

A formulation is *correct for d* if (1.30) holds. This reflects our desire to encode our ability into the machine and is our ideal in formulating an energy function.

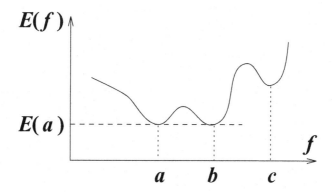

Figure 1.3: Two global minima a and b and a local minimum c.

For the time being, we do not expect the machine to surpass our ability, and therefore we desire that the minimum of an energy function, or the machine's solution, be consistent with ours. Whenever there is a difference, we assume that our own perception is correct and the difference is due to problems such as incorrect assumptions made in the modeling stage.

Assume that there are N observation data sets $d^{(n)}$ $(n = 1, 2, \ldots, N)$. Supposing that $f_{correct}[d^{(n)}]$ is the correct solution for $d^{(n)}$ in our own perception, we say that the functional form of the energy E and the chosen parameters θ are correct for the N observations if

$$f_{correct}[d] = f^* = \arg\min_f E(f \mid d, \theta) \qquad (1.31)$$

where f^* is the energy minimum, holds for each $d = d^{(n)}$. Let $\mathbb{D}$ be the space of all possible observations. If (1.31) holds for all $d \in \mathbb{D}$, we say that E and θ, and hence the formulation, are *entirely correct*.

A formulation may not be entirely correct. For example, a formulation for edge detection may give solutions having some missing or false edges or having some inaccurate edge locations. The quality of a formulation may be measured by some distance between the global energy minimum f^* and the correct solution $f_{correct}$. Ideally, it should be zero.

What about multiple global minima? Suppose there are two global minima for which $E(a) = E(b)$, as in Fig. 1.3. Both solutions a and b should be equally good if the formulation is correct. If, in the designer's opinion, one of them is better than the other in our perception, then there is still room to improve the formulation of $E(f)$. The improvement is made by adjusting either the functional form or the parameters so that the improved energy function embeds the better solution as its unique global minimum.

What if we run a program that correctly implements an energy-minimization-based system but find that the result is not what we expect? First of all, we may want to identify whether the fault is due to the local minimum problem caused by the minimization algorithm or the incorrect formulation of the energy function. The local minimum problem is identified if a better solution can be obtained with a different initialization (for an iterative algorithm). Otherwise, if the algorithm is good enough to find the global minimum or to approximate by a good-quality local minimum but such a solution is still far from our subjective judgment, we may need to check the formulation, i.e., the form of the energy function and the parameters involved.

We may try different combinations of parameters to see if we can get the desired results – though this is not a clever way. Assume that we have exhausted all possible combinations but none of the solutions (again, global minima) are close enough to the solution desired. Then there must be something fundamentally wrong. We have to take a critical look at the problem's formulation.

One possible reason may be that some constraints that may be important for solving the problem have not been properly embedded into the energy function – current imaging models are often formulated based on over-simplified assumptions (Pavlidis 1992; Rosenfeld 1994) for tractability reasons. Another possibility is that some assumptions about the prior knowledge and the observed data are made incorrectly. An example is the over-smoothing problem in the MAP restoration and reconstruction. The MAP was blamed for the problem; however, what really caused the problem was the wrong prior assumption of the quadratic smoothness rather than the MAP principle.

Chapter 2

Mathematical MRF Models

This chapter introduces foundations of MRF theory and describes important mathematical MRF models for modeling image properties. The MRF models will be used in the subsequent chapters to derive MAP-MRF image analysis models and for MRF parameter estimation.

2.1 Markov Random Fields and Gibbs Distributions

Markov random field theory is a branch of probability theory for analyzing the spatial or contextual dependencies of physical phenomena. It is used in visual labeling to establish probabilistic distributions of interacting labels. This section introduces notations and results related to MRF's.

2.1.1 Neighborhood System and Cliques

The sites in $\mathcal{S}$ are related to one another via a neighborhood system (Section 2.12). A neighborhood system for $\mathcal{S}$ is defined as

$$\mathcal{N} = \{\mathcal{N}_i \mid \forall i \in \mathcal{S}\} \tag{2.1}$$

where $\mathcal{N}_i$ is the set of sites neighboring i. The neighboring relationship has the following properties:

(1) A site is not neighboring to itself: $i \notin \mathcal{N}_i$.

(2) The neighboring relationship is mutual: $i \in \mathcal{N}_{i'} \Longleftrightarrow i' \in \mathcal{N}_i$.

S.Z. Li, *Markov Random Field Modeling in Image Analysis*,
Advances in Pattern Recognition, DOI: 10.1007/978-1-84800-279-1_2,
© Springer-Verlag London Limited 2009

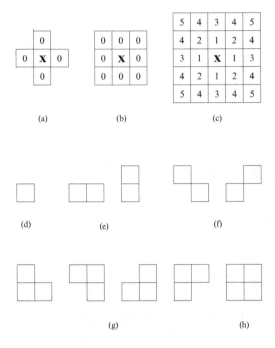

Figure 2.1: Neighborhood and cliques on a lattice of regular sites.

For a regular lattice $\mathcal{S}$, the set of neighbors of i is defined as the set of sites within a radius of $\sqrt{r}$ from i

$$\mathcal{N}_i = \{i' \in \mathcal{S} \mid [\text{dist}(\text{pixel}_{i'}, \text{pixel}_i)]^2 \leq r, \ i' \neq i\} \tag{2.2}$$

where $\text{dist}(A, B)$ denotes the Euclidean distance between A and B, and r takes an integer value. Note that sites at or near the boundaries have fewer neighbors.

In the first-order neighborhood system, also called the 4-neighborhood system, every (interior) site has four neighbors, as shown in Fig. 2.1(a) where x denotes the site considered and zeros its neighbors. In the second-order neighborhood system, also called the 8-neighborhood system, there are eight neighbors for every (interior) site, as shown in Fig. 2.1(b). The numbers $n = 1, \ldots, 5$ shown in Fig. 2.1(c) indicate the outermost neighboring sites in the nth-order neighborhood system. The shape of a neighbor set may be described as the hull enclosing all the sites in the set.

When the ordering of the elements in $\mathcal{S}$ is specified, the neighbor set can be determined more explicitly. For example, when $\mathcal{S} = \{1, \ldots, m\}$ is an ordered set of sites and its elements index the pixels of a 1D image, an interior site $i \in \{2, \ldots, m-1\}$ has two nearest neighbors, $\mathcal{N}_i = \{i-1, i+1\}$, and a site at the boundaries (the two ends) has one neighbor each, $\mathcal{N}_1 = \{2\}$ and $\mathcal{N}_m = \{m - 1\}$. When the sites in a regular rectangular lattice

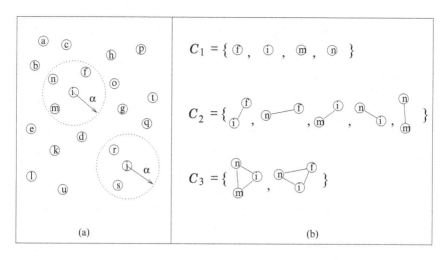

Figure 2.2: Neighborhood and cliques on a set of irregular sites.

$\mathcal{S} = \{(i,j) \mid 1 \leq i,j \leq n\}$ correspond to the pixels of an $n \times n$ image in the 2D plane, an internal site (i,j) has four nearest neighbors as $\mathcal{N}_{i,j} = \{(i-1,j),(i+1,j),(i,j-1),(i,j+1)\}$, a site at a boundary has three, and a site at the corners has two.

For an irregular $\mathcal{S}$, the neighbor set $\mathcal{N}_i$ of i is defined in the same way as (2.2) to comprise nearby sites within the radius of $\sqrt{r}$

$$\mathcal{N}_i = \{i' \in \mathcal{S} \mid [\mathrm{dist}(\mathrm{feature}_{i'}, \mathrm{feature}_i)]^2 \leq r, \; i' \neq i\} \qquad (2.3)$$

The $\mathrm{dist}(A,B)$ function needs to be defined appropriately for non-point features. Alternatively, the neighborhood may be defined by the Delaunay triangulation,[1] or its dual, the Voronoi polygon, of the sites (Besag 1975). In general, the neighbor sets $\mathcal{N}_i$ for an irregular $\mathcal{S}$ have varying shapes and sizes. Irregular sites and their neighborhoods are illustrated in Fig. 2.2(a). The neighborhood areas for sites i and j are marked by the dotted circles. The sizes of the two neighbor sets are $\#\mathcal{N}_i = 3$ and $\#\mathcal{N}_j = 2$.

The pair $(\mathcal{S}, \mathcal{N}) \overset{\triangle}{=} \mathcal{G}$ constitutes a graph in the usual sense; $\mathcal{S}$ contains the nodes and $\mathcal{N}$ determines the links between the nodes according to the neighboring relationship. A *clique* c for $(\mathcal{S}, \mathcal{N})$ is defined as a subset of sites in $\mathcal{S}$. It consists of either a single-site $c = \{i\}$, a pair of neighboring sites $c = \{i, i'\}$, a triple of neighboring sites $c = \{i, i', i''\}$, and so on. The collections of single-site, pair-site, and triple-site cliques will be denoted by $\mathcal{C}_1$, $\mathcal{C}_2$, and $\mathcal{C}_3$, respectively, where

$$\mathcal{C}_1 = \{i \mid i \in \mathcal{S}\} \qquad (2.4)$$

---

[1]Algorithms for constructing a Delaunay triangulation in $k \geq 2$ dimensional space can be found in (Bowyer 1981; Watson 1981).

$$\mathcal{C}_2 = \{\{i, i'\} \mid i' \in \mathcal{N}_i, \ i \in \mathcal{S}\} \tag{2.5}$$

and

$$\mathcal{C}_3 = \{\{i, i', i''\} \mid i, i', i'' \in \mathcal{S} \text{ are neighbors to one another}\} \tag{2.6}$$

Note that the sites in a clique are *ordered* and $\{i, i'\}$ is not the same clique as $\{i', i\}$, and so on. The collection of all cliques for $(\mathcal{S}, \mathcal{N})$ is

$$\mathcal{C} = \mathcal{C}_1 \cup \mathcal{C}_2 \cup \mathcal{C}_3 \cdots \tag{2.7}$$

where "$\cdots$" denotes possible sets of larger cliques.

The type of a clique for $(\mathcal{S}, \mathcal{N})$ of a regular lattice is determined by its size, shape, and orientation. Figures 2.1(d)–(h) show clique types for the first- and second-order neighborhood systems for a lattice. The single-site and horizontal and vertical pair-site cliques in (d) and (e) are all those for the first-order neighborhood system (a). The clique types for the second-order neighborhood system (b) include not only those in (d) and (e) but also diagonal pair-site cliques (f) and triple-site (g) and quadruple-site (h) cliques. As the order of the neighborhood system increases, the number of cliques grows rapidly and so do the computational expenses involved.

Cliques for irregular sites do not have fixed shapes like those for a regular lattice. Therefore, their types are essentially depicted by the number of sites involved. Consider the four sites f, i, m, and n within the circle in Fig. 2.2(a), in which m and n are supposed to be neighbors to each other and so are n and f. Then the single-site, pair-site, and triple-site cliques associated with this set of sites are shown in Fig. 2.2(b). The set $\{m, i, f\}$ does not form a clique because f and m are not neighbors.

2.1.2 Markov Random Fields

Let $F = \{F_1, \ldots, F_m\}$ be a family of random variables defined on the set $\mathcal{S}$ in which each random variable F_i takes a value f_i in $\mathcal{L}$. The family F is called a random field. We use the notation $F_i = f_i$ to denote the event that F_i takes the value f_i and the notation $(F_1 = f_1, \ldots, F_m = f_m)$ to denote the joint event. For simplicity, a joint event is abbreviated as $F = f$, where $f = \{f_1, \ldots, f_m\}$ is a *configuration* of F corresponding to a realization of the field. For a discrete label set $\mathcal{L}$, the probability that random variable F_i takes the value f_i is denoted $P(F_i = f_i)$, abbreviated $P(f_i)$ unless there is a need to elaborate the expressions, and the joint probability is denoted $P(F = f) = P(F_1 = f_1, \ldots, F_m = f_m)$ and abbreviated $P(f)$. For a continuous $\mathcal{L}$, we have probability density functions (p.d.f.s) $p(F_i = f_i)$ and $p(F = f)$.

F is said to be a Markov random field on $\mathcal{S}$ w.r.t. a neighborhood system $\mathcal{N}$ if and only if the following two conditions are satisfied:

$$P(f) > 0, \quad \forall f \in \mathbb{F} \qquad\qquad \text{(positivity)} \tag{2.8}$$

$$P(f_i \mid f_{\mathcal{S}-\{i\}}) = P(f_i \mid f_{\mathcal{N}_i}) \qquad \text{(Markovianity)} \qquad (2.9)$$

where $\mathcal{S} - \{i\}$ is the set difference, $f_{\mathcal{S}-\{i\}}$ denotes the set of labels at the sites in $\mathcal{S} - \{i\}$, and

$$f_{\mathcal{N}_i} = \{f_{i'} \mid i' \in \mathcal{N}_i\} \qquad (2.10)$$

stands for the set of labels at the sites neighboring i. The positivity is assumed for some technical reasons and can usually be satisfied in practice. For example, when the positivity condition is satisfied, the joint probability $P(f)$ of any random field is uniquely determined by its local conditional probabilities (Besag 1974). The Markovianity depicts the local characteristics of F. In MRF's, only neighboring labels have direct interactions with each other. If we choose the largest neighborhood in which the neighbors of any sites include all other sites, then any F is an MRF w.r.t. such a neighborhood system.

An MRF can have other properties, such as homogeneity and isotropy. It is said to be homogeneous if $P(f_i \mid f_{\mathcal{N}_i})$ is independent of the relative location of the site i in $\mathcal{S}$. So, for a homogeneous MRF, if $f_i = f_j$ and $f_{\mathcal{N}_i} = f_{\mathcal{N}_j}$, there will be $P(f_i|f_{\mathcal{N}_i}) = P(f_j|f_{\mathcal{N}_j})$ even if $i \neq j$. The isotropy will be illustrated in the next subsection with clique potentials.

In modeling some problems, we may need to use several *coupled* MRF's; each of the MRF's is defined on one set of sites, and the sites due to different MRF's are spatially interwoven. For example, in the related tasks of image restoration and edge detection, two MRF's, one for pixel values ($\{f_i\}$) and the other for edge values ($\{l_{i,i'}\}$), can be defined on the image lattice and its dual lattice, respectively. They are coupled to each other, for example, via conditional probability $P(f_i \mid f_{i'}, l_{i,i'})$ (see Section 3.3.1).

The concept of MRF's is a generalization of that of Markov processes (MPs), which are widely used in sequence analysis. An MP is defined on a domain of time rather than space. It is a sequence (chain) of random variables $\ldots, F_1, \ldots, F_m, \ldots$ defined on the time indices $\{\ldots, 1, \ldots, m, \ldots\}$. An nth-order unilateral MP satisfies

$$P(f_i \mid \ldots, f_{i-2}, f_{i-1}) = P(f_i \mid f_{i-1}, \ldots, f_{i-n}) \qquad (2.11)$$

A bilateral or noncausal MP depends not only on the past but also on the future. An nth-order bilateral MP satisfies

$$P(f_i \mid \ldots, f_{i-2}, f_{i-1}, f_{i+1}, f_{i+2}, \ldots) = P(f_i \mid f_{i+n}, \ldots, f_{i+1}, f_{i-1}, \ldots, f_{i-n}) \qquad (2.12)$$

It is generalized into MRF's when the time indices are considered as spatial indices.

There are two approaches for specifying an MRF, that in terms of the conditional probabilities $P(f_i \mid f_{\mathcal{N}_i})$ and that in terms of the joint probability $P(f)$. Besag (1974) argued for the joint probability approach in view of the disadvantages of the conditional probability approach. First, no obvious method is available for deducing the joint probability from the associated conditional probabilities. Second, the conditional probabilities themselves are subject to some non obvious and highly restrictive consistency conditions. Third, the natural specification of an equilibrium of a statistical process is in terms of the joint probability rather than the conditional distribution of the variables. Fortunately, a theoretical result about the equivalence between Markov random fields and Gibbs distributions (Hammersley and Clifford 1971; Besag 1974) provides a mathematically tractable means of specifying the joint probability of an MRF.

2.1.3   Gibbs Random Fields

A set of random variables F is said to be a *Gibbs random field* (GRF) on $\mathcal{S}$ w.r.t. $\mathcal{N}$ if and only if its configurations obey a *Gibbs distribution*. A Gibbs distribution takes the form

$$P(f) = Z^{-1} \times e^{-\frac{1}{T}U(f)} \tag{2.13}$$

where

$$Z = \sum_{f \in \mathbb{F}} e^{-\frac{1}{T}U(f)} \tag{2.14}$$

is a normalizing constant called the *partition function*, T is a constant called the *temperature*, which shall be assumed to be 1 unless otherwise stated, and $U(f)$ is the *energy function*. The energy

$$U(f) = \sum_{c \in \mathcal{C}} V_c(f) \tag{2.15}$$

is a sum of *clique potentials* $V_c(f)$ over all possible cliques $\mathcal{C}$. The value of $V_c(f)$ depends on the local configuration on the clique c. Obviously, the Gaussian distribution is a special member of this Gibbs distribution family.

A GRF is said to be homogeneous if $V_c(f)$ is independent of the relative position of the clique c in $\mathcal{S}$. It is said to be isotropic if V_c is independent of the orientation of c. It is considerably simpler to specify a GRF distribution that is homogeneous or isotropic than one without such properties. The homogeneity is assumed in most MRF vision models for mathematical and computational convenience. The isotropy is a property of direction-independent blob-like regions.

To calculate a Gibbs distribution, it is necessary to evaluate the partition function Z, which is the sum over all possible configurations in $\mathbb{F}$. Since there are a combinatorial number of elements in $\mathbb{F}$ for a discrete $\mathcal{L}$, as illustrated in Section 1.1.2, the evaluation is prohibitive even for problems of moderate size. Several approximation methods exist for solving this problem (see Chapter 8).

$P(f)$ measures the probability of the occurrence of a particular configuration, or "pattern", f. The more probable configurations are those with lower energies. The temperature T controls the sharpness of the distribution. When the temperature is high, all configurations tend to be equally distributed. Near zero temperature, the distribution concentrates around the global energy minima. Given T and $U(f)$, we can generate a class of "patterns" by sampling the configuration space $\mathbb{F}$ according to $P(f)$; see Section 3.4.1.

For discrete labeling problems, a clique potential $V_c(f)$ can be specified by a number of *parameters*. For example, letting $f_c = (f_i, f_{i'}, f_{i''})$ be the local configuration on a triple clique $c = \{i, i', i''\}$, f_c takes a finite number of states and therefore $V_c(f)$ takes a finite number of values. For continuous labeling problems, f_c can vary continuously. In this case, $V_c(f)$ is a (possibly piecewise) continuous function of f_c.

Sometimes, it may be convenient to express the energy of a Gibbs distribution as the sum of several terms, each ascribed to cliques of a certain size, that is,

$$
\begin{aligned}
U(f) \;=\; & \sum_{\{i\}\in\mathcal{C}_1} V_1(f_i) + \sum_{\{i,i'\}\in\mathcal{C}_2} V_2(f_i, f_{i'}) + \\
& \sum_{\{i,i',i''\}\in\mathcal{C}_3} V_3(f_i, f_{i'}, f_{i''}) + \cdots
\end{aligned}
\tag{2.16}
$$

The above implies a homogeneous Gibbs distribution because V_1, V_2, and V_3 are independent of the locations of i, i' and i''. For nonhomogeneous Gibbs distributions, the clique functions should be written as $V_1(i, f_i)$, $V_2(i, i', f_i, f_{i'})$, and so on.

An important special case is when only cliques of size up to two are considered. In this case, the energy can also be written as

$$
U(f) \;=\; \sum_{i\in\mathcal{S}} V_1(f_i) + \sum_{i\in\mathcal{S}} \sum_{i'\in\mathcal{N}_i} V_2(f_i, f_{i'})
\tag{2.17}
$$

Note that in the second term on the right-hand side, $\{i, i'\}$ and $\{i', i\}$ are two distinct cliques in $\mathcal{C}_2$ because the sites in a clique are *ordered*. The conditional probability can be written as

$$
P(f_i \mid f_{\mathcal{N}_i}) = \frac{e^{-\left[V_1(f_i)+\sum_{i'\in\mathcal{N}_i} V_2(f_i, f_{i'})\right]}}{\sum_{f_i\in\mathcal{L}} e^{-\left[V_1(f_i)+\sum_{i'\in\mathcal{N}_i} V_2(f_i, f_{i'})\right]}}
\tag{2.18}
$$

By incorporating (2.17) into (2.13), we can write the joint probability as the product

$$
P(f) \;=\; Z^{-1} \prod_{i\in\mathcal{S}} r_i(f_i) \prod_{i\in\mathcal{S}} \prod_{i'\in\mathcal{N}_i} r_{i,i'}(f_i, f_{i'})
\tag{2.19}
$$

where $r_i(f_i) = e^{-\frac{1}{T}V_1(f_i)}$ and $r_{i,i'}(f_i, f_{i'}) = e^{-\frac{1}{T}V_2(f_i, f_{i'})}$.

2.1.4   Markov-Gibbs Equivalence

An MRF is characterized by its local property (the Markovianity) whereas a GRF is characterized by its global property (the Gibbs distribution). The Hammersley-Clifford theorem (Hammersley and Clifford 1971) establishes the equivalence of these two types of properties. The theorem states that *F is an MRF on S w.r.t. $\mathcal{N}$ if and only if F is a GRF on S w.r.t. $\mathcal{N}$.* Many proofs of the theorem exist, e.g., in (Besag 1974), (Moussouris 1974) and (Kindermann and Snell 1980).

A proof that a GRF is an MRF is given as follows. Let $P(f)$ be a Gibbs distribution on S w.r.t. the neighborhood system $\mathcal{N}$. Consider the conditional probability

$$P(f_i \mid f_{S-\{i\}}) = \frac{P(f_i, f_{S-\{i\}})}{P(f_{S-\{i\}})} = \frac{P(f)}{\sum_{f_i' \in \mathcal{L}} P(f')} \qquad (2.20)$$

where $f' = \{f_1, \ldots, f_{i-1}, f_i', \ldots, f_m\}$ is any configuration that agrees with f at all sites except possibly i. Writing out $P(f) = Z^{-1} \times e^{-\sum_{c \in C} V_c(f)}$ gives[2]

$$P(f_i \mid f_{S-\{i\}}) = \frac{e^{-\sum_{c \in C} V_c(f)}}{\sum_{f_i'} e^{-\sum_{c \in C} V_c(f')}} \qquad (2.21)$$

Divide C into two sets $\mathcal{A}$ and $\mathcal{B}$ with $\mathcal{A}$ consisting of cliques containing i and $\mathcal{B}$ cliques not containing i. Then (2.21) can be written as

$$P(f_i \mid f_{S-\{i\}}) = \frac{\left[e^{-\sum_{c \in \mathcal{A}} V_c(f)}\right]\left[e^{-\sum_{c \in \mathcal{B}} V_c(f)}\right]}{\sum_{f_i'}\left\{\left[e^{-\sum_{c \in \mathcal{A}} V_c(f')}\right]\left[e^{-\sum_{c \in \mathcal{B}} V_c(f')}\right]\right\}} \qquad (2.22)$$

Because $V_c(f) = V_c(f')$ for any clique c that does not contain i, $e^{-\sum_{c \in \mathcal{B}} V_c(f)}$ cancels from both the numerator and denominator. Therefore, this probability depends only on the potentials of the cliques containing i,

$$P(f_i \mid f_{S-\{i\}}) = \frac{e^{-\sum_{c \in \mathcal{A}} V_c(f)}}{\sum_{f_i'} e^{-\sum_{c \in \mathcal{A}} V_c(f')}} \qquad (2.23)$$

that is, it depends on labels at i's neighbors. This proves that a Gibbs random field is a Markov random field. The proof that an MRF is a GRF is much more involved; a result to be described in the next subsection, which is about the uniqueness of the GRF representation (Griffeath 1976), provides such a proof.

The practical value of the theorem is that it provides a simple way of specifying the joint probability. One can specify the joint probability $P(F = f)$ by specifying the clique potential functions $V_c(f)$ and choosing appropriate

---

[2]This also provides a formula for calculating the conditional probability $P(f_i \mid f_{\mathcal{N}_i}) = P(f_i \mid f_{S-\{i\}})$ from potential functions.

potential functions for the desired system behavior. In this way, one encodes the a priori knowledge or preference about interactions between labels.

How to choose the forms and parameters of the potential functions for a proper encoding of constraints is a major topic in MRF modeling. The forms of the potential functions determine the form of the Gibbs distribution. When all the parameters involved in the potential functions are specified, the Gibbs distribution is completely defined. Defining the functional forms is the theme in Chapters 3 and 4, while estimating parameters is the subject in Chapters 7 and 8.

To calculate the joint probability of an MRF, which is a Gibbs distribution, it is necessary to evaluate the partition function (2.14). Because it is the sum over a combinatorial number of configurations in $\mathbb{F}$, the computation is usually intractable. The explicit evaluation can be avoided in maximum-probability-based MRF models when $U(f)$ contains no unknown parameters, as we will see subsequently. However, this is not true when the parameter estimation is also a part of the problem. In the latter case, the energy function $U(f) = U(f \mid \theta)$ is also a function of parameters θ and so is the partition function $Z = Z(\theta)$. The evaluation of $Z(\theta)$ is required. To circumvent the formidable difficulty therein, the joint probability is often approximated in practice. Several approximate formulae will be introduced in Chapter 7, where the problem of MRF parameter estimation is the subject.

2.1.5 Normalized and Canonical Forms

It is known that the choices of clique potential functions for a specific MRF are not unique; there may exist many equivalent choices that specify the same Gibbs distribution. However, there exists a unique normalized potential, called the *canonical potential*, for every MRF (Griffeath 1976).

Let $\mathcal{L}$ be a countable label set. A clique potential function $V_c(f)$ is said to be *normalized* if $V_c(f) = 0$, whenever for some $i \in c$, f_i takes a particular value in $\mathcal{L}$. The particular value can be any element in $\mathcal{L}$, e.g., 0 in $\mathcal{L} = \{0, 1, \ldots, M\}$. Griffeath (1976) established the mathematical relationship between an MRF distribution $P(f)$ and the unique canonical representation of clique potentials V_c in the corresponding Gibbs distribution (Griffeath 1976; Kindermann and Snell 1980). The result is described below.

Let F be a random field on a finite set S with local characteristics $P(f_i \mid f_{S-\{i\}}) = P(f_i \mid f_{\mathcal{N}_i})$. Then F is a Gibbs field with *canonical potential function* defined by

$$V_c(f) = \begin{cases} 0 & c = \phi \\ \sum_{b \subset c} (-1)^{|c-b|} \ln P(f^b) & c \neq \phi \end{cases} \tag{2.24}$$

where ϕ denotes the empty set, $|c - b|$ is the number of elements in the set $c - b$, and

$$f_i^b = \begin{cases} f_i & \text{if } i \in b \\ 0 & \text{otherwise} \end{cases} \tag{2.25}$$

is the configuration that agrees with f on set b but assigns the value 0 to all sites outside of b. For nonempty c, the potential can also be obtained as

$$V_c(f) = \sum_{b \subset c} (-1)^{|c-b|} \ln P(f_i^b \mid f_{\mathcal{N}_i}^b) \qquad (2.26)$$

where i is any element in b. Such a canonical potential function is *unique* for the corresponding MRF. Using this result, the canonical $V_c(f)$ can be computed if $P(f)$ is known.

However, in MRF modeling using Gibbs distributions, $P(f)$ is defined after $V_c(f)$ is determined, and therefore it is difficult to compute the canonical $V_c(f)$ from $P(f)$ directly. Nonetheless, there is an indirect way: Use a noncanonical representation to derive $P(f)$ and then canonicalize it using Griffeath's result to obtain the unique canonical representation.

The normalized potential functions appear to be immediately useful. For instance, for the sake of economy, one would use the minimal number of clique potentials or parameters to represent an MRF for a given neighborhood system. The concept of normalized potential functions can be used to reduce the number of nonzero clique parameters (see Chapter 7).

2.2   Auto-models

Contextual constraints on two labels are the lowest order constraints to convey contextual information. They are widely used because of their simple form and low computational cost. They are encoded in the Gibbs energy as pair-site clique potentials. With clique potentials of up to two sites, the energy takes the form

$$U(f) = \sum_{i \in \mathcal{S}} V_1(f_i) + \sum_{i \in \mathcal{S}} \sum_{i' \in \mathcal{N}_i} V_2(f_i, f_{i'}) \qquad (2.27)$$

where "$\sum_{i \in \mathcal{S}}$" is equivalent to "$\sum_{\{i\} \in \mathcal{C}_1}$" and "$\sum_{i \in \mathcal{S}} \sum_{i' \in \mathcal{N}_i}$" equivalent to "$\sum_{\{i,i'\} \in \mathcal{C}_2}$". Equation (2.27) is a special case of (2.16), which we call a second-order energy because it involves up to pair-site cliques. It the most frequently used form because it is the simplest in form but conveys contextual information. A specific GRF or MRF can be specified by properly selecting V_1 and V_2. Some important such GRF models will be described subsequently. Derin and Kelly (1989) presented a systematic study and categorization of Markov random processes and fields in terms of what they call strict-sense Markov and wide-sense Markov properties.

When $V_1(f_i) = f_i G_i(f_i)$ and $V_2(f_i, f_{i'}) = \beta_{i,i'} f_i f_{i'}$, where $G_i(\cdot)$ are arbitrary functions and $\beta_{i,i'}$ are constants reflecting the pair-site interaction between i and i', the energy is

$$U(f) = \sum_{\{i\} \in \mathcal{C}_1} f_i G_i(f_i) + \sum_{\{i,i'\} \in \mathcal{C}_2} \beta_{i,i'} f_i f_{i'} \qquad (2.28)$$

Such models are called *auto-models* (Besag 1974). The auto-models can be further classified according to assumptions made about individual f_i.

An auto-model is said to be an *auto-logistic* model if the f_i's take on values in the discrete label set $\mathcal{L} = \{0, 1\}$ (or $\mathcal{L} = \{-1, +1\}$). The corresponding energy is of the form

$$U(f) = \sum_{\{i\}\in\mathcal{C}_1} \alpha_i f_i + \sum_{\{i,i'\}\in\mathcal{C}_2} \beta_{i,i'} f_i f_{i'} \tag{2.29}$$

where $\beta_{i,i'}$ can be viewed as the *interaction coefficients*. When $\mathcal{N}$ is the nearest neighborhood system on a lattice (the four nearest neighbors on a 2D lattice or the two nearest neighbors on a 1D lattice), the auto-logistic model is reduced to the *Ising model*. The conditional probability for the auto-logistic model with $\mathcal{L} = \{0, 1\}$ is

$$P(f_i \mid f_{\mathcal{N}_i}) = \frac{e^{\alpha_i f_i + \sum_{i'\in\mathcal{N}_i} \beta_{i,i'} f_i f_{i'}}}{\sum_{f_i\in\{0,1\}} e^{\alpha_i f_i + \sum_{i'\in\mathcal{N}_i} \beta_{i,i'} f_i f_{i'}}} = \frac{e^{\alpha_i f_i + \sum_{i'\in\mathcal{N}_i} \beta_{i,i'} f_i f_{i'}}}{1 + e^{\alpha_i + \sum_{i'\in\mathcal{N}_i} \beta_{i,i'} f_{i'}}} \tag{2.30}$$

When the distribution is homogeneous, we have $\alpha_i = \alpha$ and $\beta_{i,i'} = \beta$, regardless of i and i'.

An auto-model is said to be an *auto-binomial* model if the f_i's take on values in $\{0, 1, \ldots, M-1\}$ and every f_i has a conditionally binomial distribution of M trials and probability of success q

$$P(f_i \mid f_{\mathcal{N}_i}) = \binom{M-1}{f_i} q^{f_i} (1-q)^{M-1-f_i} \tag{2.31}$$

where

$$q = \frac{e^{\alpha_i + \sum_{i'\in\mathcal{N}_i} \beta_{i,i'} f_{i'}}}{1 + e^{\alpha_i + \sum_{i'\in\mathcal{N}_i} \beta_{i,i'} f_{i'}}} \tag{2.32}$$

The corresponding energy takes the form

$$U(f) = -\sum_{\{i\}\in\mathcal{C}_1} \ln\binom{M-1}{f_i} - \sum_{\{i\}\in\mathcal{C}_1} \alpha_i f_i - \sum_{\{i,i'\}\in\mathcal{C}_2} \beta_{i,i'} f_i f_{i'} \tag{2.33}$$

It reduces to the auto-logistic model when $M = 1$.

An auto-model is said to be an *auto-normal model*, also called a Gaussian MRF (Chellappa 1985), if the label set $\mathcal{L}$ is the real line and the joint distribution is multivariate normal. Its conditional p.d.f. is

$$p(f_i \mid f_{\mathcal{N}_i}) = \frac{1}{\sqrt{2\pi\sigma^2}} e^{-\frac{1}{2\sigma^2}[f_i - \mu_i - \sum_{i'\in\mathcal{N}_i} \beta_{i,i'}(f_{i'} - \mu_{i'})]^2} \tag{2.34}$$

It is the normal distribution with conditional mean

$$E(f_i \mid f_{\mathcal{N}_i}) = \mu_i - \sum_{i'\in\mathcal{N}_i} \beta_{i,i'}(f_{i'} - \mu_{i'}) \tag{2.35}$$

and conditional variance

$$\text{var}(f_i \mid f_{\mathcal{N}_i}) = \sigma^2 \tag{2.36}$$

The joint probability is a Gibbs distribution

$$p(f) = \frac{\sqrt{\det(B)}}{\sqrt{(2\pi\sigma^2)^m}} \, \mathrm{e}^{\frac{(f-\mu)^{\mathrm{T}} B (f-\mu)}{2\sigma^2}} \tag{2.37}$$

where f is viewed as a vector, μ is the $m \times 1$ vector of the conditional means, and $B = [b_{i,i'}]$ is the $m \times m$ *interaction matrix* whose elements are unity and the off-diagonal element at (i, i') is $-\beta_{i,i'}$, i.e., $b_{i,i'} = \delta_{i,i'} - \beta_{i,i'}$ with $\beta_{i,i} = 0$. Therefore, the single-site and pair-site clique potential functions for the auto-normal model are

$$V_1(f_i) = (f_i - \mu_i)^2 / 2\sigma^2 \tag{2.38}$$

and

$$V_2(f_i, f_{i'}) = \beta_{i,i'}(f_i - \mu_i)(f_{i'} - \mu_{i'}) / 2\sigma^2 \tag{2.39}$$

respectively. A field of independent Gaussian noise is a special MRF whose Gibbs energy consists of only single-site clique potentials. Because all higher-order clique potentials are zero, there is no contextual interaction in the independent Gaussian noise. B is related to the covariance matrix Σ by $B = \Sigma^{-1}$. The necessary and sufficient condition for (2.37) to be a valid p.d.f. is that B be symmetric and positive definite.

A related but different model is the simultaneous auto-regression (SAR) model (Woods 1972) Unlike the auto-normal model, which is defined by the m conditional p.d.f.s, this model is defined by a set of m equations

$$f_i = \mu_i + \sum \beta_{i,i'}(f_{i'} - \mu_{i'}) + e_i \tag{2.40}$$

where e_i are independent Gaussian, $e_i \sim N(0, \sigma^2)$. It also generates the class of all multivariate normal distributions, but with joint p.d.f.s, as

$$p(f) = \frac{\det(B)}{\sqrt{(2\pi\sigma^2)^m}} \, \mathrm{e}^{\frac{(f-\mu)^{\mathrm{T}} B^{\mathrm{T}} B (f-\mu)}{2\sigma^2}} \tag{2.41}$$

where B is defined as before. Any SAR model is an auto-normal model with the B matrix in (2.37) being $B = B_2 + B_2^{\mathrm{T}} - B_2^{\mathrm{T}} B_2$, where $B_2 = B_{\text{autoregressive}}$. The reverse can also be done, though in a rather unnatural way, via Cholesky decomposition (Ripley 1981). Therefore, both models can have their p.d.f.s in the form of (2.37). However, for (2.41) to be a valid p.d.f. requires only that $B_{\text{autoregressive}}$ be nonsingular.

2.3 Multi-level Logistic Model

The auto-logistic model can be generalized to a *multilevel logistic* (MLL) model (Elliott et al. 1984; Derin and Cole 1986; Derin and Elliott 1987),

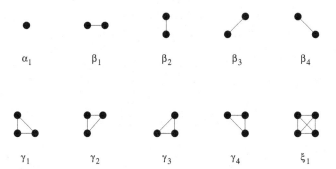

Figure 2.3: Clique types and associated potential parameters for the second-order neighborhood system. Sites are shown as dots and neighboring relationships as joining lines.

also called a Strauss process (Strauss 1977) and generalized Ising model (Geman and Geman 1984). There are M (> 2) discrete labels in the label set $\mathcal{L} = \{1, \ldots, M\}$. In this type of model, a clique potential depends on the type c (related to size, shape, and possibly orientation) of the clique and the local configuration $f_c \triangleq \{f_i \mid i \in c\}$. For cliques containing more than one site ($\#c > 1$), the MLL clique potentials are defined by

$$V_c(f) = \begin{cases} \zeta_c & \text{if all sites on } c \text{ have the same label} \\ -\zeta_c & \text{otherwise} \end{cases} \tag{2.42}$$

where ζ_c is the potential for type c cliques; for single-site cliques, they depend on the label assigned to the site

$$V_c(f) = V_c(f_i) = \alpha_I \quad \text{if } f_i = I \in \mathcal{L}_d \tag{2.43}$$

where α_I is the potential for label value I. Figure 2.3 shows the clique types and the associated parameters in the second-order (8-neighbor) neighborhood system.

Assume that an MLL model is of second order as in (2.27), so that only the α (for single-site cliques) and β (for pair-site cliques) parameters are nonzero. The potential function for pairwise cliques is written as

$$V_2(f_i, f_{i'}) = \begin{cases} \beta_c & \text{if sites on clique } \{i, i'\} = c \in \mathcal{C}_2 \text{ have the same label} \\ -\beta_c & \text{otherwise} \end{cases} \tag{2.44}$$

where β_c is the β parameter for type c cliques and $\mathcal{C}_2$ is set of pair-site cliques. For the 4-neighborhood system, there are four types of pairwise cliques (see Fig. 2.3), and so there can be four different β_c. When the model is isotropic, all the four neighbors take the same value. Owing to its simplicity, the pairwise MLL model (2.44) has been widely used for modeling regions and textures (Elliott et al. 1984; Geman and Geman 1984; Derin and Cole 1986; Derin and Elliott 1987; Murray and Buxton 1987; Lakshmanan and Derin 1989; Won and Derin 1992).

When the MLL model is isotropic, it depicts blob-like regions. In this case, the conditional probability can be expressed as (Strauss 1977)

$$P(f_i = I \mid f_{\mathcal{N}_i}) = \frac{\mathrm{e}^{-\alpha_I - \beta n_i(I)}}{\sum_{I=1}^{M} \mathrm{e}^{-\alpha_I - \beta n_i(I)}} \tag{2.45}$$

where $n_i(I)$ are the number of sites in $\mathcal{N}_i$ that are labeled I. It reduces to (2.30) when there are only two labels, 0 and 1. In contrast, an anisotropic model tends to generate texture-like patterns. See the examples in Section 3.4.

A hierarchical two-level Gibbs model has been proposed to represent both noise-contaminated and textured images (Derin and Cole 1986; Derin and Elliott 1987). The higher-level Gibbs distribution uses an isotropic random field (e.g., MLL) to characterize the blob-like region formation process. A lower-level Gibbs distribution describes the filling-in in each region. The filling-in may be independent noise or a type of texture, both of which can be characterized by Gibbs distributions. This provides a convenient approach for MAP-MRF modeling. In segmenting noisy and textured images (Derin and Cole 1986; Derin and Elliott 1987; Lakshmanan and Derin 1989; Hu and Fahmy 1987; Won and Derin 1992), for example, the higher-level model determines the prior of f for the region process, while the lower-level Gibbs model contributes to the conditional probability of the data given f. Note that different levels of MRF's in the hierarchy can have different neighborhood systems.

2.4 The Smoothness Prior

A generic contextual constraint on this world is the *smoothness*. It assumes that physical properties in a neighborhood of space or in an interval of time present some coherence and generally do not change abruptly. For example, the surface of a table is flat, a meadow presents a texture of grass, and a temporal event does not change abruptly over a short period of time. Indeed, we can always find regularities of a physical phenomenon w.r.t. certain properties. Since its early applications in vision (Grimson 1981; Horn and Schunck 1981; Ikeuchi and Horn 1981) aimed at imposing constraints (in addition to those from the data) on the computation of image properties, the smoothness prior has been one of the most popular prior assumptions in low-Level vision. It has been developed into a general framework, called regularization (Poggio et al. 85a; Bertero et al. 1988), for a variety of low-Level vision problems.

Smoothness constraints are often expressed as the prior probability or equivalently an energy term $U(f)$, measuring the extent to which the smoothness assumption is violated by f. There are two basic forms of such smoothness terms corresponding to situations with discrete and continuous labels, respectively.

Equations (2.42) and (2.44) of the MLL model with negative ζ and β coefficients provide method for constructing smoothness terms for unordered, discrete labels. Whenever all labels f_c on a clique c take the same value, which means the solution f is locally smooth on c, they incur a negative clique potential (cost); otherwise, if they are not all the same, they incur a positive potential. Such an MLL model tends to give a smooth solution that prefers uniform labels.

For spatially (and also temporally in image sequence analysis) continuous MRF's, the smoothness prior often involves derivatives. This is the case with the analytical regularization (to be introduced in Section 1.3.3). There, the potential at a point is in the form of $[f^{(n)}(x)]^2$. The order n determines the number of sites in the cliques involved; for example, $[f'(x)]^2$, where $n = 1$ corresponds to a pair-site smoothness potential. Different orders imply different classes of smoothness.

Let us take continuous restoration or reconstruction of nontexture surfaces as an example. Let $f = \{f_1, \ldots, f_m\}$ be the sampling of an underlying "surface" $f(x)$ on $x \in [a, b]$, where the surface is one-dimensional for simplicity. The Gibbs distribution $P(f)$, or equivalently the energy $U(f)$, depends on the type of surface f we expect to reconstruct. Assume that the surface is flat a priori. A flat surface that has equation $f(x) = a_0$ should have zero first-order derivative, $f'(x) = 0$. Therefore, we may choose the prior energy as

$$U(f) = \int [f'(x)]^2 dx \qquad (2.46)$$

which is called a *string*. The energy takes the minimum value of zero only if f is absolutely flat, or a positive value otherwise. Therefore, the surface which minimizes (2.46) alone has a constant height (gray value for an image).

In the discrete case where the surface is sampled at discrete points $a \leq x_i \leq b$, $i \in \mathcal{S}$, we use the first-order difference to approximate the first derivative and use a summation to approximate the integral, so (2.46) becomes

$$U(f) = \sum_i [f_i - f_{i-1}]^2 \qquad (2.47)$$

where $f_i = f(x_i)$. Expressed as the sum of clique potentials, we have

$$U(f) = \sum_{c \in \mathcal{C}} V_c(f) = \sum_{i \in \mathcal{S}} \sum_{i' \in \mathcal{N}_i} V_2(f_i, f_{i'}) \qquad (2.48)$$

where $\mathcal{C} = \{(1,2), (2,1), (2,3), \cdots, (m-2, m-1), (m, m-1), (m-1, m)\}$ consists of only pair-site cliques and

$$V_c(f) = V_2(f_i, f_{i'}) = \frac{1}{2}(f_i - f_{i'})^2 \qquad (2.49)$$

Its 2D equivalent is

$$\int\int \{[f_x(x,y)]^2 + [f_y(x,y)]^2\}\mathrm{d}x\mathrm{d}y \qquad (2.50)$$

and is called a *membrane*.

Similarly, the prior energy $U(f)$ can be designed for planar or quadratic surfaces. A planar surface, $f(x) = a_0 + a_1 x$, has zero second-order derivative, $f''(x) = 0$. Therefore, one may choose

$$U(f) = \int [f''(x)]^2 \mathrm{d}x \qquad (2.51)$$

which is called a *rod*. The surface that minimizes (2.51) alone has a constant gradient. In the discrete case, we use the second-order difference to approximate the second-order derivative, and (2.51) becomes

$$U(f) = \sum_i [f_{i+1} - 2f_i + f_{i-1}]^2 \qquad (2.52)$$

For a quadratic surface, $f(x) = a_0 + a_1 x + a_2 x^2$, the third-order derivative is zero, $f'''(x) = 0$, and the prior energy may be

$$U(f) = \int [f'''(x)]^2 \mathrm{d}x \qquad (2.53)$$

The surface that minimizes (2.53) alone has a constant curvature. In the discrete case, we use the third-order difference to approximate the second-order derivative and (2.53) becomes

$$U(f) = \sum_i [f_{i+1} - 3f_i + 3f_{i-1} - f_{i-2}]^2 \qquad (2.54)$$

The above smoothness models can be extended to 2D. For example, the 2D equivalent of the rod, called a plate, comes in two varieties, the quadratic variation

$$\int\int \{[f_{xx}(x,y)]^2 + 2[f_{xy}(x,y)]^2 + [f_{yy}(x,y)]^2\}\mathrm{d}x\mathrm{d}y \qquad (2.55)$$

and the squared Laplacian

$$\int\int \{f_{xx}(x,y) + f_{yy}(x,y)\}^2 \mathrm{d}x\mathrm{d}y \qquad (2.56)$$

The surface that minimizes one of the smoothness prior energies alone has either a constant gray level, a constant gradient, or a constant curvature. This is undesirable because constraints from other sources such as the data are not used. Therefore, a smoothness term $U(f)$ is usually utilized in conjunction with other energy terms. In regularization, an energy consists of a smoothness term and a closeness term, and the minimal solution is a compromise between the two constraints; refer to Section 1.3.3.

The encodings of the smoothness prior in terms of derivatives usually lead to *isotropic* potential functions. This is due to the assumption that the underlying surface is nontextured. *Anisotropic* priors have to be used for texture patterns. This can be done, for example, by choosing (2.27) with direction-dependent V_2. This will be discussed in Section 3.4.

2.5   Hierarchical GRF Model

A hierarchical two-level Gibbs model has been proposed to represent both noise-contaminated and textured images (Derin and Cole 1986; Derin and Elliott 1987). The higher-level Gibbs distribution uses an isotropic random field (e.g., MLL) to characterize the blob-like region's formation process. A lower-level Gibbs distribution describes the filling-in in each region. The filling-in may be independent noise or a type of texture, both of which can be characterized by Gibbs distributions. This provides a convenient approach for MAP-MRF modeling. In segmenting noisy and textured images (Derin and Cole 1986; Derin and Elliott 1987; Lakshmanan and Derin 1989; Hu and Fahmy 1987; Won and Derin 1992), for example, the higher-level model determines the prior of f for the region process, while the lower-level Gibbs model contributes to the conditional probability of the data given f. Note that different levels of MRF's in the hierarchy can have different neighborhood systems.

Various hierarchical Gibbs models result, according to what are chosen for the regions and for the filling-ins. For example, each region may be filled in by an auto-normal texture (Manjunath et al. 1990; Won and Derin 1992) or an auto-binomial texture (Hu and Fahmy 1987); the MLL for the region formation may be substituted by another appropriate MRF. The hierarchical MRF model for textured regions will be further discussed in Section 3.4.1.

A drawback of the hierarchical model is that the conditional probability $P(d_i \mid f_i = I)$ for regions given by $\{i \in \mathcal{S} \mid f_i = I\}$ cannot always be written exactly. For example, when the lower-level MRF is a texture modeled as an auto-normal field, its joint distribution over an irregularly shaped region is not known. This difficulty may be overcome by using approximate schemes such as pseudo-likelihood (to be introduced in Section 7.1) or by using the eigenanalysis method (Wu and Leahy 1993).

2.6   The FRAME Model

The FRAME (filter, random fields and maximum entropy) model, proposed in (Zhu et al. 1997),(Zhu and Mumford 1997) and (Zhu et al. 1998), is a generalized MRF model that fuses the essence of filtering theory and MRF modeling through the maximum entropy principle. It is generalized in the following two aspects: (1) The FRAME model is defined in terms of statistics (i.e., potential functions) calculated from the output of a filter bank by which the image is filtered, instead of the clique potentials of the image itself. Given an image (a realization of an MRF), the image is filtered by a bank of filters, giving a set of output images. Some statistics are then calculated from the output images. (2) The FRAME model provides a means of learning the model parameters from a set of samples (example images) representative of the MRF to be modeled. Besides, it also gives an algorithm for filter selection.

The joint distribution of the FRAME model is constrained in such a way that the model can reproduce the statistics of the example images. It is found by solving a constrained maximum entropy problem. Let $G^{(k)}$ ($k = 1, \ldots, K$) be a bank of K filters (such as Gabor filters), $f^{(k)} = G^{(k)} * f$ the output of filtering f by $G^{(k)}$, and $H^{(k)} \in \mathcal{L}^S$ (the $\mathcal{L}$ is assumed to be the same for all the K filter outputs) the histogram of $f^{(k)}$ defined by

$$H^{(k)}(I) = \frac{1}{|\mathcal{S}|} \sum_{i \in \mathcal{S}} \delta(I - f_i^{(k)}) \tag{2.57}$$

where $\delta(t) = 1$ if $t = 0$ or 0 otherwise. For the filtered sample images, we denote the averaged histogram of the kth filter output by $\overline{H_{samp}^{(k)}}$ (averaged across all example images). Now, the joint distribution of the FRAME is defined as:

$$p(f) = \arg \max_{p} \left\{ - \int p(f) \log(p(f)) \mathrm{d}f \right\} \tag{2.58}$$

subject to

$$\overline{H_{p(f)}^{(k)}}(I) = \overline{H_{samp}^{(k)}}(I) \qquad \forall k, \forall I \tag{2.59}$$

$$\int p(f) \mathrm{d}f = 1 \tag{2.60}$$

where

$$\overline{H_{p(f)}^{(k)}} = \int H^{(k)}(f) p(f) \mathrm{d}f \tag{2.61}$$

is the expectation of $H^{(k)}$ w.r.t. $p(f)$. By using Lagrange multipliers $\theta_I^{(k)}$ for the constraints of (2.59), we get the Lagrangian

$$L(p, \theta) = - \int p(f) \log(p(f)) \mathrm{d}f \tag{2.62}$$

$$+ \int_I \sum_k \theta_I^{(k)} \left\{ \int_f p(f) \sum_i \delta(I - f_i^{(k)}) \mathrm{d}f - |\mathcal{S}| \, \overline{H_{samp}^{(k)}}(I) \right\} \mathrm{d}I$$

(Note that the constraints are multiplied by the factor of $|\mathcal{S}|$.) By setting $\frac{\partial L(p,\theta)}{\partial p} = 0$, the solution to the constrained optimization (ME) problem can be derived as (consider $p(f) = p(f \mid \theta)$ when θ is given)

$$
\begin{aligned}
p(f \mid \theta) &= \frac{1}{Z(\theta)} e^{-\sum_{k=1}^{K} \sum_{i \in \mathcal{S}} \{\int \theta^{(k)}(I)\delta(I - f_i^{(k)})dI\}} \\
&= \frac{1}{Z(\theta)} e^{-\sum_{k=1}^{K} \sum_{i \in \mathcal{S}} \{\theta^{(k)}(f_i^{(k)})\}}
\end{aligned}
\tag{2.63}
$$

where $\theta^{(k)}(\cdot)$ are the potential functions of the FRAME model and Z the normalizing factor.

In the discrete form, assume that $I^{(k)} = f_i^{(k)}$ is quantized into L discrete values $I_1^{(k)}, \dots, I_L^{(k)}$. The solution in (2.63) can be written as

$$
\begin{aligned}
p(f \mid \theta) &= \frac{1}{Z(\theta)} e^{-\sum_{k=1}^{K} \sum_{i \in \mathcal{S}} \sum_{\ell=1}^{L} \{\theta_\ell^{(k)} \delta(I_i^{(k)} - f_i^{(k)})\}} \\
&= \frac{1}{Z(\theta)} e^{-\sum_{k=1}^{K} \sum_{\ell=1}^{L} \theta_\ell^{(k)} H_\ell^{(k)}} \\
&= \frac{1}{Z} e^{-<\theta, H>}
\end{aligned}
\tag{2.64}
$$

where $\theta_\ell^{(k)} = \theta^{(k)}(I_\ell^{(k)})$, $H_\ell^{(k)} = H^{(k)}(I_\ell^{(k)})$, and $< a, b >$ is the inner product of a and b.

The FRAME model provides a means of modeling complicated high-order patterns in a tractable way. In the traditional MRF model, the neighborhood is usually small to keep the model tractable, and therefore it is difficult to model patterns in which interaction in a large neighborhood is necessary. In contrast, the FRAME model is able to model more complicated patterns by incorporating larger neighborhood and potential functions of higher-order cliques implicitly determined by the filter windows; moreover, it uses an accompanying learning procedure to estimate high-order potential functions from the filter outputs. This makes the high-order model tractable in formulation, albeit expensive in computation. There are two things to learn in the FRAME model: (1) the potential functions $\theta_I^{(k)}$; and (2) the types of filters $G^{(k)}$ to use. These will be described in Section 7.1.7.

Zhu and his colleagues (Wu et al. 2000) have established an equivalence between the FRAME model and another mathematical model of texture, called Julesz ensembles (Julesz 1962), when the size of the image lattice goes to infinity. On the other hand, they also propose fast MCMC algorithms for sampling $p(f \mid \theta)$ which involves hundreds of parameters to estimate in a large neighborhood (Zhu and Liu 2000).

2.7   Multiresolution MRF Modeling

The motivation for multiresolution MRF (MRMRF) modeling here is similar to that of FRAME modeling: to model complex and macro patterns by incorporating interactions in a large neighborhood. The approach used in MRMRF modeling is to build an MRF model based on the outputs of multiresolution filters.

From orthogonal wavelet decomposition, such as in Haar or Daubechies wavelets, nonredundant subbands can be obtained in different scales and directions. They can be used to represent the original image completely. On the other hand, these subbands are downsampled with the discrete wavelet transform. Therefore, the texture structure represented by the information of two faraway pixels in the original image may become that of immediate neighbors in the subband images on the higher-levels. Figure 2.4 shows an example of multiresolution wavelet decomposition. We can see that with the decomposition and downsampling, the subbands of different scales and directions can reveal different characteristics of the original image. The pixel relationship at different scales is not the same, even in the same direction.

Let f be an image defined on a lattice of sites $\mathcal{S}$ indexed by $i \in \mathcal{S}$, $G = \{G^{(1)}, G^{(2)}, ..., G^{(K)}\}$ a set of multiresolution filters such as wavelet filters, and $f^{(k)} = G^{(k)} * f$ the kth subband output of filtering f with $G^{(k)}$. Assume that the pixel values of the K filter outputs are quantized into M levels, giving the label set $\mathcal{L} = \{1, \dots, M\}$, which is the same for all the K subbands. Then the distribution of image f can be written in the form

$$P(f \mid G) = \frac{1}{Z(G)} \exp(-U(f \mid G)) \qquad (2.65)$$

where $Z(G)$ is the normalizing partition function. $U(f \mid G)$ is the energy function, which takes the form

$$U(f \mid G) = \sum_{k=1}^{K} \sum_{c \in \mathcal{C}} V_c^{(k)}(f) \qquad (2.66)$$

where $\mathcal{C}$ is the set of all cliques in a neighborhood system and $V_c^{(k)}(f)$ is the clique potential associated with the filter output $f^{(k)}$.

Consider cliques of up to two sites. Let $\theta = \{\theta^{(k)}\} = \{\alpha^{(k)}, \beta^{(k)}\}$ be the set of MRMRF parameters, where $\alpha^{(k)} = \{\alpha^{(k)}(I)\}$ consists of M components for the eight quantized pixel values $I = f_i^{(k)}$, and $\beta^{(k)} = \{\beta_c^{(k)}\}$ consists of four components for cliques $c = (i, i')$ in the 4-neighborhood system. For homogeneous MRMRF's, the potential functions are location independent, though the pair-site clique potentials are direction dependent. The following energy function is used to include cliques of up to two sites:

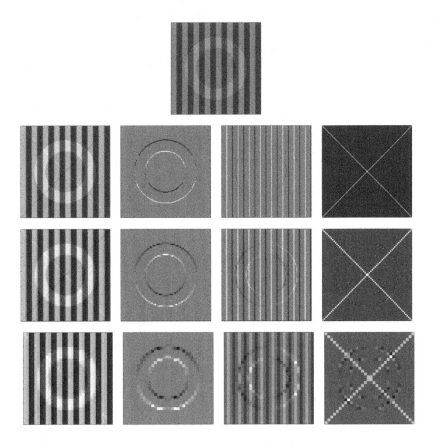

Figure 2.4: The original image (row 0). Haar wavelet decomposition at level 1 (row 1), level 2 (row 2), and level 3 (row 3). From left to right of rows 1–3 are the low-pass, horizontal, vertical, and diagonal subbands.

$$U(f \mid G, \theta) = \hspace{6cm} (2.67)$$

$$\sum_{k=1}^{K} \sum_{i \in \mathcal{S}} \left\{ \alpha_i^{(k)}(f_i^{(k)}) + \sum_{i' \in \mathcal{N}_i, c=(i,i')} \beta_c^{(k)} \left[1 - 2\exp(-(f_i^{(k)} - f_{i'}^{(k)})^2) \right] \right\}$$

The corresponding potential functions are $V_1^{(k)}(i) = \alpha_i^{(k)}(f_i^{(k)})$ for the single-site clique and $V_2^{(k)}(i, i') = \beta_c^{(k)} \left[1 - 2\exp(-(f_i^{(k)} - f_{i'}^{(k)})^2) \right]$ for $c = (i, i')$.

In this model, the exponential form of the pair-site clique potential is similar to that of the multilevel logistic (MLL) model (see Section 2.3). But it is easier to process and more meaningful in texture representation than the MLL model (Liu and Wang 1999). In MLL, if the two labels are not exactly the same, they contribute nothing to the potential even if they are similar; in contrast, the use of the exponential term in MRMRF incorporates the label similarity into the potential in a soft way.

Before giving the conditional probability $P(f_i|f_{\mathcal{N}_i}, G, \theta)$, we first define a generalized energy of f_i as

$$U(f_i \mid G, \theta) = \tag{2.68}$$

$$\sum_{k=1}^{K} \left\{ \alpha^{(k)}(f_i^{(k)}) + \sum_{i' \in \mathcal{N}_i, c=(i,i')} \beta_c^{(k)} \left[1 - 2\exp(-(f_i^{(k)} - f_{i'}^{(k)})^2) \right] \right\}$$

Thus

$$P(f_i|f_{\mathcal{N}_i}, G, \theta) = \frac{\sum_{k=1}^{K} \exp(-U(f_i^{(k)} \mid \theta^{(k)}))}{\sum_{k=1}^{K} \sum_{I^{(k)}=1}^{M} \exp(-U(I^{(k)} \mid \theta^{(k)}))} \tag{2.69}$$

where $I^{(k)} = f_i^{(k)}$ is the value of pixel i in the kth subband image. To define a pseudo-likelihood, we simplify it by

$$P(f_i \mid f_{\mathcal{N}_i}, G, \theta) \approx \prod_{k=1}^{K} \frac{\exp(-U(f_i^{(k)} \mid \theta^{(k)}))}{\sum_{I^{(k)}=1}^{M} \exp(-U(I^{(k)} \mid \theta^{(k)}))} \tag{2.70}$$

Then the pseudo-likelihood can be written as

$$\begin{aligned}
PL(f \mid G, \theta) &= \prod_{i \in S} P(f_i \mid f_{\mathcal{N}_i}, G, \theta) \\
&\approx \prod_{i \in S} \prod_{k=1}^{K} \frac{\exp(-U(f_i^{(k)} \mid \theta^{(k)}))}{\sum_{I^{(k)}=1}^{M} \exp(-U(I^{(k)} \mid \theta^{(k)}))} \\
&= \prod_{k=1}^{K} PL(f^{(k)} \mid \theta^{(k)}) \tag{2.71}
\end{aligned}$$

where

$$PL(f^{(k)} \mid \theta^{(k)}) = \prod_{i \in S} \frac{\exp(-U(f_i^{(k)} \mid \theta^{(k)}))}{\sum_{I^{(k)}=1}^{M} \exp(-U(I^{(k)} \mid \theta^{(k)}))} \tag{2.72}$$

Thus the pseudo-likelihood $PL(f \mid G, \theta)$ can be approximated by the product of individual pseudo-likelihoods of the subband outputs $f^{(k)}$, in which an assumption is made that the parameters at different subbands are independent of each other. With this simplification, the parameter of the model can be estimated easily.

The parameters are estimated from sample data f_{samp}, which is given. The estimation can be done by maximum likelihood $P(f_{samp}^{(k)} \mid \theta^{(k)})$ or by maximizing the pseudo-likelihood $PL(f_{samp}^{(k)} \mid \theta^{(k)})$ defined in (2.71). According to that definition, each subband can be considered as an independent MRF model,and hence the parameters of each subband can be estimated independently without considering the other subbands. Any method for MRF parameter estimation can be used to estimate parameters of each subband. The Markov chain Monte Carlo (MCMC) method (Section 7.1.6) used in the parameter estimation for the FRAME model (Section 7.1.7) would be a proper choice.

The MRMRF model attempts to incorporate information in a large neighborhood by fusing filtering theory and MRF models, which is similar to the FRAME model (Zhu et al. 1998). Compared with the traditional MRF model, the MRMRF model can reveal more information contained in the textures since the original images are decomposed into subbands of different scales and directions and downsampled. For this reason, the MRMRF model is more powerful than the traditional MRF model; however, it seems less powerful than the FRAME model since only up to pair-site interactions are considered in MRMRF. Computationally, it is also between the traditional MRF and the FRAME, the FRAME being very expensive.

2.8 Conditional Random Fields

In the MAP-MRF framework, the optimal configuration is the optimum of the posterior probability $P(f \mid d)$, or equivalently that of the joint probability $P(f, d) = p(d \mid f)P(f)$. The prior is formulated as an MRF, and the likelihood is due to the observation model. Usually, for tractability reasons $p(d \mid f)$ is assumed to have the factorized form (Besag 1974)

$$p(d \mid f) = \prod_{i \in \mathcal{S}} p(d_i \mid f_i) \qquad (2.73)$$

even though the underlying observation model is not as simple.

The conditional random field (CRF) models the posterior probability $P(f \mid d)$ directly as an MRF without modeling the prior and likelihood individually. The label set f is said to be a CRF, given d, if every f_i satisfies the Markovianity (with positivity assumed) (Lafferty et al. 2001)

$$P(f_i \mid d, f_{\mathcal{S}-\{i\}}) = P(f_i \mid d, f_{\mathcal{N}_i}) \qquad (2.74)$$

According to the Markov-Gibbs equivalence, we have

$$P(f \mid d) = \frac{1}{Z} \exp\left(-\frac{1}{T}E(f \mid d)\right) \qquad (2.75)$$

where Z is the partition function and $E(f \mid d)$ the energy function. If only up to pairwise clique potentials are nonzero, the posterior probability $P(f \mid d)$ has the form

$$P(f \mid d) = \frac{1}{Z} \exp \left\{ -\sum_{i \in \mathcal{S}} V_1(f_i \mid d) - \sum_{i \in \mathcal{S}} \sum_{i' \in \mathcal{N}_i} V_2(f_i, \ f_{i'} \mid d) \right\} \qquad (2.76)$$

where $-V_1$ and $-V_2$ are called the association and interaction potentials, respectively, in the CRF literature (Lafferty et al. 2001). Generally, these potentials are computed as a linear combination of some feature attributes extracted from the observation.

There are two main differences between the CRF and MRF. First, in a CRF, the unary (or association) potential at site i is a function of all the observation data $d_1, \ldots, d_n$ as well as that of the label f_i; in an MRF, however, the unary potential is a function of the observations f_i and d_i only. Second, in an MRF, the pairwise (or interaction) potential for each pair of sites i and i' is independent of the observation; however, in a CRF, it is also a function of all $d_1, \ldots, d_n$ as well as that of the labels f_i and $f_{i'}$ (Lafferty et al. 2001; Ng and Jordan 2002; Rubinstein and Hastie 1997).

Therefore, a CRF may be suitable for dealing with situations where the likelihood of an MRF is not of a factorized form such that all the d_i ($\forall i \in \mathcal{S}$) can explicitly exist in both unary and pairwise potentials. Moreover, unlike in an MRF, where $d_{i'}$ can influence f_i ($i \neq i'$) indirectly through the neighborhood system, in a CRF, this is done directly by the link between $d_{i'}$ and i. The CRF has so far been used mainly in speech (1D signal) analysis. It can be extended to discriminative random fields (DRF) for image analysis as follows.

2.9 Discriminative Random Fields

While the MRF is a generative model for modeling a spatial pattern such as an image, the discriminative random field (DRF) (Kumar and Hebert 2003; Kumar 2005; Kumar and Hebert 2006) is a discriminative model that has been used for classifying patterns directly (e.g., target vs. non target classification) in images. The DRF is a special type of CRF with two extensions to it. First, a DRF is defined over 2D lattices (such as the image grid), as illustrated in Fig. 2.5. Second, the unary (association) and pairwise (interaction) potentials therein are designed using local discriminative classifiers.

The DRF of Kumar and Hebert (2003) and Kumar and Hebert (2006) defines the potentials in terms of generalized linear models as

$$V_1(f_i \mid d) \;=\; -\log\left(\sigma[f_i T_i(d)]\right) \qquad (2.77)$$

$$V_2(f_i, f_{i'} \mid d) \;=\; \alpha f_i f_{i'} + \beta(2\sigma[\delta(f_i, f_{i'}) T_{i,i'}(d)] - 1) \qquad (2.78)$$

where $\sigma[x]$ is the logistic function (e.g., $1/(1 + e^{-x})$), $T_i(d)$ and $T_{i,i'}(d)$ are functions that transform d into the unary and binary feature attributes and then linearly combine them into scalars, α and β are parameters to be learned from training examples, and $\delta(f_i, f_{i'})$ is an indication function

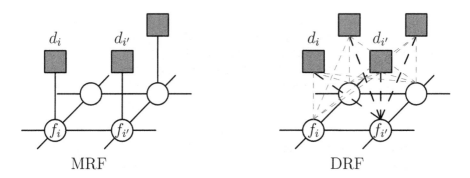

Figure 2.5: MRF vs. DRF. In an MRF, each site is connected to one observation datum. In a DRF, each site is connected to all the observation data.

$$\delta(f_i, f_{i'}) = \left\{ \begin{array}{ll} 1 & \text{if } f_i = f_{i'} \\ -1 & \text{otherwise} \end{array} \right. \tag{2.79}$$

In their work, the solution to the target detection problem is based on the maximum posterior marginal principle. Belief propagation and sampling algorithms are employed to find the MPM estimate.

2.10 Strong MRF Model

In addition to the (local) Markovianity (2.9) introduced in Section 2.1.2, there are two other variants of Markov properties, namely pairwise Markovianity and global Markovianity (Lauritzen 1996).

Pairwise Markovianity. A Markovianity is pairwise if for any non-adjacent sites i and i', it satisfies $P(f_i \mid f_{i'}) = P(f_i \mid f_{\mathcal{S}-\{i\}-\{i'\}})$. This means that the labels of two nonadjacent sites are independent given the labels of the other sites.

Global Markovianity. A Markovianity is global if for any disjoint subsets $\mathcal{A}$, $\mathcal{B}$, and $\mathcal{C}$ of $\mathcal{S}$, $\mathcal{C}$ separating $\mathcal{A}$ from $\mathcal{B}$, it satisfies $P(f_{\mathcal{A}} \mid f_{\mathcal{B}}) = P(f_{\mathcal{A}} \mid f_{\mathcal{C}})$; that is, given a set of sites, the labels of any two separated subsets are independent.

Generally, pairwise Markovianity can be deduced from local Markovianity, and local Markovianity can be deduced from global Markovianity, but the reverse does not always hold (Lauritzen 1996). From this viewpoint, the local Markovianity is stronger than the pairwise Markovianity and weaker than the global Markovianity.

A strong MRF is a special case of the standard MRF (Moussouris 1974). Let $\mathcal{G} = (\mathcal{S}, \mathcal{N})$ represent a graph, and suppose F is an MRF defined on $\mathcal{G}$ w.r.t. $\mathcal{N}$. Assuming that $\mathcal{D} \subseteq \mathcal{S}$ is a subset of $\mathcal{S}$, an MRF is a strong MRF, then it satisfies

$$P(f_i \mid f_{\mathcal{D}-\{i\}}) = P(f_i \mid f_{\mathcal{N}_i \cap \mathcal{D}}) \quad \forall \mathcal{A} \subseteq \mathcal{S} \tag{2.80}$$

which is the global Markovianity for i. That is, an MRF is strong if the Markovianity holds not only w.r.t. the neighborhood system but also any of the subsets $\mathcal{D} \subseteq \mathcal{S}$ (Moussouris 1974; Paget 2004). In such a case, if the label of a neighboring site i is undefined, the label of the site i is still conditionally dependent on the labels of its neighboring sites in $\mathcal{N}_i$ that have been labeled.

While in the standard MRF the conditional distribution $P(f_i \mid f_{\mathcal{S}-\{i\}})$ is a Gibbs distribution of clique potentials, the strong MRF, based on the strong Markovianity (2.80) models $P(f_i \mid f_{\mathcal{D}-\{i\}})$ without the potentials. Therefore, it can be used to develop a nonparametric model. It has been used for texture classification in which images contain other textures of unknown origins.

2.11 $\mathcal{K}$-MRF and Nakagami-MRF Models

In a GMRF model, the joint prior distribution $p(f)$ is multivariate normal. In an analysis of ultrasound envelopes of backscattered echo and spatial interaction, the prior $p(f)$ takes the form of a $\mathcal{K}$-distribution or Nakagami distribution. Therefore, the $\mathcal{K}$-MRF (Bouhlel et al. 2004) and Nakagami-MRF (Bouhlel et al. 06a) have been proposed for the modeling problems therein.

A $\mathcal{K}$-distribution (Jakeman and Pusey 1976) with parameters (α, β) has the form

$$\mathcal{K}_{\alpha,\beta}(x) = \frac{2\beta}{\Gamma(\alpha)} \left(\frac{\beta x}{2} \right)^2 B_{\alpha-1}(\beta x) \quad \forall x \in R_+ \tag{2.81}$$

where $\Gamma(\cdot)$ is the Gamma function, α is the shape parameter, $B_{\alpha-1}(\cdot)$ is a modified Bessel function of the second kind of order $(\alpha - 1)$, and β is the scaling parameter of the $\mathcal{K}$-distribution.

The conditional density of a $\mathcal{K}$-MRF model is also a $\mathcal{K}$-distribution (Bouhlel et al. 2004; Bouhlel et al. 06b)

$$p(f_i \mid f_{\mathcal{N}_i}) \propto \mathcal{K}_{\alpha_i,\beta}(f_i) \tag{2.82}$$

where the parameter α_i is given by

$$\alpha_i = a_i + 1 + \sum_{i' \in \mathcal{N}_i} b_{i,i'} \ln f_{i'} \tag{2.83}$$

where the real valued a_i, $b_{i,i'}$, and β can be estimated from examples by solving the following system of equations (Bouhlel et al. 06b)

$$E[f_i \mid f_{\mathcal{N}_i}] = \frac{2\Gamma(\alpha_i + 0.5)}{\beta\Gamma(\alpha_i)}\Gamma(1.5) \tag{2.84}$$

$$E[f_i^2 \mid f_{\mathcal{N}_i}] = 4\frac{\alpha_i}{\beta^2} \tag{2.85}$$

where $E[\cdot]$ is the mathematical expectation.

The Nakagami distribution, with parameters (α, β), has the form

$$\mathcal{N}_{\alpha,\beta}(x) = \frac{2\beta^\alpha}{\Gamma(\alpha)} x^{2\alpha-1} \exp(-\beta x^2) \quad \forall x \in R_+ \tag{2.86}$$

where $\Gamma(\cdot)$ is the Gamma function. The conditional density of a Nakagami-MRF model is also a Nakagami distribution (Bouhlel et al. 06a)

$$p(f_i \mid f_{\mathcal{N}_i}) \propto \mathcal{N}_{\alpha_i,\beta}(f_i) \tag{2.87}$$

where the parameter α_i is given by

$$\alpha_i = \frac{1}{2}\left(a_i + 1 + \sum_{j \in \mathcal{N}_i} b_{i,i'} \ln f_{i'} \right) \tag{2.88}$$

where the parameters a_i, $b_{i,i'}$, and β can be estimated from examples by solving the system of equations

$$E[f_i^2 \mid f_{\mathcal{N}_i}] = \frac{\alpha_i}{\beta} \tag{2.89}$$

$$D[f_i^2 \mid f_{\mathcal{N}_i}] = \frac{\alpha_i}{\beta^2} \tag{2.90}$$

where $E[\cdot]$ is the variance.

2.12 Graphical Models: MRF's versus Bayesian Networks

The MAP-MRF approach models MRF problems defined on undirected graphs. The graphical model (GM) (or probabilistic graphical model) approach incorporates the probability theory in the manipulation of more general graphs (Pearl 1988; Jordan 1998; Jensen 2001). The graph theory part represents a complex system by a graph built on many simpler parts linked by relations and provides the data structure required by efficient algorithms. The probability theory part manipulates on the graph, provides interfaces between the model and data, and ensures consistency therein. A GM can be undirected or directed.

An undirected GM, also called a Markov network, is equivalent to a pairwise or second-order MRF. It can be denoted as $\mathcal{G} = (\mathcal{S}, \mathcal{N})$, where each node (site) is associated with a label, with or without an observation on the node, and the relationships between nodes are modeled via the neighborhood system $\mathcal{N}$.

A directed GM is denoted as $\mathcal{G} = (\mathcal{S}, \mathcal{M})$, where $\mathcal{S}$ is the set of nodes and $\mathcal{M}$ is the "parent system". If $i' \in \mathcal{S}$ is a parent node of i, then there is a directed edge from i' to i. All the nodes that i is dependent on constitute the parent set $\mathcal{M}_i$. All the $\mathcal{M}_i$s constitute the parent system $\mathcal{M}$. Such a GM can used to depict causal relationships.

A directed GM is a Bayesian network (BN) or belief network when the graph is acyclic, meaning there are no loops in the directed graph. In a BN, a node i is associated with a random variable taking a discrete or continuous value f_i. The labels and observations can be defined on disjoint subsets of nodes and related through the parent system $\mathcal{M}$. Figure 2.6 illustrates differences between an MRF and a Bayesian network. Both are referred to as inference network in machine learning literature.

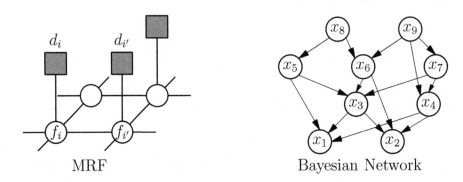

MRF   Bayesian Network

Figure 2.6: MRF vs. BN. Left: A part of an MRF, with sites (circles), labels (f_i's), and observations (squares). Right: A simple instance of a BN, where nodes x_1 and x_2 are the observation variables and the other nodes are latent variables whose values are to be inferred. The arrows show the dependency of the nodes (note that the observation nodes x_1 and x_2 depend on no other nodes).

The relationships in a BN can be described by local conditional probabilities $P(x_i \mid \mathcal{M}_i)$; if i has no parents, as for observation nodes, its local probability is considered unconditional as $P(x_i \mid \mathcal{M}_i) = P(x_i)$. The joint distribution for a BN can be expressed as the product of the local conditional probabilities

$$P(f) = P(f_1, \ldots, f_M) = \prod_{i=1}^{M} P(f_i \mid f_{\mathcal{M}_i}) \qquad (2.91)$$

Similar to MRF's, issues in BNs include representation, inference (finding the optimal solution), learning (parameter estimation), decision, and application. The inference can be formulated as a maximum posterior probability or maximum marginal probability problem. Efficient algorithms exist for inference (such as belief propagation; see Section 9.3.3) and learning in BNs (Jordan 1998).

Chapter 3

Low-Level MRF Models

Low-level processing is performed on regular lattices of images. The set of sites $\mathcal{S} = \{1, \ldots, m\}$ indexes image pixels in the image plane. The set $\mathcal{L}$ contains continuous or discrete labels for the pixels. Therefore the problems fall into categories LP1 and LP2. Most existing MRF vision models are for low-Level processing. MRF models for image restoration and segmentation have been studied most comprehensively (Hassner and Slansky 1980; Hansen and Elliott 1982; Chellappa and Kashyap 1982; Derin et al. 1984; Geman and Geman 1984; Chellappa 1985; Cohen and Cooper 1987; Dubes et al. 1990). Surface reconstruction can be viewed as a more general case than restoration in that the data can be sparse; i.e., available at certain locations of the image lattice (Marroquin 1985; Blake and Zisserman 1987; Marroquin et al. 1987; Chou and Brown 1990; Geiger and Girosi 1991). MRF's can play a fuller role in texture analysis because textured images present anisotropic properties (Chellappa and Kashyap 1982; Cross and Jain 1983; Elliott et al. 1984; Derin and Cole 1986; Derin and Elliott 1987; Lakshmanan and Derin 1989; Hu and Fahmy 1987; Pappas 1992; Won and Derin 1992). The treatment of an optical flow as an MRF (Koch 1988; Shulman and Herve 1989; Harris et al. 1990) is similar to that of restoration and reconstruction. Edge detection (Geman and Geman 1984; Blake and Zisserman 1987; Geman et al. 1990) is often addressed along with restoration, reconstruction, and analysis of other image properties such as texture, flow, and motion. We can also find low-Level applications of MRF's such as active contours, (Kass et al. 1987; Amini et al. 1988; Storvik 1994) deformable templates, (Mardia et al. 1991; Mardia et al. 1992) data fusion, and visual integration (Szeliski 1989; Clark and Yuille 1990; Nadabar and Jain 1995).

In this chapter, we formulate various MAP-MRF models for low-Level vision following the procedure summarized in Section 1.3.4. We begin with the prototypical MAP-MRF models for image restoration. The presentation therein introduces the most important concepts in MRF modeling. After that, the formulations for the image restoration are extended to a closely related

S.Z. Li, *Markov Random Field Modeling in Image Analysis*, Advances in Pattern Recognition, DOI: 10.1007/978-1-84800-279-1_3, © Springer-Verlag London Limited 2009

problem, surface reconstruction, in which the observation may be sparser. The MRF models for boundary detection, texture and optical flow will be described subsequently. How to impose the smoothness constraint while allowing discontinuities is an important issue in computer vision (Terzopoulos 1983b; Geman and Geman 1984; Blake and Zisserman 1987) that deserves a thorough investigation; it is the topic of Chapter 5. Another important issue, MRF parameter estimation in low-Level vision, will be discussed in Chapter 7.

3.1   Observation Models

In low-Level vision, an observation $d = \{d_1, \ldots, d_m\}$ is a rectangular array of pixel values. In some cases, the observation may be sparse; that is, it is available at $i \in \mathcal{A} \subset \mathcal{S}$. Every pixel takes a value d_i in a set $\mathcal{D}$. In practice, $\mathcal{D}$ is often the set of integers encoded by a byte (8 bits), as a result of normalization and quantization, so that $\mathcal{D} = \{0, 1, \ldots, 255\}$.

An observation d can be considered as a transformed and degraded version of an MRF realization f. The transformation may include geometric transformation and blurring, and the degradation is due to random factors such as noise. These determine the conditional distribution $p(d \mid f)$ or the likelihood of f. A general observation model can be expressed as

$$d = \varphi(B(f)) \odot e \qquad (3.1)$$

where B is a blurring effect, φ is a transformation which can be linear or nonlinear and deterministic or probabilistic, e is the sensor noise, and $\odot$ is an operator of addition or multiplication. In practice, a simple observation model of no blurring, linear transformation, and independent additive Gaussian noise is often assumed. Each observed pixel value is assumed to be the sum of the true gray value and independent Gaussian noise

$$d_i = \varphi(f_i) + e_i \qquad (3.2)$$

where $\varphi(\cdot)$ is a linear function and $e_i \sim N(0, \sigma_i^2)$. The probability distribution of d conditional on f, or the likelihood of f, is

$$p(d \mid f) = \frac{1}{\prod_i^m \sqrt{2\pi\sigma_i^2}} e^{-U(d \mid f)} \qquad (3.3)$$

where

$$U(d \mid f) = \sum_{i \in \mathcal{S}} (\varphi(f_i) - d_i)^2 / [2\sigma_i^2] \qquad (3.4)$$

is the likelihood energy. Obviously, (3.3) is a special form of Gibbs distribution whose energy is purely due to single-site cliques in the zero-th order neighborhood system (where the radius $r = 0$) with clique potentials being

$[\varphi(f_i) - d_i]^2/[2\sigma_i^2]$. If the noise distribution is also homogeneous, then the deviations are the same for all $i \in \mathcal{S}$; i.e., $\sigma_i = \sigma$.

The function $\varphi(\cdots)$ maps a label f_i to a real gray value, where f_i may be numerical or symbolic and continuous or discrete. When f_i is symbolic, for example, representing a texture type, the ordering of the MRF labels is usually not defined unless artificially. Without loss of generality, we can consider that there is a unique numerical value for a label f_i and denote $d_i = \varphi(f_i) + e_i$ simply as $d_i = f_i + e_i$. Then the likelihood energy becomes

$$U(d \mid f) = \sum_{i \in \mathcal{S}} (f_i - d_i)^2/[2\sigma^2] \tag{3.5}$$

for i.i.d. Gaussian noise.

A special observation model for a discrete label set is random replacement. An f_i is transformed into d_i according to the likelihood probability

$$P(d_i = k \mid f_i = k') = \begin{cases} p & = e^{\ln p} & \text{if } k = k' \\ \frac{1-p}{M-1} & = e^{\ln \frac{1-p}{M-1}} & \text{otherwise} \end{cases} \tag{3.6}$$

where $\mathcal{L} = \mathcal{D} = \{0, \ldots, M-1\}$ contains unordered labels for f_i and d_i, $0 \le p \le 1$, $k \in \mathcal{D}$, and $k' \in \mathcal{L}$. In this model, a label value remains unchanged with probability p and changes to any other value with equal probability $\frac{1-p}{M-1}$. This describes the transition from one state to another. The simplest case of this is the random flipover of binary values.

A useful and convenient model for both the underlying MRF and the observation is the hierarchical GRF model (Derin and Cole 1986; Derin and Elliott 1987). There two levels of Gibbs distributions are used to represent noisy or textured regions. The higher-level Gibbs distribution, which is usually an MLL, characterizes the blob-like regions formation process, while the lower-level Gibbs distribution describes the filling-in, such as noise or texture, in each region. This will be described in detail in Section 3.4.1.

3.2 Image Restoration and Reconstruction

The purpose of image restoration is to recover the true pixel values, f, from the observed (noisy) image pixel values, d. When the image $d(x, y)$ is considered as a noise-added surface, the problem is to restore the underlying surface $f(x, y)$.

3.2.1 MRF Priors for Image Surfaces

The underlying surface from which f is sampled is a graph surface $\{f(x, y)\}$ defined on a continuous domain, where $f(x, y)$ is the height of the surface at (x, y). The first factor affecting the specification of the MRF prior distributions is whether $f(x, y)$ takes a continuous or discrete value. We are interested in $\{f(x,y)\}$, which is either piecewise continuous or piecewise constant; i.e., being piecewise because of the discontinuities involved. In the image representation, $\{f(x,y)\}$ is sampled at an image lattice, giving sampled surface heights $\{f_{i,j}\}$. In the subsequent presentation, the double subscript (i,j) is replaced by the single subscript i and the MRF sample is denoted as $f = \{f_1, \ldots, f_m\}$ unless there is a need for the elaboration.

The set of the sampled heights $f = \{f_1, \ldots, f_m\}$ is assumed to be a realization of an MRF; that is, f_i at a particular location i depends on those in the neighborhood. According to MRF-Gibbs equivalence, specifying the prior distribution of an MRF amounts to specifying the clique potential functions V_c in the corresponding Gibbs prior distribution (2.13). Depending on whether $\{f(x,y)\}$ is continuous, piecewise continuous, or piecewise constant, there will be different Gibbs distributions for the MRF.

MRF Prior for Piecewise Constant Surfaces

Piecewise constant surfaces, or homogeneous blob-like regions, can be properly characterized by the MLL (multilevel logistic), more specifically the homogeneous and isotropic MLL model described in Section 2.3. For cliques containing more than one site, the clique potentials are defined as[1]

$$V_c(f) = \begin{cases} 0 & \text{if all sites in } c \text{ have the same label} \\ -\zeta_c & \text{otherwise} \end{cases} \quad (3.7)$$

where $\zeta_c < 0$ is a constant dependent on c. That "all sites in c have the same label" (that is, all $\{f_i \mid i \in c\}$ are the same) means the entire smoothness of labels f on the clique c. Any violation of the entire smoothness incurs a penalty of the positive number $-\zeta_c > 0$. Because the more probable configurations are those with higher $P(f)$ or lower $U(f)$ values, the MLL model (3.7) favors smooth f.

For single-site cliques, the clique potentials depend on the label assigned to the site

$$V_c(f) = V_1(f_i) = \alpha_l \quad \text{if } f_i = l \in \mathcal{L}_d \quad (3.8)$$

where α_l is the penalty against which f_i is labeled l; see (2.43). The higher α_l is, the fewer pixels will be assigned the value l. This has the effect of controlling the percentage of sites labeled l.

A special case of (3.7) is where v_c is nonzero only for the pair-site cliques and zero for all the other types. In this case, $\zeta_c = 0$ for all c of size $\#c > 2$; when $\#c = 2$, it is

$$V_c(f) = V_2(f_i, f_{i'}) = v_{20}[1 - \delta(f_i - f_{i'})] \quad (3.9)$$

where $\delta(\cdot)$ is the Kronecker delta function and v_{20} is the penalty against nonequal labels on two-site cliques. The prior energy is the sum of all the clique potentials

---

[1]The definition in (3.7) is effectively the same as (2.42) for the restoration purpose.

$$U(f) = \sum_{i \in \mathcal{S}} \sum_{i' \in \mathcal{N}_i} v_{20}[1 - \delta(f_i - f_{i'})] \qquad (3.10)$$

where "$\sum_{i \in \mathcal{S}} \sum_{i' \in \mathcal{N}_i}$" is equivalent to "$\sum_{\{i,i'\} \in \mathcal{C}_2}$". This simple form has been used for the restoration of piecewise constant images by Geman and Geman (1984), Elliott, Derin, Cristi, and Geman (1984), Derin and Cole (1986), Derin and Elliott (1987), Leclerc (1989) and Li (1990a).

MRF Prior for Piecewise Continuous Surfaces

In specifying prior clique potentials for continuous surfaces, only pair-site clique potentials are normally used for piecewise continuous surfaces. In the simplest case, that of a flat surface, they can be defined by

$$V_2(f_i, f_{i'}) = g(f_i - f_{i'}) \qquad (3.11)$$

where $g(f_i - f_{i'})$ is a function penalizing the violation of smoothness caused by the difference $f_i - f_{i'}$. For the purpose of restoration, the function g is generally even

$$g(\eta) = g(-\eta) \qquad (3.12)$$

and nondecreasin

$$g'(\eta) \geq 0 \qquad (3.13)$$

on $[0, +\infty)$. The derivative of g can be expressed

$$g'(\eta) = 2\eta h(\eta) \qquad (3.14)$$

One may choose appropriate g functions to impose either complete smoothness or piecewise smoothness.

When the underlying $\{f(x, y)\}$ contains no discontinuities (i.e., continuous everywhere) $g(\cdot)$ is usually a quadratic function

$$g(\eta) = \eta^2 \qquad (3.15)$$

Under the quadratic g, the penalty to the violation of smoothness is proportional to $(f_i - f_{i'})^2$. The quadratic pair-site potential function is not suitable for prior surface models in which the underlying $\{f(x, y)\}$ is only piecewise continuous. At discontinuities, $\|f_i - f_{i'}\|$ tends to be very large and the quadratic function brings about a large smoothing force when the energy is being minimized, giving an oversmoothed result.

To encode piecewise smoothness, g has to satisfy a necessary condition[2]

$$\lim_{\eta \to \infty} |g'(\eta)| = C < \infty \qquad (3.16)$$

where $C \in [0, \infty)$ is a constant. This condition means that g should saturate at its asymptotic upper bound when $\eta \to \infty$ to allow discontinuities. A

---

[2]This equation will be referred to again as (5.26).

possible choice is the truncated quadratic used in the line process model
(Geman and Geman 1984; Marroquin 1985; Blake and Zisserman 1987), in
which

$$g(\eta) = \min\{\eta^2, \alpha\} \qquad (3.17)$$

Equation (3.17) satisfies condition (3.16). How to properly choose the g (or
h) function for discontinuity-adaptive image restoration is the subject to be
studied in Chapter 5.

A significant difference between (3.11) and (3.9) is due to the nature of
label set $\mathcal{L}$. For piecewise constant restoration where labels are considered
unordered, the difference between any two labels f_i and $f_{i'}$ is symbolic; e.g.,
in the set of $\{yes, no\}$. In the continuous case where labels are ordered by the
relation (e.g., "smaller than"), the difference takes a continuous quantitative
value. We may define a "softer" version of (3.9) to provide a connection between continuous and discrete restoration. For piecewise constant restoration
of a two-level image, for example, we may define

$$V_2(f_i, f_{i'}) = v_{20}(1 - e^{-\eta^2/\gamma}) \qquad (3.18)$$

which in the limit is $\lim_{\gamma \to 0} V_2(f_i, f_{i'}) = v_{20}[1 - \delta(f_i - f_{i'})]$.

In the prior models above, the clique potential is a function of the *first-
order* difference $f_i - f_{i'}$, which is an approximation of f_i'. They have a tendency
to produce surfaces of constant or piecewise constant height, though in the
continuous restoration continuous variation in height is allowed. They are not
very appropriate models for situations where the underlying surface is not
flat or piecewise flat.

According to the discussions in Section 2.4, we may design potential functions $g(f_i')$, $g(f_i'')$, and $g(f_i''')$ for surfaces of constant gray level (horizontally
flat), constant gradient (planar but maybe slanted) and constant curvature,
respectively. It is demonstrated (e.g., in (Geman and Reynolds 1992)), that
potentials involving f'' give better results for surfaces of constant gradient.
Higher-order models are also necessary for image reconstruction from sparse
data (Grimson 1981; Terzopoulos 1983a; Poggio et al. 85a; Blake and Zisserman 1987).

3.2.2   Piecewise Constant Restoration

In piecewise constant restoration, $\mathcal{L}$ consists of discrete values. The task is
to recover the true configuration f from the observed image d. This is akin
to region segmentation because f in effect partitions the set of lattice sites
into mutually exclusive regions. In the MAP-MRF framework, the optimal f
is the one that minimizes the posterior energy.

Deriving Posterior Energy

Let us derive the posterior energy using the four-step procedure summarized
in Section 1.3.4:

1. Define a neighborhood system and the set of cliques for it. The neighborhood system is defined according to (2.2). The set of cliques for the 4-neighborhood system is shown in Fig. 2.1(d)-(e). For the 8-neighborhood system, cliques in Fig. 2.1(f)–(h) are also included.

2. Define the prior clique potential functions in the Gibbs prior distribution (2.13)–(2.15). For the MLL prior, it is (3.9).

3. Derive the likelihood energy from the observation model. Assume the i.i.d. additive Gaussian model. The likelihood function $p(f \mid d)$ takes the form of (3.5), with f_i taking a discrete value.

4. Add the prior energy $U(f)$ and the likelihood energy $U(d \mid f)$ to yield the posterior energy

$$E(f) = U(f \mid d) = \sum_{i \in \mathcal{S}} (f_i - d_i)^2 / [2\sigma^2] + \sum_{i \in \mathcal{S}} \sum_{i' \in \mathcal{N}_i} v_{20}[1 - \delta(f_i - f_{i'})]$$

(3.19)

Note that

$$\sum_{i' \in \mathcal{N}_i} [1 - \delta(f_i - f_{i'})] = \#\{f_{i'} \neq f_i \mid i' \in \mathcal{N}_i\}$$

(3.20)

is simply the number of neighboring sites whose label $f_{i'}$ is different from f_i.

In this problem, as well as in other labeling problems in this chapter, the parameters, such as σ^2 and v_{20} here, are assumed known. The more advanced topic of labeling with unknown noise and MRF parameters will be discussed in Section 7.2.1.

Energy Minimization

Because $\mathcal{L}_d$ is discrete, minimizing (3.19) is a combinatorial problem. The minimal solution f^* is the optimally restored image in the configuration space $\mathbb{F} = \mathcal{L}_d^m$. The simplest algorithm is the steepest local energy descent or the "greedy" method. It proceeds as follows: Start with an initial configuration $f^{(0)} \in \mathbb{F}$. For site i, choose the new label $f_i^{(t+1)}$ among all $f_i^{(t+1)} \in \mathcal{L}_d$, $f_i^{(t+1)} \neq f_i^{(t)}$, that minimizes $E(f^{(t+1)}) - E(f^{(t)})$ locally. The iteration continues until no further energy descent is possible. An example of such greedy algorithms is the iterative conditional modes (ICM) (Besag 1986); see Section 9.3.1 for details.

This simple algorithm finds a local energy minimum whose quality depends on the initial estimate $f^{(0)}$. Global minimization algorithms such as simulated annealing (Geman and Geman 1984) need to be used if global solutions are required. This is the subject of Chapter 10.

3.2.3 Piecewise Continuous Restoration

In piecewise continuous restoration, $\mathcal{L}$ consists of a real interval. The task is to recover the underlying $\{f(x, y)\}$, which is a piecewise continuous graph surface, being piecewise because discontinuities may exist in $\{f(x, y)\}$.

Deriving the Posterior Energy

The first step is the same as that for piecewise constant restoration. The second step is to define the prior clique potential functions. Here, the single-site clique potentials are set to zero; only the pair-site cliques are considered, with potential function (3.11) rewritten as

$$V_2(f_i, f_{i'}) = g(f_i - f_{i'}) \tag{3.21}$$

To impose the piecewise smoothness prior, the g function is required to satisfy condition (3.16).

The third step is to derive the likelihood energy. When the noise is i.i.d. additive Gaussian, $e_i \sim N(0, \sigma^2)$, the likelihood energy is given in (3.5) with continuously valued f_i.

In the fourth step, the prior energy $U(f)$ and the likelihood energy $U(d \mid f)$ are added to obtain the posterior energy

$$E(f) = \sum_{i \in \mathcal{S}} (f_i - d_i)^2 / [2\sigma^2] + \sum_{i \in \mathcal{S}} \sum_{i' \in \mathcal{N}_i} g(f_i - f_{i'}) \tag{3.22}$$

or equivalently

$$E(f) = \sum_{i \in \mathcal{S}} (f_i - d_i)^2 + \lambda \sum_{i \in \mathcal{S}} \sum_{i' \in \mathcal{N}_i} g(f_i - f_{i'}) \tag{3.23}$$

where $\lambda = 2\sigma^2$. When no noise is present ($\lambda = 0$), only the first term is effective, so the MAP solution is exactly the same as the data, $f^* = r$, which is also the maximum likelihood solution. As λ becomes larger, the solution becomes more influenced by the second (i.e., the smoothness) term. Figure 3.1 shows the effect of λ on f^* with the quadratic g function. It can be seen that the larger λ is, the smoother f^* is.

In the 1D case where $\mathcal{S} = \{1, \dots, m\}$ can be an ordered set such that the nearest neighbor set for each interior site i is $\mathcal{N}_i = \{i - 1, i + 1\}$, (3.22) can be written as

$$E(f) = \sum_{i=1}^{m} (f_i - d_i)^2 + 2\lambda \sum_{i=2}^{m} g(f_i - f_{i-1}) \tag{3.24}$$

(The terms involving the two boundary points are omitted.) When $f = \{f_1, \dots, f_m\}$ is the sample of a continuous function $f = f(x)$, $a \leq x \leq b$, (3.24) is an approximation to the regularization energy

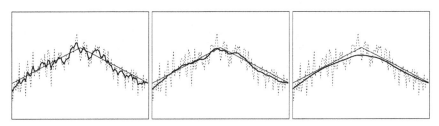

Figure 3.1: Solution f^* obtained with $\lambda = 2$ (left), 20 (middle), and 200 (right). The original signals are the thinner solid lines, noisy data the dotted lines, and the minimal solutions the thicker solid lines.

$$E(f) = \sum_{i=1}^{m}(f_i - d_i)^2 + 2\lambda \int_a^b g(f'(x))\mathrm{d}x \qquad (3.25)$$

(see Section 1.3.3). When $g(\eta)$ is as in (3.15), then (3.25) is the energy for the standard quadratic regularization (Tikhonov and Arsenin 1977; Poggio et al. 85a). When it is defined as (3.17), the energy is called a *weak string* and its two-dimensional counterpart is called a *weak membrane* (Blake and Zisserman 1987). Regularization of ill-posed problems will be discussed further in surface reconstruction.

In the 2D case where $\mathcal{S} = \{(i,j) \mid 1 \le i, j \le n\}$ is a lattice, each site (i,j) (except at the boundaries) has four nearest neighbors $\mathcal{N}_{i,j} = \{(i-1,j), (i+1,j), (i,j-1), (i,j+1)\}$. The corresponding posterior energy is

$$E(f) \;=\; \sum_{i,j}(d_{i,j} - f_{i,j})^2 + \lambda \sum_{i,j} \sum_{(i',j') \in \mathcal{N}_{i,j}} g(f_{i,j} - f_{i',j'}) \qquad (3.26)$$

This is a direct extension of (3.24).

Energy Minimization

The MAP solution is defined as $f^* = \arg\min_f E(f)$. The simplest way to find the f^* is to perform gradient descent. Start with an initial configuration and iterate with

$$f^{(t+1)} \leftarrow f^{(t)} - \mu \nabla E(f^{(t)}) \qquad (3.27)$$

where $\mu > 0$ is a step size and $\nabla E(f)$ is the gradient of the energy function until the algorithm converges to a point f for which $\nabla E(f) = \mathbf{0}$. Consider $E(f) = \sum_{i=1}^{m}(f_i - d_i)^2 + \lambda \sum_i \sum_{i' \in \mathcal{N}_i} g(f_i - f_{i-1})$. The gradient is composed of the components

$$\frac{\partial E(f)}{\partial f_i} \;=\; 2(f_i - d_i) + \lambda \sum_{i' \in \mathcal{N}_i} g'(f_i - f_{i'})$$

$$\;=\; 2(f_i - d_i) + 2\lambda \sum_{i' \in \mathcal{N}_i} (f_i - f_{i'})h(f_i - f_{i'}) \qquad (3.28)$$

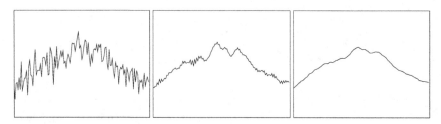

Figure 3.2: Minimizing $E(f)$. From left to right: the initial f set to the observed data d, f after five iterations and the final result f^*.

where $g'(\eta) = 2\eta h(\eta)$ as assumed in (3.14). Figure 3.2 shows restored curves at three stages of an iterative minimization process. In the 2D case, it is

$$\frac{\partial E(f)}{\partial f_{i,j}} \;=\; 2(f_{i,j} - d_{i,j}) + 2\lambda \sum_{(i',j')\in\mathcal{N}_{i,j}} (f_{i,j} - f_{i',j'})\, h(f_{i,j} - f_{i',j'}) \quad (3.29)$$

Alternatively, one may directly solve the system of equations

$$\nabla E(f^*) = \mathbf{0} \qquad\qquad (3.30)$$

which is composed of m simultaneous equations $\frac{\partial E(f)}{\partial f_i} = 0$ $(i = 1, \ldots, m)$.

When $E(f)$ is convex w.r.t. f, there is a unique global energy minimum f^* and the convergence to the global is guaranteed by the gradient-descent algorithm. When it is not, as with the truncated quadratic potential function (3.17), there are local minima in the energy landscape and the gradient-descent algorithm only gives a local minimum. Some global optimization techniques may be used to overcome this difficulty, such as simulated annealing (Geman and Geman 1984) and graduated nonconvexity (Blake and Zisserman 1987) (see Chapter 10).

3.2.4   Surface Reconstruction

Surface reconstruction also recovers surface values but from sparse data. It consists of two main components: restoration and interpolation. The situation for sparse data are typical of stereopsis (Marr and Poggio 1979; Grimson 1981; Mayhew and Frisby 1981). There, a pair of images taken from two slightly different viewpoints are compared and matched to give the corresponding points. The depth values at the matched points are computed using triangulation. Since the matched points are sparsely distributed, so are the computed depth values. The sparseness brings about more uncertainties, which have to be resolved by interpolation.

Let $d = \{d_i \mid i \in \mathcal{A}\}$ be a set of available depth values, where $d_i = d(x_i, y_i)$ is the depth value at location (x_i, y_i) and $\mathcal{A} \subset \mathcal{S}$ is the set of subscripts for

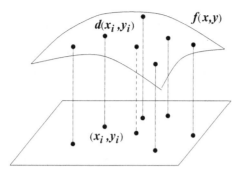

Figure 3.3: Surface reconstruction from sparse data.

the available data. Assume the observation model

$$d(x_i, y_i) = f(x_i, y_i) + e_i \tag{3.31}$$

where e_i is identical independent Gaussian noise and $f(x_i, y_i)$ is the true value of the underlying surface. Figure 3.3 illustrates (x_i, y_i), $d(x_i, y_i)$, and $f(x, y)$. Our aim is to recover the underlying surface $\{f(x, y)\}$ from the sparse data d. When continuous solutions are considered, it may be reasonable to look at the problem from the analytic viewpoint.

Regularization Solution

From the analytic viewpoint, the reconstruction problem is mathematically *ill-posed* (Tikhonov and Arsenin 1977; Poggio et al. 85a) at least in that the solution is not unique (see Section 1.3.3). Regularization methods can be used to convert the problem into a well-posed one.

In reconstruction from sparse data, the regularizer (i.e., the smoothness term) has to involve the second- or higher-order derivatives. This is because regularizers with first-order derivatives can be creased (e.g., in the absence of data at lattice points) without any increase in the associated energy. Regularizers with second- or higher-order derivatives are sensitive to creases. The smoothness term with the second-order derivative is

$$U(f) = \int [f''(x)]^2 \mathrm{d}x \tag{3.32}$$

Its 2D counterpart, the plate with quadratic variation, is

$$U(f) = \int \{[f_{xx}(x, y)]^2 + 2[f_{xy}(x, y)]^2 + [f_{yy}(x, y)]^2\} \mathrm{d}x\mathrm{d}y \tag{3.33}$$

For spatially discrete computation where the continuous domain is sampled at regular lattice sites, the derivatives are approximated by the finite differences. A discrete form for (3.33) is

$$
\begin{aligned}
U(f) \; = \; & \sum_u \sum_v [f_{i+1,j} - 2f_{i,j} + f_{i-1,j}]^2 + \\
& 2\sum_u \sum_v [f_{i,j} - f_{i-1,j} - f_{i,j-1} + f_{i-1,j-1}]^2 + \\
& \sum_u \sum_v [f_{i,j+1} - 2f_{i,j} + f_{i,j-1}]^2
\end{aligned}
\tag{3.34}
$$

For piecewise smoothness, the prior terms may be defined with a g function having the property (3.16). In the 1D case,

$$
U(f) = \int g(f''(x))\mathrm{d}x
\tag{3.35}
$$

In the 2D case,

$$
U(f) = \int \left[g(f_{xx}^2(x,y)) + g(f_{xy}^2(x,y)) + g(f_{yy}^2(x,y)) \right] \mathrm{d}x\mathrm{d}y
\tag{3.36}
$$

They are called the weak rod and weak plate, respectively. The latter has been used by Geman and Reynolds (1992) to restore damaged movie film.

Compared with the first-order models such as strings and membranes, the second-order models are more capable of interpolating under sparse data, but they are less flexible in dealing with discontinuities and more expensive to compute. It is also reported that the mixture of first- and second-order prior energies, which is akin to a "spring under tension" (Terzopoulos 1983b; Terzopoulos 1986b), performs poorly (Blake and Zisserman 1987).

3.3   Edge Detection

Edges correspond to abrupt changes or discontinuities in certain image properties between neighboring areas. The image properties may be nontexture or texture. In this section, we are interested in nontexture edges due to changes in image intensity, such as jump edges and roof edges. Jump edges correspond to the discontinuities in the underlying surface $\{f(x,y)\}$ or to the maxima and minima[3] of its first (directional) derivative, f'. Roof edges correspond to the discontinuities in the first derivative, f', or to the maxima and minima of its second derivative, f''. However, when pixel sites are spatially quantized (in the x–y plane), the pixel values are subject to noise. In this case, discontinuities, maxima, and minima are not well defined (their definitions become part of the solution to the edge detection problem). These cause problems for edge detection.

---

[3]More exactly, positive and negative impulses.

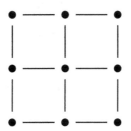

Figure 3.4: A lattice of sites (dots) and its dual sites (bars) in the 4-neighborhood system.

The first step in edge detection (Torre and Poggio 1986; Canny 1986) is to estimate the derivatives from noisy and spatially quantized data. The second step is to detect the zeros and extrema of the estimated derivative function. The final step is to link edges, which are detected based on local changes, to form boundaries which are coherent in a more global sense. The final step also relates to perceptual organization (Lowe 1985). This section concerns only the first two steps, by which pixel locations where sharp changes occur are marked.

Edge detection is closely related to image restoration and surface reconstruction involving discontinuities. There, we are mainly interested in removing noise and getting (piecewise) smooth surfaces; although discontinuities are also taken care of to avoid oversmoothing, they are not required to be marked explicitly. In this subsection, we modify the piecewise continuous restoration model to obtain explicit labeling of edges.

3.3.1 Edge Labeling Using Line Process

Rewrite the posterior energy (3.23) for the piecewise continuous restoration

$$E(f) = \sum_{i \in S} (f_i - d_i)^2 + \lambda \sum_{i \in S} \sum_{i' \in \mathcal{N}_i} g(f_i - f_{i'}) \qquad (3.37)$$

To impose the piecewise smoothness prior, the g function must have the property (3.16). The $g(f_i - f_{i'})$ is modified for the purpose of explicitly marking edges.

In addition to the existing MRF for pixel values, we introduce another coupled MRF, called *line process* (LP) (Geman and Geman 1984; Marroquin 1985), in which each label takes a value on $\{0,1\}$ to signal the occurrence of edges. The two coupled MRF's are defined on two spatially interwoven sets as illustrated in Fig. 3.4 (assuming the 4-neighborhood system). One set is the existing lattice for the intensity field, and the other is the dual lattice for the introduced edge field. A possible edge may exist between each pair of neighboring pixel sites.

We use the following notations. Denote the lattice of the m pixel sites by

$$\mathcal{S}^P = \{i \mid i = 1, \ldots, m\} \tag{3.38}$$

(corresponding to the dots in Fig. 3.4) and the dual lattice (corresponding to the bars) by

$$\mathcal{S}^E = \{(i, i') \mid i, i' \in \mathcal{S}^P; i \bowtie i'\} \tag{3.39}$$

where $i \bowtie i'$ means i and i' are neighbors. Let f_i^P, $i \in \mathcal{S}^P$, be an intensity label taking a value in a real interval $\mathcal{L}^P$. Let $f_{i,i'}^E$, $(i, i') \in \mathcal{S}^E$, be an edge label, also called a line process variable (Geman and Geman 1984), taking a value in $\mathcal{L}^E = \{0, 1\}$, with 0 and 1 representing the absence or presence of an edge, respectively.

The interaction between the two coupled MRF's is determined by the joint probability $P(f^P, f^E)$ or the prior energy $U(f^P, f^E)$. Consider the energy (3.23) with the potential function $g(f_i^P - f_{i'}^P) = \min\{(f_i^P - f_{i'}^P)^2, \alpha\}$ defined in (3.17). It was for piecewise continuous restoration without explicit labeling of the edge. We modify it into a function of both intensity and edge variables

$$g(f_i^P, f_{i'}^P, f_{i,i'}^E) = (f_i^P - f_{i'}^P)^2 (1 - f_{i,i'}^E) + \alpha f_{i,i'}^E \tag{3.40}$$

where i and i' are neighbors. To minimize the $g(f_i^P, f_{i'}^P, f_{i,i'}^E)$ above alone, the intensity labels and the edge labels are determined as follows: If $[f_i^P - f_{i'}^P]^2 < \alpha$, then it is cheaper to pay the price $[f_i^P - f_{i'}^P]^2$ and set $f_{i,i'}^E = 0$; otherwise it is more economical to set $f_{i,i'}^E = 1$ to insert an edge at the cost of α.

The prior energy with the g function in (3.40) is

$$U(f^P, f^E) = \sum_{i \in \mathcal{S}^P} \sum_{i' \in \mathcal{N}_i} g(f_i^P, f_{i'}^P, f_{i,i'}^E) \tag{3.41}$$

It can be expressed as

$$U(f^P, f^E) = \sum_{i \in \mathcal{S}^P} \sum_{i' \in \mathcal{N}_i} V(f_i^P, f_{i'}^P \mid f_{i,i'}^E) + \sum_{i \in \mathcal{S}^P} \sum_{i' \in \mathcal{N}_i} V(f_{i,i'}^E) \tag{3.42}$$

where $V(f_i^P, f_{i'}^P \mid f_{i,i'}^E) = (f_i^P - f_{i'}^P)^2 (1 - f_{i,i'}^E)$ and $V(f_{i,i'}^E) = \alpha f_{i,i'}^E$. In terms of probability, (3.42) corresponds to

$$P(f^P, f^E) = P(f^P \mid f^E) \, P(f^E) \tag{3.43}$$

Adding the prior energy and the likelihood energy yields the posterior energy

$$U(f^P, f^E \mid d) = \sum_{i \in \mathcal{S}^P} (f_i^P - d_i)^2 + \sum_{i \in \mathcal{S}^P} \sum_{i' \in \mathcal{N}_i} \lambda g(f_i^P, f_{i'}^P, f_{i,i'}^E) \tag{3.44}$$

Minimization of equation (3.44) has to be performed over all f^P and f^E. Since the latter field assumes discrete configurations, the minimization is a

```
    0                 1                  1
  0  1  0          1  0  1           1  1  1            1  1
                                        1               1  1
```

Figure 3.5: Forbidden edge patterns.

combination of real and combinatorial problems. This can be converted to a simpler real minimization with the energy (Blake and Zisserman 1987)

$$E(f^P) = \sum_{i \in \mathcal{S}} (f_i^P - d_i)^2 + \sum_{i \in \mathcal{S}} \sum_{i' \in \mathcal{N}_i} \lambda_i g(f_i^P - f_{i'}^P) \qquad (3.45)$$

where g is the truncated quadratic potential function (3.40). Equation (3.45) is exactly the same as (3.22). Its minimization can be performed using algorithms working on real numbers, such as the graduate nonconvexity (GNC) (Blake and Zisserman 1987). By minimizing it only w.r.t. f^P, we obtain the restored image $(f^P)^*$, in which the restored pixel values near the edges are properly preserved (not over-smoothed). The edge field f^E is then determined by thresholding $(f^P)^*$

$$f_{i,i'}^E = \left\{ \begin{array}{ll} 1 & \text{if } (f_i^P - f_{i'}^P)^2 > \alpha \\ 0 & \text{otherwise} \end{array} \right. \qquad (3.46)$$

where i and i' are neighbors.

Do we need the edge field f^E explicitly? The current trend seems to use a g function without the explicit edge labels (i.e., performing minimization after eliminating the line process as illustrated previously). This is reflected, for example, by (Koch et al. 1986; Blake and Zisserman 1987; Geiger and Girosi 1989; Shulman and Herve 1989; Lange 1990; Li 1990b; Rangarajan and Chellappa 1990; Geman and Reynolds 1992; Geman et al. 1992; Bouman and Sauer 1993; Stevenson et al. 1994). It is less difficult to perform the real minimization than the combinatorial minimization. However, an explicit edge field may be useful for modeling contextual interactions of complex edge configurations, such as forbidden edge patterns, which are naturally described by a discrete field.

3.3.2 Forbidden Edge Patterns

There are essentially two ways to impose constraints: exact and inexact. In the inexact way, any violation of the constraints incurs a *finite* cost. The use of inexact constraints is seen in works on edge and boundary detection using relaxation labeling, for example, in (Zucker 1976; Hanson and Riseman 1978), and Faugeras and Berthod (1981). This is also what we have done so

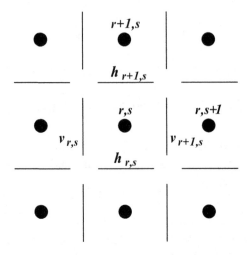

Figure 3.6: Horizontal and vertical edges.

far; e.g., with the g function. On the other hand, when the constraints are imposed exactly, any violation is forbidden. For example, if edges are required to form a boundary of single-pixel width, the adjacent parallel edges are not allowed. A family of four forbidden edge configurations is given in Fig. 3.5; they correspond, respectively, to an isolated edge (a terminated boundary), sharp turn, quadruple junction, and "small structure".

Let $U^E(f^E)$ be the total number of occurrences of the forbidden configurations in f^E. When there are no forbidden edge configurations, $U^E(f^E) = 0$ must hold exactly. Under this constraint, the MAP edge detection problem may be posed as (Geman et al. 1990; Heitz and Bouthemy 1993)

$$(f^P, f^E)^* = \arg \min_{(f^P, f^E):U^E(f^E)=0} E(f^P, f^E) \qquad (3.47)$$

where $E(f^P, f^E)$ is the posterior energy due to the inexact constraints. The minimization is performed over all configurations for which $U^E(f^E) = 0$. This is a problem of constrained minimization (see Section 9.4).

The constrained minimization may be computed by using the penalty method. Any forbidden local configuration is finally penalized by using an infinitely large clique potential; the exact constraints are imposed by gradually increasing the penalties against the violation of inexact constraints. An example is given below that rules out adjacent, horizontally parallel edges. Divide $f^E = \{f^E_{i,i'}\}$ into a horizontal edge field $\{h_{i,j}\}$ and a vertical edge field $\{v_{i,j}\}$ $((i,j) \in \mathcal{S}^E)$, as in Fig. 3.6. Each $h_{i,j}$ and $v_{i,j}$ variable takes a value in $\{0, 1\}$ as $f^E_{i,i'}$. We may define the energy to penalize adjacent, horizontally parallel edges (Koch et al. 1986) as

$$U_h(h) = \lambda_h \sum_{i,j} h_{i,j} h_{i+1,j} \qquad (3.48)$$

where $\lambda_h > 0$ is a constant. It is nonzero when both $h_{i,j}$ and $h_{i+1,j}$ take the value 1, indicating the occurrence of a pair of horizontally parallel edges. This incurs a penalty λ_h. To prohibit the formation of such edge configurations, we can gradually increase λ_h to a sufficiently large value ($\rightarrow +\infty$) so that the final minimal solution does not contain any horizontally parallel edges as a result of energy minimization. In the same way, $U_v(v) = \lambda_v \sum_{i,j} v_{i,j} v_{i,j+1}$ can be added to rule out adjacent, vertically parallel edges.

3.4 Texture Synthesis and Analysis

Three important issues in MRF texture analysis are texture modeling, texture classification, and texture segmentation. In MRF-based texture modeling, a texture is assumed to be an MRF, and to model a texture is to specify the corresponding conditional probabilities or Gibbs clique potential parameters. Texture classification is an application of pattern recognition techniques. It concerns the extraction of texture features and the design of a decision rule or classifier for classification. In MRF modeling, texture features correspond to the MRF texture parameters and feature extraction is equivalent to parameter estimation. In the supervised case, the estimation is performed using training data. This establishes reference parameters. Textures are classified by comparing texture feature vectors, extracted from the image and the reference feature vectors, which is basically a pattern recognition problem. Texture segmentation partitions a textured image into regions such that each region contains a single texture and no two adjacent regions have the same texture. To do this, one has to deal with issues in texture modeling, parameter estimation, and labeling. This chapter discusses problems in MRF texture modeling and segmentation.

3.4.1 MRF Texture Modeling

An MRF model of texture can be specified by the joint probability $P(f)$. The probability determines how likely a texture pattern f is to occur. Several MRF models, such as the auto-binomial (Cross and Jain 1983), auto-normal (or GMRF) (Chellappa 1985; Cohen and Cooper 1987), and multilevel logistic (MLL) models (Derin et al. 1984; Derin and Elliott 1987), have been used to model textures. A particular MRF model tends to favor the corresponding class of textures by associating it with a larger probability than others. Here, "a particular MRF model" is a particular probability function $P(f)$, which is specified by its form and the parameters in it.

In the MLL model, for example, the probability of a texture pattern f is defined based on the MLL clique potential function (2.42). In the second-order (8-neighbor) neighborhood system, there are ten types of cliques, as

---

Sampling Algorithm 1
Begin Algorithm
(1)   randomly initialize f to a point in $\mathcal{L}^S$;
(2)   for $i \in \mathcal{S}$ do
(2.1)   let $f'_{i'} = f_{i'}$ for all $i' \neq i$;
                        choose $f_i \in \mathcal{L}$ at random;
(2.2)   let $p = \min\{1, P(f')/P(f)\}$
                            where P is the given Gibbs distribution;
(2.3)   replace f by f' with probability p;
(3)   repeat (2) for N times;
End Algorithm

Figure 3.7: Generating a texture using a Metropolis sampler.

illustrated in Fig. 2.1, with the parameters associated with them shown in Fig. 2.3. When only the pairwise clique potentials are nonzero, equation (2.42) reduces to (2.44), rewritten as

$$V_2(f_i, f_{i'}) = \begin{cases} \beta_c & \text{if sites on } \{i, i'\} = c \in \mathcal{C}_2 \text{ have the same label} \\ -\beta_c & \text{otherwise} \end{cases}$$

$$(3.49)$$

where $f_i \in \mathcal{L} = \{1, \ldots, M\}$ and $\mathcal{C}_2$ is a set of pair-site cliques, and β_c is a parameter associated with the type c pair-site cliques. After the MLL is chosen, to specify an MRF model is to specify the β_c's. When the model is anisotropic (i.e., with all $\beta_c = \beta$), the model tends to generate texture-like patterns; otherwise it generates blob-like regions. The probability $P(f)$ can be calculated from $V_2(f_i, f_{i'})$'s using the corresponding Gibbs distribution.

A textured pattern corresponding to a realization of a Gibbs distribution $P(f) = \frac{1}{Z} e^{-U(f)}$ can be generated by sampling the distribution. Two often-used sampling algorithms are the Metropolis sampler (Metropolis et al. 1953) and the Gibbs sampler (Geman and Geman 1984) (see Section 7.1.6). Figure 3.7 and 3.8 list two sampling algorithms (Chen 1988). Algorithm 1 is based on the Metropolis sampler (Hammersley and Handscomb 1964) whereas Algorithm 2 is based on the Gibbs sampler of Geman and Geman (1984). A sampling algorithm generates a texture f with probability $P(f)$.

The differences between the two sampling algorithms are in step (2). Algorithm 1 needs only to evaluate one exponential function because the update is based on the ratio $P(f')/P(f)$ in step (2.2). Algorithm 2 needs to compute M exponential functions, and when the exponents are very large, this computation can be inaccurate. The following conclusions on the two algorithms were made by Chen (1988): (1) $N = 50$ iterations are enough for both

---

Sampling Algorithm 2
Begin Algorithm
(1)   randomly initialize f to a point in $\mathcal{L}^{\mathcal{S}}$;
(2)   for $i \in \mathcal{S}$ do
(2.1)   compute $p_l = P\{f_i = l \mid f_{\mathcal{N}_i}\}$ for all $l \in \mathcal{L}$;
                  where $f_{\mathcal{N}_i}$ are the pixel values at neighboring sites;
(2.2)   set f_i to l with probability p_l;
(3)   repeat (2) for N times;
End Algorithm

Figure 3.8: Generating a texture using a Gibbs sampler.

algorithms, (2) Algorithm 2 tends to generate a realization with a lower energy than Algorithm 1 if N is fixed, and (3) Algorithm 1 is faster than Algorithm 2.

Figure 3.9 shows four texture images generated using the MLL model (2.44) and Algorithm 1. It can be seen that when all the β_c parameters are the same such that the model is isotropic, the regions formed are blob-like; when the model is anisotropic, the pattern generated looks textured. So, blob regions are modeled as a special type of texture whose MRF parameters tend to be isotropic.

The method above for modeling a single texture provides the basis for modeling images composed of multiple textured regions. In the so-called hierarchical model (Derin and Cole 1986; Derin and Elliott 1987; Won and Derin 1992), blob-like regions are modeled by a high-level MRF that is an isotropic MLL; these regions are filled in by patterns generated according to MRF's at the lower-level. A filling-in may be either additive noise or a type of texture characterized by a Gibbs distribution. Let f be a realization of the region process with probability $P(f)$. Let d be the textured image data with probability $P(d \mid f)$. A texture label f_i indicates which of the M possible textures in the texture label set $\mathcal{L}$ pixel i belongs to.

The MLL model is often used for the higher-level region process. Given f, the filling-ins are generated using the observation model

$$d = \varphi(f) \qquad (3.50)$$

where φ is a random function that implements the mechanism of a noise or texture generator. To generate noisy regions, the observation model is simply $d_i = \varphi(f_i) = \ell(f_i) + e_i$, where $\ell(f_i)$ denotes the true gray level for region type f_i, as in (3.2). When the noise is also Gaussian, it follows a special Gibbs distribution, that in the zeroth order neighborhood system

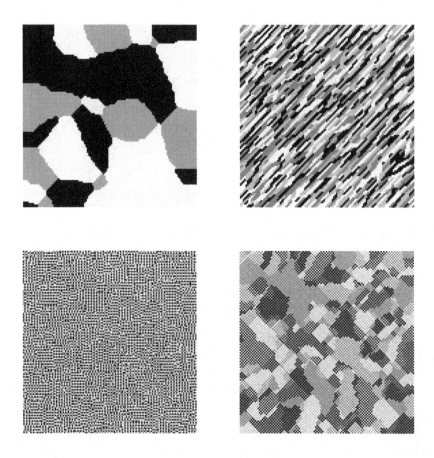

Figure 3.9: Textured images (128 × 128) generated by using MLL models. Upper left: Number of pixel levels $M = 3$ and parameters $\beta_1 = \beta_2 = \beta_3 = \beta_4 = 1$. Upper right: $M = 4$, $\beta_1 = \beta_2 = \beta_3 = 1$, $\beta_4 = -1$. Lower left: $M = 4$, $\beta_1 = \beta_2 = \beta_3 = \beta_4 = -1$. Lower right: $M = 4$, $\beta_1 = \beta_2 = -1$, $\beta_3 = \beta_4 = 2$.

(where the radius of the neighborhood is zero). When the noise is an identical, independent, and zero-mean distribution, the likelihood potential function is $V(d_i \mid f_i) = [\ell(f_i) - d_i]^2/2\sigma^2$, where σ^2 is the variance.

For textured regions, $\varphi(f)$ implements anisotropic MRF's. Because of the contextual dependence of textures, the value for d_i depends on the neighboring values $d_{\mathcal{N}_i}$. Given f_i and $d_{\mathcal{N}_i}$, its conditional probability is $P(d_i \mid d_{\mathcal{N}_i}, f)$, where all the sites in $\{i, \mathcal{N}_i\}$ are assumed to belong to the same texture type, f_i, and hence some treatment may be needed at and near boundaries. Note that the texture configuration is d rather than f; f is the region configuration.

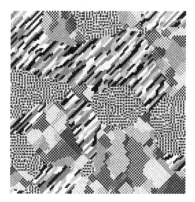

Figure 3.10: Multiple texture regions generated using the hierarchical Gibbs model of textures.

The configuration d for a particular texture can be generated using an appropriate MRF model (such as MLL used previously for generating a single texture). That is to say, d is generated according to $P(d \mid f)$; see further discussions in Section 3.4.2. Figure 3.10 shows an image of textured regions generated using the hierarchical Gibbs model. There, the region process is the same as that for the upper left of Fig. 3.9, and the texture filling-ins correspond to the three other realizations.

3.4.2 Texture Segmentation

Texture segmentation partitions an image into regions according to the texture properties of the regions. In the supervised texture segmentation, it is assumed that all the parameters for the textures, and for the noise if it is present, are specified (Hansen and Elliott 1982; Elliott et al. 1984; Geman and Geman 1984; Besag 1986; Derin and Cole 1986; Cohen and Cooper 1987; Derin and Elliott 1987), and the segmentation is to partition the image in terms of the textures whose distribution functions have been completely specified.

Unsupervised segmentation not only does the partition but also needs to estimate the parameters involved (Lakshmanan and Derin 1989; Manjunath and Chellappa 1991; Cohen and Fan 1992; Hu and Fahmy 1987; Won and Derin 1992). Obviously, this is more practical in applications and is also more challenging. There is a chicken-and-egg problem. The estimation should be performed by using realizations of a single class of MRF's (i.e., a single type of texture) (see Chapter 7), either noisy or noise-free. This requires that the segmentation be done. However, the segmentation depends on the parameters of the underlying textures. A strategy for solving this is to use an iterative algorithm alternating between segmentation and estimation. While the more advanced topic of unsupervised segmentation will be discussed in Section 7.2.2, we focus here on the supervised MRF segmentation with all parameters known.

Texture segmentation, like other labeling problems, is usually performed in an optimization sense, such as MAP. A main step is to formulate the posterior energy. In MAP texture segmentation with the hierarchical texture model, the higher-level MRF determines the prior probability $P(f)$, where the segmentation f partitions $\mathcal{S}$ into regions each of which is assigned a texture type from the label set $\mathcal{L}$; the lower-level field contributes to the likelihood function $P(d \mid f)$, where d is the image data composed of multiple textures.

Let the label set be $\mathcal{L} \in \{1, \ldots, M\}$ and f represent a segmentation in which $f_i \in \mathcal{L}$ is the indicator of the texture type for pixel i. Denote the set of all sites labeled I by

$$\mathcal{S}^{(I)}(f) = \{i \in \mathcal{S} \mid I = f_i\} \qquad (3.51)$$

Then

$$\mathcal{S} = \bigcup_{I \in \mathcal{L}} \mathcal{S}^{(I)}(f) \qquad (3.52)$$

and

$$\mathcal{S}^{(I)}(f) \bigcap \mathcal{S}^{(J)}(f) = \phi, \quad I \neq J \qquad (3.53)$$

The likelihood energy function can be expressed as

$$U(d \mid f) = \sum_{c \in \mathcal{C}} V_c(d \mid f) = \sum_{I \in \mathcal{L}} \sum_{\forall c \subset \mathcal{S}^{(I)}} V_c^{(I)}(d \mid f) \qquad (3.54)$$

where $V_c^{(I)}(d \mid f)$ is the potential function for the data d on c labeled as type I.

Suppose that a type I texture is modeled as an MLL with parameters

$$\theta^{(I)} = \{\alpha^{(I)}, \beta^{(I)}, \cdots\} \qquad (3.55)$$

Then, according to (2.43), the single-site clique potentials are

$$V_c^{(I)}(d \mid f) = \alpha^{(I)} \qquad (3.56)$$

where $I = f_i$ and, according to (2.42), the multi-site clique potentials are

$$V_c^{(I)}(d \mid f) = \begin{cases} \beta_c^{(I)} & \text{if all } d_i, \ i \in c, \text{ are the same} \\ -\beta_c^{(I)} & \text{otherwise} \end{cases} \qquad (3.57)$$

This equation is for the cliques in the interior of $\mathcal{S}^{(I)}$. At and near region boundaries, a clique c may ride across two or more $\mathcal{S}^{(I)}$'s. In this case, the following rule may be used to determine the type of texture for generating the data: If c sits mostly in a particular $\mathcal{S}^{(I)}$, then choose model parameters $\theta^{(I)}$; if it sits equally in all the involved $\mathcal{S}^{(I)}$'s, choose an I at random from the involved labels. When the gray levels for type I texture image data are known as $\mathcal{D}_I = \{\ell_1^{(I)}, \dots, \ell_{M_I}^{(I)}\}$, where M_I is the number of gray levels for type I texture, more constraints are imposed on the texture segmentation and better results can be expected. When the gray levels are also subject to noise, then the constraints become inexact.

After $U(f)$ is also defined for the region process, the posterior energy can be obtained as

$$U(f \mid d) = U(f) + U(d \mid f) \qquad (3.58)$$

Minimizing (3.58) w.r.t. f is more complicated than the minimization for the restoration and reconstruction formulated in the previous sections because neither term on the right-hand side can be decomposed into independent subterms. in (Derin and Elliott 1987), some assumptions of independence are made to simplify the formulation and a recursive dynamic programming algorithm (see Section 9.3.6) is used to find a suboptimal solution.

3.5 Optical Flow

Optical flow is the velocity field in the image plane caused by the motion of the observer and objects in the scene. It contains important information about cues for region and boundary segmentation, shape recovery, and so on. This section describes methods for computing optical flow and motion from a sequence of time-varying images.

Two major paradigms exist for determining visual motion: feature-based (e.g., edge-based) (Nagel 1983; Paquin and Dubios 1983; Bouthemy 1989) and gradient-based (Horn and Schunck 1981). In the former paradigm, features are extracted from the sequence of images, matched between two neighboring frames, and tracked over time. This gives a sparse flow for which the information is available at the sparse set of the extracted image features. In the gradient-based paradigm, the flow is recovered based on local spatio-temporal changes in image intensity. This gives a dense flow field whose values are available throughout the image plane. The focus of this section will be on finding flow field using gradient-based methods.

3.5.1 Variational Approach

Gradient-based methods, advocated by Horn and Schunck (1981), utilize the constraints of intensity constancy, and spatial and temporal coherence. Let $d(x, y, t)$ denote the intensity at (x, y) of the image plane at time t. Consider a brightness pattern that is displaced in the direction $(\delta x, \delta y)$ in time δt. A constraint is that the intensity of the brightness remain constant; i.e.,

$d(x, y, t) = d(x + \delta x, y + \delta y, t + \delta t)$. This gives the equation of *intensity constancy*

$$\frac{\partial d}{\partial x}\frac{\mathrm{d}x}{\mathrm{d}t} + \frac{\partial d}{\partial y}\frac{\mathrm{d}y}{\mathrm{d}t} + \frac{\partial d}{\partial t} = 0 \qquad (3.59)$$

This equation relates the change in image intensity at a point to the motion of the intensity pattern (the distribution of pixel intensities). Let

$$f(x, y) = [u(x, y), v(x, y)]^T = (dx/dt, dy/dt)^T \qquad (3.60)$$

designate the optical flow vector at (x, y). Then $f = \{f(x, y)\}$ is the optical flow field we want to compute. The intensity constancy constraint is a single equation with the two unknowns u and v

$$ud_x + vd_y + d_t = 0 \qquad (3.61)$$

This corresponds to a straight line in the velocity space. The equation shows that only the velocity components parallel to the spatial image gradient can be recovered through the local computation. This is termed the "aperture problem" by Marr (1982). When noise is taken into consideration, the flow may be computed by minimizing

$$V(d(x, y) \mid f(x, y)) = (ud_x + vd_y + d_t)^2 \qquad (3.62)$$

However, optical flow f thus computed under the intensity constancy constraint alone is not unique. Additional constraints must be imposed on the flow.

An important constraint is the smoothness; that is, the flow at nearby places in the image will be similar unless discontinuities exist there. The smoothness is imposed in a similar way as in the restoration and reconstruction discussed earlier. The movement of intensity points should present some coherence and so the flow in a neighborhood should change smoothly unless there are discontinuities. One way to impose this constraint is to minimize the squared magnitude of the gradient of the optical flow (Horn and Schunck 1981)

$$V(f(x, y)) = \|\nabla u\|^2 + \|\nabla v\|^2 = \left(\tfrac{\partial u}{\partial x}\right)^2 + \left(\tfrac{\partial u}{\partial y}\right)^2 + \left(\tfrac{\partial v}{\partial x}\right)^2 + \left(\tfrac{\partial v}{\partial y}\right)^2 \qquad (3.63)$$

See also (Snyder 1991) for a study of smoothness constraints for optical flow.

In the MAP-MRF framework, (3.62) corresponds to the likelihood potential due to independent Gaussian noise in the image intensities and (3.63) corresponds to the prior potential due to the prior distribution of the MRF, f. The posterior potential is obtained by combining (3.62) and (3.63) into a weighted sum

$$V(f(x, y) \mid d(x, y)) = V(f(x, y) \mid d(x, y)) + \lambda V(f(x, y)) \qquad (3.64)$$

where λ is a weighting factor. Integrating it over all (x, y) gives the posterior energy

$$E(f) = \int \int [V(f(x,y) \mid d(x,y)) + \lambda V(f(x,y))] \, dx dy \qquad (3.65)$$

This formulation (Horn and Schunck 1981) not only provides a method for computing optical flow, but also is a pioneer work on the variational approach in low-Level vision that was later developed into the regularization framework (Poggio et al. 85a; Bertero et al. 1988). Various MAP-MRF formulations for flow estimation can be found, for example in (Murray and Buxton 1987; Konrad and Dubois 1988a; Black and Anandan 1990; Konrad and Dubois 1992; Heitz and Bouthemy 1993).

In discrete computation of the variational solution, it is important that the spatial and temporal intensity partial derivatives of the input image sequence $d = \{d_{i,j,k}\}$, which appear in (3.62), be consistent. They may be estimated by averaging the four first-order differences taken over the adjacent measurements (Horn and Schunck 1981)

$$
\begin{aligned}
d_x &\approx \tfrac{1}{4} \; [(d_{i,j+1,k} - d_{i,j,k}) + (d_{i+1,j+1,k} - d_{i+1,j,k}) + \\
&\quad (d_{i,j+1,k+1} - d_{i,j,k+1}) + (d_{i+1,j+1,k+1} - d_{i+1,j,k+1})] \\
d_y &\approx \tfrac{1}{4} \; [(d_{i+1,j,k} - d_{i,j,k}) + (d_{i+1,j+1,k} - d_{i,j+1,k}) + \\
&\quad (d_{i+1,j,k+1} - d_{i,j,k+1}) + (d_{i+1,j+1,k+1} - d_{i,j+1,k+1})] \\
d_t &\approx \tfrac{1}{4} \; [(d_{i,j,k+1} - d_{i,j,k}) + (d_{i+1,j,k+1} - d_{i+1,j,k}) + \\
&\quad (d_{i,j+1,k+1} - d_{i,j+1,k}) + (d_{i+1,j+1,k+1} - d_{i+1,j+1,k})]
\end{aligned}
\qquad (3.66)
$$

Another formula calculates them using a three-point approximation (Battiti et al. 1991). The spatial derivatives of the flow $f_{i,j,k}$ in (3.68) are usually approximated by the simple differences

$$
\begin{aligned}
u_x &= u_{i,j+1,k} - u_{i,j,k} \\
u_y &= u_{i+1,j,k} - u_{i,j,k} \\
v_x &= v_{i,j+1,k} - v_{i,j,k} \\
v_y &= v_{i+1,j,k} - v_{i,j,k}
\end{aligned}
\qquad (3.67)
$$

3.5.2 Flow Discontinuities

The quadratic smoothness in (3.63) is unable to deal with discontinuities (i.e., boundaries between regions moving differently). For the discontinuity-preserving computation of optical flow, the smoothness is more appropriately imposed as minimizing

$$V(f(x,y)) = g\left(\left[\tfrac{du}{dx}\right]^2 + \left[\tfrac{dv}{dx}\right]^2\right) + g\left(\left[\tfrac{du}{dy}\right]^2 + \left[\tfrac{dv}{dy}\right]^2\right) \qquad (3.68)$$

where the g function satisfies a necessary condition (3.16). When g is a truncated quadratic, (3.68) gives the line process model (Koch et al. 1986; Murray and Buxton 1987). A more careful treatment, the "oriented smoothness" constraint (Nagel and Enkelmann 1986), may be used to avoid smoothing across intensity, rather than flow, discontinuities. In (Shulman and Herve 1989), g is chosen to be the Huber robust penalty function (Huber 1981). In (Black and Anandan 1993), such a robust function g is also applied to the data term, giving $V(d(x,y) \mid f(x,y)) = g(ud_x + vd_y + d_t)$.

Heitz and Bouthemy (1993) propose an MRF interaction model that combines constraints from both gradient-based and edge based paradigms. It is assumed that motion discontinuities appear with a rather low probability when there is no intensity edge at the same location (Gamble and Poggio 1987). This is implemented as an energy term that imposes a constraint on the interaction between motion discontinuities and intensity edges. The energy is designed to prevent motion discontinuities from appearing at points where there are no intensity edges.

The computation of optical flow discussed so far is based on the assumption of constant intensity. The assumption is valid only for very small displacements, or short-range motion. For long-range motion, there is a significant time gap between different frames. In this case, the analysis is usually feature-based. This now requires us to resolve the correspondence between features in successive frames.

3.6 Stereo Vision

Assume that images obtained from different viewpoints are available. A goal of multiview stereo vision is to reconstruct (the visible part of) the shape of a 3D object through finding the camera–surface distances from information contained in the multiple calibrated images. This is illustrated in Fig. 3.11. Note that the distance from any C_i to the corresponding surface point has a unique depth value (as referenced w.r.t. C_1) for calibrated cameras. MRF formulations for stereo vision have been studied in connection with graph cuts (Roy and Cox 1998; Kolmogorov and Zabih 2002; Boykov and Kolmogorov 2004; Vogiatzis et al. 2005; Sinha and Pollefeys 2005).

Assume that K calibrated images are taken from K different viewpoints. Let $\mathcal{S}_k = \{(x,y,k)\}$ be the set of image pixel index3es for camera k, and $\mathcal{S} = \mathcal{S}_1 \cup \cdots \cup \mathcal{S}_K$ be the set of all pixel locations where $\mathcal{S}_k$ is the set of sites for image $k \in \{1, \cdots, K\}$. A pixel $i \in \mathcal{S}$ corresponds to a ray in the 3D space that connects between the viewpoint of the camera to which i corresponds and the location of site i in the image (e.g., the line connecting C_1 and p_i in Fig. 3.11). All points along the ray are projected at point i.

Let $\mathcal{N}_k$ be the neighborhood system defined on the set $\mathcal{S}_k$, and thus $\mathcal{N} = \{\mathcal{N}_1 \cup \cdots \cup \mathcal{N}_K\}$ on the whole $\mathcal{S}$. Pixels belonging to different images are considered not to be neighboring each other.

Let $\mathcal{L} = \{1, \cdots, M\}$ be the set of depth labels whose values are relative to the reference camera C_1 in Fig. 3.11. Let $p_i = (x_i, y_i, f_i)$ be a point in the 3D

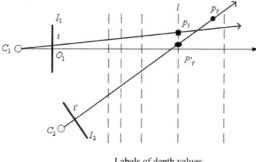

Labels of depth values

Figure 3.11: Two-camera stereo vision. C_1 and C_2 in 3D are the viewpoints of cameras 1 and 2. I_1 and I_2 are the two images viewed from C_1 and C_2. Consider camera 1 as the reference camera with C_1 as the reference point, and the center O_1 of I_1 as the reference pixel without loss of generality, the reference axis being C_1-O_1. The dashed lines, corresponding to the planes perpendicular to C_1-O_1, represent the depths. The goal is to find the depth labels.

space, where (x_i, y_i) is known from the site index in $\mathcal{S}$ with the calibration information for pixel i. According to the stereo vision theory, the depth labels f_i can be inferred from information about pixels $i \in \mathcal{S}$ that are visible from at least two cameras.

The problem can be formulated as finding the MAP-MRF labeling $f :$ $\mathcal{S} \to \mathcal{L}$ that minimizes the posterior energy function $U(f \mid d) = U(d \mid f) + U(f)$. In the following formulation, the likelihood (data) term imposes the photo consistency constraint, whereas the prior term imposes the smoothness and visibility constraints (Kolmogorov and Zabih 2002).

The photo consistency states that if pixels i and i' come from the same point (i.e., having the same depth value) on the surface, they should have similar intensity values; Denote the intensity of pixel i by $I(i)$, and define a lower-bounded intensity discrepancy of any two pixels in a form similar to the line process (see Chapter 5)

$$D(i, i') = \min\{0, [I(i) - I(i')]^2 - \alpha\} \tag{3.69}$$

where $\alpha > 0$ is a constant. Consider a set $\mathcal{A}$ of 3D point pairs that are close to each other on the surface and have the same label $f_i = f_{i'}$

$$\mathcal{A} = \{\langle p_i, p_{i'} \rangle \mid i, i' \in \mathcal{S}, f_i, f_{i'} \in \mathcal{L}, f_i = f_{i'}, \|p_i - p_{i'}\| < \beta\} \tag{3.70}$$

where $\| \cdot \|$ is some norm (such as L^2) and $\beta > 0$ is a predefined threshold. Because $f_i = f_{i'}$ is satisfied, $\|p_i - p_{i'}\| < \beta$ implies that the points are close in the 2D plane. The photo consistency is encoded in the likelihood energy

$$U(d \mid f) = \sum_{\langle p_i, p_{i'} \rangle \in \mathcal{A}} D(i, i') \tag{3.71}$$

The smoothness prior could be defined by line processes for discontinuity preservation (see Chapter 5)

$$U_s(f) = \sum_{\{i,i'\} \in \mathcal{N}} \min\{|f_i - f_{i'}|, \gamma\} \tag{3.72}$$

where $\gamma > 0$ is a constant. The visibility prior imposes the opaqueness in that only the outermost surface is visible. Let p_i and $p_{i'}$ be any two points in 3D with $f_i \neq f_{i'}$. Let $p'_{i'}$ be the intersection of ray i' and the f_i plane. If pair $< p_i, p'_{i'} > \in \mathcal{A}$, then p_i and $p'_{i'}$ are in the same depth. This contradicts the condition $f_i \neq f_{i'}$ because the coexistence of p_i and $p_{i'}$ would violate the visibility constraint, and such occurrences should be penalized. Define the set of pairs that violate the visibility constraint as

$$\mathcal{B} = \{\langle p_i, p_{i'} \rangle \mid \langle (i, \ell), (i', \ell) \rangle \in \mathcal{A}, \ \ell = \min\{f_i, f_{i'}\}\} \tag{3.73}$$

The visibility prior term can be defined as

$$U_v(f) = \sum_{\langle p_i, p_{i'} \rangle \in \mathcal{B}} C \tag{3.74}$$

where $C \in (0, \infty)$ is a large penalty constant.

The above analysis leads to the posterior energy $U(f \mid d) = U(d \mid f) + U_s(f) + U_v(f)$ for stereo vision. For the energy minimization, the graph cuts approach has become popular in recent years (Roy and Cox 1998; Kolmogorov and Zabih 2002; Boykov and Kolmogorov 2004; Vogiatzis et al. 2005; Sinha and Pollefeys 2005).

3.7 Spatio-temporal Models

Video processing and analysis takes into consideration not only the pixel values in a single static frame but also the temporal relations between frames. Thus the neighborhood system should be defined on both spatial and temporal domains. Typical applications of spatio-temporal MRF models include stitching (Pérez et al. 2003; Agarwala et al. 2004; Rother et al. 2006), background cut (Sun et al. 2006), object tracking (Nguyen et al. 2007), and video synopsis (Rav-Acha et al. 2006; Pritch et al. 2007). Here we consider the problem of video synopsis for fast video browsing and indexing.

A video synopsis condenses an original video into a short period representation, while preserving the essential activities in the input video (see Fig.3.12). It can show multiple activities simultaneously, even if they originally happened at different times. The indexing into the original video functionality is provided by pointing each activity toward the original time frames.

Input Video

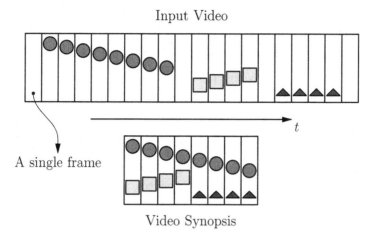

A single frame

Video Synopsis

Figure 3.12: Video synopsis: input video (top) and video synopsis (bottom). In the input video, three different events happen sequentially, denoted by filled circles, squares, and triangles in 20 frames. By rearranging the events in the time coordinates, the output video synopsis contains only eight frames without losing the information on the activities.

Let I denote the input video consisting of N_I frames and $I(x,y,t)$ represent the intensity (or RGB values for color video) of pixel (x,y) in frame t. I can be represented as a space-time volume (or tube) in 3D. Let O be the output video synopsis of $N_O \ll N_I$ frames. It should contain as much information as possible about the activities in I and preserve the motion dynamics of the activities. Moreover, visible seams and fragmented objects should be avoided.

Define the set of sites S to consist of pixels in the synopsis video O, $S = \{(x,y,t) \mid 1 \le t \le N_O\}$ and the labels to consist of all the time indices of the input video I, $\mathcal{L} = \{1, 2, \cdots, N_I\}$. The synopsis video can be obtained from the labeling $f : S \to \mathcal{L}$. If site $i = (x,y,t)$ is labeled $f_i \in \mathcal{L}$, then $O(x,y,t) = I(x,y,f_i)$. Define a neighborhood system $\mathcal{N}_i$ to consist of the six direct neighbors in the 3D volume: the four spatial neighbors in the image plane and the two temporal neighbors in time.

The problem can be formulated as one of MAP-MRF labeling minimizing $U(f \mid d) = U(d \mid f) + U(f)$. The likelihood energy is defined as the activity loss, whereas the prior energy encodes the smoothness. The loss of activity is defined as the number of active pixels in the input video I that do not appear in the synopsis video O. The activity measure of each pixel can be represented by the characteristic function indicating its difference from the background

$$\chi(x,y,t) = \|I(x,y,t) - B(x,y,t)\| \tag{3.75}$$

where B is the background video, which in (Rav-Acha et al. 2006; Pritch et al. 2007) is built using graph cuts (see Section 10.4). With this, the activity loss is given by the difference in the input and output videos

$$U(d \mid f) = \sum_{\forall (x,y,t) \text{ of } I} \chi(x,y,t) - \sum_{\forall (x,y,t) \text{ of } O} \chi(x,y,f(x,y,t)) \qquad (3.76)$$

The first term is the total activity in the input video I, and the second, which is smaller than the first, is that in the synopsis O. The smoothness prior, or discontinuity cost, $U(f)$, is defined as the sum of color differences across seams between spatio-temporal neighbors in the synopsis video and the corresponding neighbors in the input video:

$$U(f) = \sum_{i=(x,y,t) \in \mathcal{S}} \sum_{i' \in \mathcal{N}_i} \|O(i') - I(x,y,f_{i'})\|^2 \qquad (3.77)$$

In (Rav-Acha et al. 2006; Pritch et al. 2007), the posterior energy function is minimized using simulated annealing (see Section 10.1).

3.8 Bayesian Deformable Models

Correct location and extraction of objects of interest is an important step in many applications such as industrial inspection, medical image analysis, and object recognition. Deformable models, such as the snake (Terzopolous et al. 1987; Kass et al. 1988; Terzopolous et al. 1988), G-Snake (Lai and Chin 1995), active shape model (ASM) (Cootes et al. 1995) and deformable templates (Jain et al. 1996), are able to fit objects more closely and have proven to be more effective than rigid models in detecting shapes subject to local deformation. Such models can be classified into two classes (Jain et al. 1996; Jain et al. 1998): (i) free-form and (ii) parametric. In the free-form class, such as the snake, there is no global structure of the template; the template is constrained only by local continuity and smoothness constraints. Such models can be deformed to any arbitrary shape, deviating greatly from the shape of interest.

In the parametric class, a global shape model is designed and encoded by a small number of parameters (Yuille et al. 1989; Grenander et al. 1991; Staib and Duncan 1992; Lai and Chin 1995; Cootes et al. 1995; Mumford 1996; Jain et al. 1996; Blake et al. 1998). *Prior* knowledge about the structural properties is incorporated into the shape models, assuming that prior information of the geometrical shape or a set of training samples is available. Compared with the free-form class, the parametric class is more robust against irrelevant structures, occlusion, and bad initialization. A quite successful and versatile scheme in this class is that using statistical shape models in the Bayesian framework (Grenander 1976; Staib and Duncan 1992; Mumford 1996; Lai and Chin 1995; Jain et al. 1996; Blake et al. 1998). In these models, both the prior knowledge and observation statistics are used to define the optimal estimate in the Bayesian sense.

In this section, a deformable contour model, called EigenSnake, is presented in the Bayesian framework (Section 3.8.1). Three strategies are combined into the EigenSnake to make the object search more robust and accurate. (i) The prior knowledge of a specific object shape is derived from a training set of shape samples, and expressed as a prior distribution for defining a Bayesian estimate as the optimal solution for the object contour. The prior distribution can explain and represent the global and local variations in the training set. (ii) Constraints on object shapes are imposed in a "soft" manner, following the derived prior distribution. The shape encoding can adjust itself dynamically based on the up-to-date knowledge learned during the solution-finding process. This adjustment allows the solution to fit more closely to the object shape. (iii) A shape space is constructed based on a set of eigenvectors derived by principal component analysis (PCA) performed on the training set. It is used to restrict and stabilize the search for the Bayesian optimum.

The affine invariant *shape matrix* method in the G-Snake (Lai and Chin 1995) is used for encoding the global shape information. However, we noticed two drawbacks of the G-Snake model. (i) It encodes the shape information in a "hard" manner in that the shape matrix is fixed. This actually assumes that the object shape considered undergoes only *rigid* (such as affine) transformation. So, only a small amount of local deformation is tolerated, and an incorrect solution results when the shape deformation is large, which is often a problem in many applications, such as face detection. (ii) A more fundamental problem is what we call "incorrect formulation" of the G-Snake: A better solution, which is a contour more closely fitting the object in the image, may have a higher energy value than an inferior solution. The problem is due to the lack of constraints on the optimal solution. The EigenSnake overcomes these problems in two ways: (i) It modifies the shape matrix dynamically based on the up-to-date information gathered from the solution-finding process, and (ii) it encodes prior constraints specific to objects of interest in its energy function. More accurate solutions are obtained by using these methods.

The shape space constraint used in the ASM (Cootes et al. 1995; Lanitis et al. 1997) is applied to the solution-finding process. The shape space, which is a subspace of the whole arbitrary contour space, explains and represents most of the nonrigid variance in the training set, based on the eigenvectors derived through PCA performed on the training contours. Every intermediate solution (energy minimum) is projected into the subspace, and thus the solution is always restricted to be within the subspace. This proves to be effective in stabilizing the solution. However, in the ASM, no prior distribution accounting for global rigid motion is incorporated; therefore, the original ASM is unable to handle large-scale and orientational shape changes. This is overcomed in the EigenSnake.

Experiments are performed in the application domain of face extraction (Section 3.8.2). Four methods based on the G-Snake and the EigenSnake are compared with evaluate how various constraints and different prior knowledges have affected the results. The results demonstrate that the use of various prior constraints in the EigenSnake helps produce more stable and accurate solutions. The EigenSnake, as compared with the three other methods, effectively overcomes problems due to initialization, occlusion, noise, and variations in scale and rotation.

3.8.1   Formulation of EigenSnake

The EigenSnake incorporates the prior knowledge about the distribution of training contours. The global and local variations in the training set are well represented by the resulting prior distribution. Moreover, a shape space is constructed to further constrain the solution.

Joint Prior Distribution

Let $\mathbb{E} = \{(x, y) \in \mathbb{R}^2\}$ be the image plane. An arbitrary contour can be represented by a vector of m points in $\mathbb{E}$, $f = [f_1^T, f_2^T, \cdots, f_m^T]$ (a $2 \times m$ matrix), where $f_i \in \mathbb{E}$ is a point. Given a prototype contour $\overline{f}$, a shape matrix $A = A(\overline{f})$ can be determined by $A(\overline{f})\overline{f}^T = \mathbf{0}$. The shape matrix is under affine transformation: If f^{aff} is an affine-transformed version of $\overline{f}$, then $A(\overline{f}) f^{aff^T} = \mathbf{0}$ (Lai and Chin 1995).

For the convenience of the following description, we express any contour as a $2m \times 1$ vector $f = [f_1, f_2, \cdots, f_m]^T \in \mathbb{R}^{2m}$. Given a prototype contour $\overline{f}$ and the corresponding shape matrix $A(\overline{f})$, the prior distribution of f is given as

$$p(f \mid \overline{f}) \propto \exp\left\{-E(f \mid \overline{f})\right\} \tag{3.78}$$

where $E(f \mid \overline{f}) = \sum_{i=1}^{m} \frac{E(f_i \mid \overline{f})}{\sigma_i^2}$ is the internal energy (Lai and Chin 1995).

In the G-Snake, $\overline{f}$ is chosen to be the mean of the training contours and is fixed. This actually assumes that the shape of the object considered (that of $\overline{f}$) is subject to *rigid* (such as affine) transformation only. However, it ignores prior knowledge about the fact that local variations exist among the training contours. This is equivalent to the assumption that f is a distribution with zero variance, which is obviously invalid with the training set. Since $\overline{f}$, and hence $A(\overline{f})$, is fixed, it is difficult for the solution f to converge closely to the object contour in the image when the local shape deformation is large (which is often the case for many applications such as face detection). Because of this, a better solution, which is a contour more closely fitting the object in the image, may have a higher energy value than an inferior solution, as will be illustrated by experimental results. This is referred to as the incorrectness of an energy model.

The EigenSnake amends this problem by allowing $\overline{f}$ to deform according to a prior distribution derived from the training contour. In this case, the joint prior

$$P(f, \overline{f}) = p(f \mid \overline{f}) \cdot p(\overline{f}) \propto \exp\{-E(f, \overline{f})\} \qquad (3.79)$$

is considered. Hence, the shape information is encoded in a "soft" manner, and $\overline{f}$ and hence $A(\overline{f})$ can be modified dynamically during the computation. The prior $p(\overline{f})$ is derived from a set of training contours in the following.

Learning Prior from Training Set

Let a set of L face contours $\{f^1, f^2, \ldots, f^L\}$ be given as the training set where each contour f^i is represented by a sequence of m evenly placed (done manually) points in $\mathbb{E}$ (i.e., $f_k^i \in \mathbb{E}$ and $f^i \in \mathbb{R}^{2m}$). The training contours are aligned at the centers of their gravities and then scaled and rotated by using an iterative alignment algorithm so as to remove the rigid variation in the training samples as much as possible. Thus preprocessed contours constitute a random distribution. The mean contour $\overline{f}_{mean}$ is computed by averaging the corresponding contour points. See Fig. 3.13.

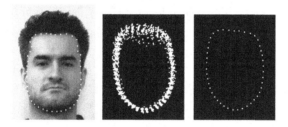

Figure 3.13: Left: A sample of a face contour with $m = 44$ points. Middle: Scatter of points from normalized and aligned set of face boundary shapes. Right: The mean (average) of the training face contours.

Assume that the training samples obey a Gaussian distribution with the mean vector $\overline{f}_{mean}$ and the covariance matrix C:

$$p(f) \propto \exp\left\{-\frac{1}{2}\left(f - \overline{f}_{mean}\right)^T C^{-1} \left(f - \overline{f}_{mean}\right)\right\} \qquad (3.80)$$

The distribution represents the prior knowledge about local variation derived from the training contours. However, the covariance matrix C is often singular, and thus C^{-1} is nonexisting, because the dimensionality of the samples is high and the number of available training contours is often small. The distribution has to be approximated by some means.

Approximation of Prior Using PCA

We adopt an optimal method proposed by Moghaddam and Pentland (1997) to approximate the Gaussian density $p(f)$ by applying a PCA to the training samples to capture the main modes of the variation in the directionally elongated (i.e., elliptical) Gaussian distribution.

Let Ψ be the eigenvector matrix of C and Λ be the corresponding diagonal matrix of eigenvalues. We have $\Lambda = \Psi^T C \Psi$. The first $M(\ll 2m)$ principal eigenvectors are selected: $\Phi = [\phi_1, \phi_2, \cdots, \phi_M]$. They are used as the basis vectors to constitute a specific *shape space* (i.e., the M-dimensional space spanned by the M orthonormal vectors). The shape space efficiently explains most of the local variation within the training set. Any contour $f \in \mathbb{R}^{2m}$ can be approximated by a linear combination of the M eigenvectors (Turk and Pentland 1991)

$$\hat{f} = \overline{f}_{mean} + \Phi\omega \tag{3.81}$$

where $\omega = \Phi^T(f - \overline{f}_{mean})$ is the vector of weights, $\hat{f}$ is the projection of f into the shape space, and ω is the coordinate of the projection point in the shape space. It provides an optimum approximation for f in the sense of least squares error. It discards the variation of f that is inconsistent with those reflected by training contours while retaining most of the variation that the shape space can explain. The projection $\hat{f}$ has the distribution (Moghaddam and Pentland 1997)

$$p(\hat{f}) = p(\omega) = \frac{\exp\left\{-\frac{1}{2}\sum_{k=1}^{M}\frac{\omega_k^2}{e_k}\right\}}{(2\pi)^{M/2}\prod_{k=1}^{M}e_k^{1/2}} \tag{3.82}$$

where e_k is the kth largest eigenvalue of C, $\omega_k = \omega_k(f) = \phi_k^T\left(f - \overline{f}_{mean}\right)$, and $\varepsilon^2(f) = \left\|f - \overline{f}_{mean}\right\|^2 - \sum_{k=1}^{M}\omega_k^2$ is the residual error.

Prior Distribution of Prototype Contour

During the solution-finding process, the prototype contour $\overline{f}$ is modified dynamically according to the up-to-date MAP solution, and is confined in the shape space. The method is described below.

Let $f' = \arg\min_{f \in \mathcal{N}(f^{(t-1)})} E(f, \overline{f} \mid d)$ be the MAP solution found at iteration t, where $E(f, \overline{f} \mid d)$ is the posterior energy (to be defined later) and $\mathcal{N}(f^{(t-1)})$ is the neighborhood of the previously found solution $f^{(t-1)}$. The MAP estimate f' is aligned with $\overline{f}_{mean}$ by minimizing $\left\|u + s \cdot R_\theta(f') - \overline{f}_{mean}\right\|$, where u (translation), s (scaling), and θ (rotation) are the rigid transformation parameters and $R_\theta(f)$ is a function that rotates the contour f by θ. The translation vector can be found as $u = (x_u, y_u) =$

$\frac{1}{m} \sum_{i=1}^{m} [\overline{f}_{mean,i} - f'_i]$. The two other parameters can be calculated using a simple alignment algorithm. The aligned contour is given as $f'' = u + s \cdot R_\theta(f')$.

The aligned contour f'' is then projected into the shape space using (3.81), resulting in the projected version $\hat{f}''$. The prototype contour $\overline{f}$ is set to

$$\overline{f} = \hat{f}'' = \overline{f}_{mean} + \Phi\omega \qquad (3.83)$$

where $\omega = \Phi^T(f'' - \overline{f}_{mean})$ are the coordinates of f'' in the shape space.

Therefore, $\overline{f}$ obtained in such a way follows the same distribution as that of $\hat{f}''$

$$p(\overline{f}) = p(\hat{f}'') = p(\omega) \qquad (3.84)$$

in (3.82). The corresponding prototype contour energy is

$$E(\overline{f}) = \frac{1}{2} \sum_{i=1}^{M} \frac{\omega_i^2}{e_i} \qquad (3.85)$$

Note that $p(f \mid \overline{f})$ is scale and rotation invariant, and $p(\overline{f})$ depends only on the local deformation parameter vector ω and is independent of the global deformation parameters such as u, s, and θ. As a consequence, $p(f, \overline{f})$ is also scale and rotation invariant, retaining the advantage of the prior distribution of the G-Snake.

To take advantage of the shape space constraint, the solution found in iteration t is correspondingly revised to

$$f^{(t)} = -u + \frac{1}{s} \cdot R_{-\theta}(\hat{f}'') \qquad (3.86)$$

The revised solution $f^{(t)}$ is in the shape space. It retains the rigid variation of f' (w.r.t. $\overline{f}_{mean}$), and has the local deformation consistent with the samples in the training set.

Two remarks about the imposition of the shape space constraints follow: (i) The imposition of the shape space constraint makes the EigenSnake more robust against spurious structures and occlusion. The effect will be demonstrated in Section 3.8.2. (ii) Because $\omega = \mathbf{0}$ when $f'' = \overline{f}_{mean}$, ω can be considered as the local deformation parameters of f'' or f'. Hence, the rigid transformation parameters (u, s, and θ) and the nonrigid deformation parameters (ω) are separated. Therefore, the prior distribution $p(f, \overline{f})$ explains not only the global variation but also the local variation, derived from a training set.

Likelihood Distribution

Edges are detected from the image $d = \{d(x, y)\}$. The normalized edge direction at each pixel location (x, y), denoted $d^e(x, y) = (d^e_x(x, y), d^e_y(x, y))$, is used to provide constraints on the data, with $\|d^e(x,y)\| \in (0,1]$ if an edge is detected at (x,y) or a zero vector $d^e(x,y) = \mathbf{0}$ if an edge is not detected there.

Denote the unit tangent directions of f by $f^e = [f_1^e, \ldots, f_m^e]$. The direction at a point $f_i = (x_i, y_i)$ can be estimated by using a finite difference formula: $\frac{f_{i+1}-f_i}{|f_{i+1}-f_i|} + \frac{f_i-f_{i-1}}{|f_i-f_{i-1}|}$. It is normalized into a unit-length vector $f_i^e = (f_{x_i}^e, f_{y_i}^e)$, with $\|f_i^e(x,y)\| = 1$.

Constraints on the image d are imposed through the differences $\|f_i^e - d_i^e\|$. Assuming the differences have the same variance σ_e^2 for all i, we define the likelihood energy function as

$$E(d \mid f) = \frac{1}{\sigma_e^2} \sum_{i=1}^{m} \|f_i^e - d_i^e\|^2 \qquad (3.87)$$

where $d_i^e = d^e(x_i + x_u, y_i + y_u)$. The likelihood distribution can be defined as

$$p(d \mid f) = \frac{1}{Z'} \exp\{-E(d \mid f)\} \qquad (3.88)$$

where Z' is the normalizing constant.

Bayesian Estimates

In the Bayesian framework, the problem of extracting a contour with unknown deformation from a given image can be posed as a problem of maximum a posteriori (MAP) estimation. The posterior distribution of the Eigen-Snake is then defined as

$$p(f, \overline{f} \mid d) = \frac{p(d \mid f) \cdot p(f, \overline{f})}{p(d)} \qquad (3.89)$$

where $p(d \mid f) = p(d \mid f, \overline{f})$. The corresponding energy is

$$E(f, \overline{f} \mid d) = E(d \mid f) + E(f \mid \overline{f}) + E(\overline{f}) \qquad (3.90)$$

The MAP estimates f_{MAP} and $\overline{f}_{MAP}$ are then defined as

$$\{f_{MAP}, \overline{f}_{MAP}\} = \arg\min_{f, \overline{f}}\{E(f, \overline{f} \mid d)\} \qquad (3.91)$$

MultiEigenSnake

The EigenSnake is expanded into the MultiEigenSnake to detect a compound object with multiple separate subobjects. Suppose a compound object consists of L simple subobjects. The shape of each subobject is represented by a simple contour, and their combination is represented by a compound contour. Therefore, the MultiEigenSnake consists of $L + 1$ contours. For example, a

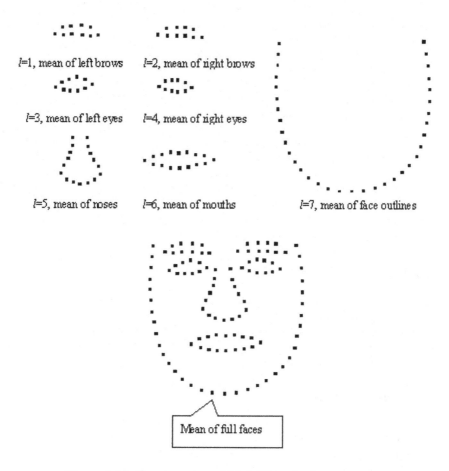

Figure 3.14: Construction of MultiEigenSnake contour.

face is considered to be composed of $L = 7$ subobjects: a face outline, a pair of brows, a pair of eyes, a nose, and a mouth. Figure 3.14 shows the construction of the MultiEigenSnake from individual training contours for the representation of a full face.

Let $\overline{f^{ME}}$ denote the prototype contours of the compound object and $\overline{f}_\ell$ denotes the prototype contour of the ℓth subobject. Corresponding to (3.90), we define the posterior energy for the MultiEigenSnake

$$E(\overline{f^{ME}}, f^{ME}, f^1, \cdots, f^L, \overline{f^1}, \cdots, \overline{f^L} \mid d) =$$
$$E(\overline{f^{ME}}) + \sum_{\ell=1}^{L}[E(d \mid f^\ell) + E(f^\ell \mid \overline{f^\ell})] \qquad (3.92)$$

The prior energy $E(\overline{f^{ME}})$ accounts for prior knowledge about local deformation of the compound object and the spatial configuration of its L subobjects.

There are two reasons why we partition a compound object into many submodels instead of using a single EigenSnake model.

1. The prior distribution of the EigenSnake, $p(\mathbf{f}\,|\overline{\mathbf{f}})$, is based on the assumption that every contour point is dependent only on its two direct neighboring points (the Markovian property). However, no neighboring relationship is defined between points belonging to different subobjects, and so each contour has to be modeled by its one energy $E_{GS}(\mathbf{f}^\ell, \overline{\mathbf{f}}^\ell, \mathbf{u}^\ell)$.

2. We assume that the shape of each subobject is stable, so the prior for each subobject can be modeled effectively by its shape model, and close fitting to the contour of a subobject can be achieved. However, the spatial configuration (i.e., relations among the subobjects) is subject to larger changes and therefore has to be modeled by using the PCA method. The effect of the strategy to design the MultiEigenSnake will be demonstrated in Section 3.8.2.

3.8.2 Experiments

Two sets of experiments are presented. The first examines the effects of using the specific prior knowledge and the shape space constraint by comparing the EigenSnake and other methods in extracting face outlines. The second set demonstrates the efficacy of the MultiEigenSnake in the extraction of the full facial features.

Comparison of EigenSnake with Other Methods

Here, four methods are compared: the original G-Snake (GS), enhanced G-Snake with the shape space constraint imposed (GS+), EigenSnake without the imposition of the shape space (ES−), and the full EigenSnake (ES). Their respective performances will tell us how the incorporation of the specific prior knowledge (by comparing ES− vs. GS, and ES vs. GS+) and the imposition of the shape space constraint (ES vs. ES−, and GS+ vs. GS) helps improve the extraction of the face outline.

The experiment data set contains 60 frontal face images of size 512×342 from the face database of Bern University. $m = 44$ landmark points are manually and evenly placed on the face outline of each training image as accurately as possible. A subset of 40 images are used as the *training set*, and all the 60 images are used as the *testing set*.

The mean contours, which were shown on the right in Fig. 3.13, is calculated from the training set. Eight eigenvectors are used to form the shape space. $M = 35$ eigenvectors are used to estimate the prior energy $E(\overline{f})$.

Contour initialization is done manually, as commonly practiced for snakes. It has the same shape as the mean contour but is rescaled randomly to 70% and 120% of the true mean scale, shifted away randomly from the true position by between ±20 pixels, and rotated randomly from the true orientation in the range of $\pm 15°$. All the methods have the same initial contour for each test image.

Table 3.1: Accuracy of the four methods.

Method	Init	GS	GS+	ES−	ES
D(method)	23.8	17.1	12.7	7.7	4.8

Table 3.1 compares the accuracy, denoted D(method), of the face outlines detected by the four methods (plus that of the initial outline). There, the accuracy is measured by the average Euclidean distance (in pixels) between corresponding points in the extracted outline and the manual marked outline, and the statistics are obtained with the results for the 60 test images. The compared results illustrate that both the prior energy $E(\bar{f})$ and the shape space constraint can improve the accuracy. The improvement in accuracy by using the shape space constraint may be calculated as $\Delta_S = [(D(\text{GS}) - D(\text{GS+})) + (D(\text{ES−}) - D(\text{ES}))]/2 = 3.65$ (the average of the two improvements, one from GS to GS+, the other from ES− to ES). The improvement in accuracy made by the use of the prior energy $E(\bar{f})$ may be calculated as $\Delta_P = [(D(\text{GS}) - D(\text{ES−})) + (D(\text{GS+}) - D(\text{ES}))]/2 = 8.65$ (the average of the two improvements, one from GS to ES−, the other from GS+ to ES). The two numbers Δ_S and Δ_P suggest that the contribution of the prior energy $E(\bar{f})$ is greater than that of the shape space constraint; in other words, more was done by imposing the specific prior knowledge than by using the shape space constraint.

Figure 3.15 shows an example of face outline extraction. It can be seen that the GS often fails to localize objects. Although GS+ can modify the results of GS so that it looks more like a "face outline", the improvement is less than that produced by ES− and ES. The two ES methods, which use the new energy functions with abundant prior knowledge incorporated, can effectively resist the disturbance from spurious structures, a noisy edge map, and improper initial scales.

The final energy values tell us the "correctness" of the four energy-minimization-based models (for a correct energy model, a better solution should have a lower energy). The GS+ produced better results than the GS; however, the final GS+ energies are higher than corresponding GS energies. This means that the GS energy does not really define the optimal solution in its minimum and is therefore "incorrect". This is due to a lack of constraints on the optimal solution. In contrast, the ES model has a proper energy function because a better solution is always associated with a lower energy.

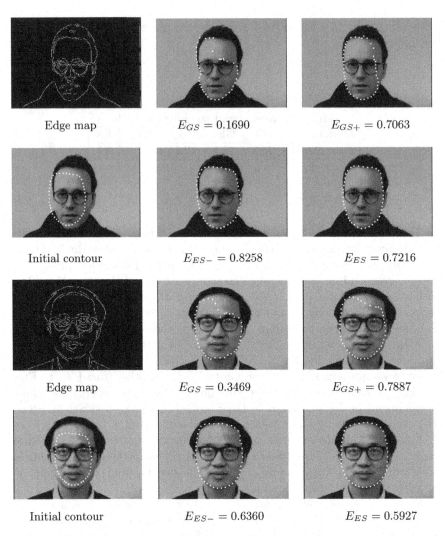

Figure 3.15: Comparison of the four methods.

Extracting Full Face Using MultiEigenSnake

This experiment demonstrates the MultiEigenSnake applied to extract a full face with $L = 7$ facial features. The training set consists of 120 frontal face images from the face database of Bern University. The testing set was chosen from the face database of Carnegie Mellon University.

A total of 108 landmark points are used to represent a full face, among which are 10 for the left brow, 10 for the right brow, 10 for the left eye, 10 for the right eye, 17 for the nose, 16 for the mouth, and 35 for the face boundary.

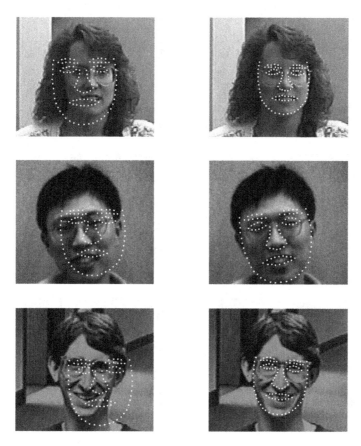

Figure 3.16: Examples of MultiEigenSnake results.

The means of the training contours (shown as Fig. 3.14) are used as initial prototype contours of L submodels and the compound model.

Some experimental results are shown in Fig. 3.16. Basically, the MultiEigenSnake can successfully extract the full face contours if the initial distance between the initial prototype contour and the target object is less than one-fourth of the initial prototype contour size. The MultiEigenSnake has certain scale- and rotation-insensitivity; for example, it allows up to $\pm 20°$ rotation of the initial prototype contour from the orientation of the true object.

Conclusion

The EigenSnake is trainable, general, and capable of representing any arbitrary shape. It incorporates prior knowledge about both rigid and nonrigid

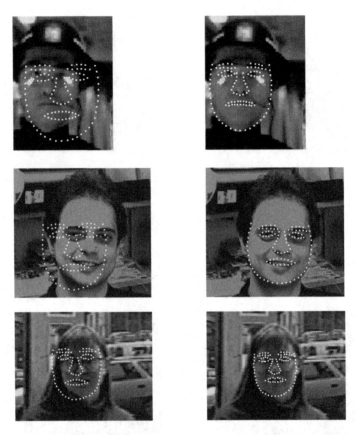

Figure 3.17: Examples of MultiEigenSnake results (continued).

deformations of expected shapes and deforms to fit the contour in a way that reflects variations in the training set. The EigenSnake has the ability not only to account for global changes due to rigid motions but also to learn local deformation of contours from examples.

Experimental results show that the EigenSnake is more robust to initialization, spurious features, and occlusion and can produce more accurate contours than existing methods. The results comparing various schemes show that both the incorporation of the prior knowledge about the specific shape of interest into the energy function and the shape space constraint can improve the accuracy. Of the two contributions, the improvement made by using the prior shape knowledge is greater than that for the shape space constraint.

Although the EigenSnake presented here is for the face contour and facial feature extraction, the method may be applied to other applications, such as tracking object in motion, as long as a set of training samples of the object shape is provided.

Chapter 4

High-Level MRF Models

High-level vision tasks, such as object matching and recognition and pose estimation, are performed on features extracted from images. The arrangements of such features are usually irregular, and hence the problems fall into categories LP3 and LP4. In this chapter, we present MAP-MRF formulations for solving these problems.

We begin with a study on the problem of object matching and recognition under contextual constraints. An MAP-MRF model is then formulated following the systematic approach summarized in Section 1.3.4. The labeling of a scene in terms of a model[1] object is considered as an MRF. The optimal labeling of the MRF is obtained by using the MAP principle. The matching of different types of features and multiple objects is discussed. A related issue, MRF parameter estimation for object matching and recognition, will be studied in Chapter 7.

We then derive two MRF models for pose computation, pose meaning the geometric transformation from one coordinate system to another. In visual matching, the transformation is from the scene (image) to the model object considered (or vice versa). In derived models, the transformation is from a set of object features to a set of image features. They minimize posterior energies derived for the MAP pose estimation, possibly together with an MRF for matching.

4.1 Matching under Relational Constraints

In high-level image analysis, we are dealing with image features, such as critical points, lines, and surface patches, that are more abstract than image pixels. Such features in a scene are not only attributed by (unary) properties about the features themselves but also related to each other by relations

---

[1] In this chapter, the word "model" is used to refer to both mathematical vision models and object models.

S.Z. Li, *Markov Random Field Modeling in Image Analysis*,
Advances in Pattern Recognition, DOI: 10.1007/978-1-84800-279-1_4,
© Springer-Verlag London Limited 2009

between them. In other words, an object or a scene is represented by features
constrained by the properties and relations. It is the bilateral or higher-order
relations that convey the contextual constraints. They play a crucial role in
visual pattern matching.

4.1.1   Relational Structure Representation

The features, properties and relations can be denoted compactly as a *relational structure* (RS) (Fischler and Elschlager 1973; Ambler et al. 1973; Cheng
and Huang 1984; Radig 1984; Li 1992c; Li 1992a). An RS describes a scene
or (part of) a model object. The problem of object recognition is reduced to
that of RS matching.

Let us start with a scene RS. Assume there are m features in the scene.
These features are indexed by a set $S = \{1, \ldots, m\}$ of sites. The sites
constitute the nodes of the RS. Each node $i \in S$ has associated with
it a vector $d_1(i)$ composed of a number of K_1 *unary properties* or *unary
relations*, $d_1(i) = [d_1^{(1)}(i), \ldots, d_1^{(K_1)}(i)]^T$. A unary property could be, for
example, the color of a region, the size of an area, or the length of a
line. Each pair of nodes $(i, i' \in S, i' \neq i)$ are related to each other by
a vector $d_2(i, i')$ composed of a number of K_2 *binary (bilateral) relations*,
$d_2(i, i') = [d_2^{(1)}(i, i'), \ldots, d_2^{(K_2)}(i, i')]^T$. A binary relation could be, for example, the distance between two points or the angle between two lines. More generally, among n features $i_1, \ldots, i_n \in S$, there may be a vector $d_n(i_1, \ldots, i_n)$
of K_n n-ary relations. This is illustrated in Fig. 4.1. An n-ary relation is also
called a relation, or constraint, of order n. The scope of relational dependencies can be determined by a neighborhood system $\mathcal{N}$ on S. Now, the RS for
the scene is defined by a triple

$$\mathcal{G} = (S, \mathcal{N}, d) \qquad (4.1)$$

where $d = \{d_1, d_2, \ldots, d_H\}$ and H is the highest-order. For $H = 2$, the RS
is also called a *relational graph (RG)*. The highest-order H cannot be lower
than 2 when contextual constraints must be considered.

The RS for a model object is similarly defined as

$$\mathcal{G}' = (\mathcal{L}, \mathcal{N}', D) \qquad (4.2)$$

where $D = \{D_1, D_2, \ldots, D_H\}$, $D_1(I) = [D_1^{(1)}(I), \ldots, D_1^{(K_1)}(I)]^T$, $D_2(I, I') = [D_2^{(1)}(I, I'), \ldots, D_2^{(K_2)}(I, I')]^T$, and so on. In this case, the set of labels $\mathcal{L}$
replaces the set of sites. Each element in $\mathcal{L}$ indexes one of the M model
features. In addition, the "neighborhood system" for $\mathcal{L}$ is defined to consist
of all the other elements, that is,

$$\mathcal{N}'_I = \{I' \mid \forall I' \in \mathcal{L}, I' \neq I\} \qquad (4.3)$$

This means each model feature is related to all the other model features. The
highest-order considered, H, in $\mathcal{G}'$ is equal to that in $\mathcal{G}$. For particular n and

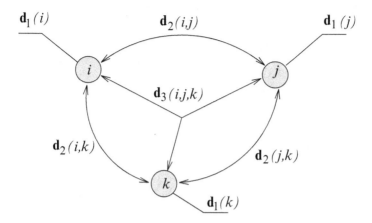

Figure 4.1: Three nodes and their unary, binary, and triple relations in the RS representation. From (Li 1992a) with permission; ©1992 Elsevier.

k $(1 \leq k \leq K_n; 1 \leq n \leq H)$, $D_n^{(k)}$ represents the same type of constraint as $d_n^{(k)}$; for example, both represent the angle between two line segments.

Relations of various orders impose unary, binary, ... , H-ary constraints on the features. Intercontextual constraints are represented by the second- or higher-order relations. Due to these constraints, a scene, an object, or a view of an object is seen as an integrated part rather than as individual features. The higher the order of relations is, the more powerful the constraints are but the higher the complication and expenses are in the computation.

There can be multiple model RSs in a model base. A model RS describes a whole model object or a part of it. When an RS is used to describe a part (for example, a view) of an object, the whole object may be described by several RSs and these RSs may be related by some inter-RS constraints.

Now introduce a virtual model composed of a single node $\mathcal{L}_0 = \{0\}$. It is called the NULL model. This special model represents everything not modeled by $\mathcal{G}'$, such as features due to all the other model objects and the noise. So the actual label set in matching the scene to the model plus the NULL contains $M + 1$ labels. It is denoted by

$$\mathcal{L}^+ = \{0, 1, \ldots, M\} \tag{4.4}$$

After the introduction of the NULL, the mapping from $\mathcal{S}$ to $\mathcal{L}$ is illustrated in Fig. 4.2.

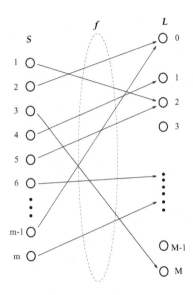

Figure 4.2: Discrete mapping involving the NULL label (numbered 0). All scene nodes (sites) not modeled by the object considered should be matched to this special label.

Figures 4.3 and 4.4 demonstrate two cup images and their segmentations based on the H-K surface curvatures (Besl and Jain 1985). Suppose we are matching the RG in Fig. 4.4(b) to that in Fig. 4.4(a). Based on the constraints from the unary properties of surface curvature and region area and the binary relations of distance between regions, the correct matching from Fig. 4.4(b) to Fig. 4.4(a) is $5 \rightarrow 1$, $1 \rightarrow 2$, and the rest to NULL.

Model-based matching can be considered as finding the optimal mapping from the image RS to the model RS (or vice versa). Such a mapping from one RS to another is called a *morphism*, written as

$$f : \mathcal{G}(\mathcal{S}, \mathcal{N}, d) \rightarrow \mathcal{G}'(\mathcal{L}, \mathcal{N}', D) \tag{4.5}$$

which maps each node in $\mathcal{S}$ to a node in $\mathcal{L}$

$$f : \mathcal{S} \rightarrow \mathcal{L} \tag{4.6}$$

and thus maps relations d_n to relations D_n

$$f : d_n \rightarrow D_n \tag{4.7}$$

A morphism is called an *isomorphism* if it is one-to-one and onto. It is called a *monomorphism* if it is one-to-one but not onto. It is called a *homomorphism* if it is many-to-one. We do not allow one-to-many mappings because

Figure 4.3: Cup images and segmentation based on *H-K* surface curvatures. Top: A cup image and its *H-K* map. Bottom: A transformed version of the cup image and the *H-K* map (note that some noise is introduced after the transformation due to quantization). From (Li 1992c) with permission; ©1992 Elsevier.

they contradict the definition of functions and, more crucially, increase the difficulties in finding the optimal solutions.

When the properties and relations in the RSs considered include numerical values, the mappings are numerical morphisms, which are more difficult to resolve than symbolic morphisms. Since such morphisms do not preserve relations in the exact, symbolic sense, they may be called *weak morphisms* – a term extended from the weak constraint models (Hinton 1978; Blake 1983; Blake and Zisserman 1987).

The goodness of a numerical morphism is usually judged by an objective function such as an energy. It is not very difficult to find a correct one-to-one mapping (isomorphism) between two *identical* RSs. The requirement that the unary properties of two nodes and the binary relations of two links must be exactly the same in order to be matched to each other provides a strong constraint for resolving the ambiguities. For the case where the two RSs have different numbers of nodes and the matched properties and relations are not

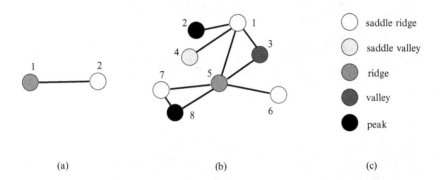

Figure 4.4: Relational graphs built from the cup *H-K* segmentation maps.
The textures of the nodes denote different surface types, and the links represent the adjacency relations. (a) The RG for the original upright cup image.
Node 1 corresponds to the body of the cup and node 2 to the handle. (b) The
RG for the transformed cup. Node 5 corresponds to the body, node 1 to the
handle, and the rest to NULL. (c) Legend for the correspondences between
texture types and *H-K* surface types. From (Li 1992c) with permission;
©1992 Elsevier.

exactly the same, the matching is more difficult because it cannot exploit
the advantage of the strong constraint of the exact equalities. Only "weak
constraints" are available. The subsequent sections use MRF's to establish
the objective function for weak homomorphisms for inexact, partial matching.

4.1.2 Work in Relational Matching

Two important computational issues in object matching are how to use contextual constraints and how to deal with uncertainties. Contextual constraints
in object recognition are often represented using the notion of relational
graphs. An object is represented by a set of features, their properties, and relations. Relative values between two image features embed context in matching
(Fischler and Elschlager 1973). Object matching is then reduced to matching between two relational graphs. On the other hand, noise is inevitably
introduced in the process of feature extraction and relation measurement.

In the maximum clique method (Ambler et al. 1973; Ghahraman et al.
1980) for relational graph matching, an associated graph is formed from the
two relational graphs, one for the scene and the other for a model object

(Ambler et al. 1973). Each node of the associated graph represents a possible match. The optimal matching is given by the maximal cliques.

A class of constraint satisfaction problems for matching was studied by Shapiro and Haralick (1981). In their work, the criterion is the weighted number of mismatched relations. Most such work involves definiting some criteria to describe the "goodness", or conversely the cost, of matching with relational constraints. Mistakes in symbolic relations are mapped into a number, and this number is then used as a criterion to decide whether a homomorphism is acceptable or not. The inexact matching problem is viewed as finding the best homomorphism (i.e., the one with minimum number of errors w.r.t. a given attribute value threshold, a missing part threshold, and a relation threshold).

Relaxation labeling (RL) (Rosenfeld et al. 1976) has been a useful method for solving the matching problem. The constraints are propagated via a compatibility function, and the ambiguity of labeling is reduced by using an iterative RL algorithm. In our view, the most important part in RL-based recognition is the definition of the compatibility function. Various RL schemes should be considered as algorithms for finding solutions. This will be further examined in Section 9.3.2.

Typical early works on relational matching using RL include (Davis 1979) and (Bhanu and Faugeras 1984; Bhanu 1984). In (Davis 1979), the objective function consists of four terms. Each term either encodes a particular constraint or penalizes unmatched features, the idea dating back to work by Fischler and Elschlager (1973). An association graph (Ambler et al. 1973) is used for matching relational structures. In the search for optimal matching, incompatible nodes for which some evaluation function is below a threshold are deleted from the graph. This generates a sequence of association graphs until a fixed point is reached. In (Bhanu and Faugeras 1984; Bhanu 1984), matching is posed as an optimization problem, and the optimization is performed by using an RL algorithm presented in (Faugeras and Berthod 1981).

A feature common to most of the matching works above is that thresholds are used to determine whether two matches are compatible. This effectively converts the weighted-graph matching into symbolic matching. While greatly reducing search space, this may rule out the potential matches. Because of the noise, the observation of the objects, which represents feature properties and relations extracted from the scene, can be considered as a set of random variables. Furthermore, some object features may be missing, and spurious features may emerge due to noise and unmodeled objects. The matching strategy has to deal with these uncertainties. It is hard to judge that in the presence of uncertainties a difference of 1.000001 is impossible while 0.999999 is possible.

In the weak constraint satisfaction paradigm, "hard" constraints are allowed to be violated without causing the failure of constraint satisfaction. However, each such violation is penalized by adding an amount to a cost function that measures the imperfection of the matching. This is usually implemented by using the line process in low-Level vision (Geman and Geman 1984; Marroquin 1985). In a weak notion of graph matching, Bienenstock (1988) proposed a scheme for an approximation of graph isomorphism in which relation-preserving characteristics of isomorphism can be violated but each violation incurs a small penalty. This is a transplant of the idea of the line process at the lower-level to the higher-level perception. Nevertheless, at a higher-level where more abstract representations are used, the weak constraint satisfaction problem becomes more complicated.

Li makes use of contextual constraints not only on the prior configuration of labelings but also on the observed data into the labeling process (Li 1992c; Li 1992a; Li 1992b; Li et al. 1993). He proposes an energy function, on a heuristic basis, that combines contextual constraints from both the prior knowledge and the observation. Kittler et al. (1993) later derive from probabilistic viewpoint the same compatibility used in Li's energy function.

MRF's provide a formal basis for matching and recognition under contextual constraints. Modestino and Zhang (1989) describe an MRF model for image interpretation. They consider an interpretation of a scene as an MRF and define the optimal matching as the MAP estimate of the MRF. Unfortunately, the posterior probability therein is derived not by using the laws of probability but designed directly by using some heuristic rules. This contradicts the promises of MAP-MRF modeling. Cooper (1990) describes a coupled network for simultaneous object recognition and segmentation. MRF is used to encode prior qualitative and possibly quantitative knowledge in the nonhomogeneous and anisotropic situations. The network is applied to recognize Tinkertoy objects. An interesting development are Markov processes of objects proposed by Baddeley and van Lieshout (1993). Other works in MRF-based recognition can be found in (Grenander et al. 1991; Baddeley and van Lieshout 1992; Friedland and Rosenfeld 1992; Kim and Yang 1992; Cooper et al. 1993). The MRF model described in the next section is based on (Li 1994a).

4.2   Feature-Based Matching

The labeling of a scene in terms of a model object is denoted by $f = \{f_i \in \mathcal{L}^+ \mid i \in \mathcal{S}\}$, where elements in $\mathcal{S}$ index image features and those in $\mathcal{L}$ model object features plus the NULL. It is also interpreted as a relational mapping from $\mathcal{G}(\mathcal{S}, \mathcal{N}, d)$ to $\mathcal{G}'(\mathcal{L}, \mathcal{N}', D)$; see (4.5). We assume that f is a realization of an MRF w.r.t. $\mathcal{N}$. Below, we derive its posterior probability using the MAP-MRF approach, in which contextual constraints not only on the prior configuration but also the observation are considered.

4.2.1 Posterior Probability and Energy

The prior distribution of f is defined using MRF's. In RS matching, the neighborhood covers all other related sites (scene features). One may restrict the scope of interaction by defining $\mathcal{N}_i$ as the set of other features that are within a distance r from i (see (2.3))

$$\mathcal{N}_i = \{i' \in \mathcal{S} \mid [\text{dist}(\text{feature}_{i'}, \text{feature}_i)]^2 \leq r, \ i' \neq i\} \quad (4.8)$$

where the function dist is a suitably defined distance function for features. The distance threshold r may be reasonably related to the size of the model object. The set of first-order cliques is

$$\mathcal{C}_1 = \{\{i\} \mid i \in \mathcal{S}\} \quad (4.9)$$

The set of second-order cliques is

$$\mathcal{C}_2 = \{\{i, i'\} \mid i' \in \mathcal{N}_i, i \in \mathcal{S}\} \quad (4.10)$$

Here, only cliques of up to order two are considered.
 The single-site potential is defined as

$$V_1(f_i) = \begin{cases} v_{10} & \text{if } f_i = 0 \\ 0 & \text{otherwise} \end{cases} \quad (4.11)$$

where v_{10} is a constant. If f_i is the NULL label, it incurs a penalty of v_{10}; otherwise the nil penalty is imposed . The pair-site potential is defined as

$$V_2(f_i, f_{i'}) = \begin{cases} v_{20} & \text{if } f_i = 0 \text{ or } \ f_{i'} = 0 \\ 0 & \text{otherwise} \end{cases} \quad (4.12)$$

where v_{20} is a constant. If either f_i or $f_{i'}$ is the NULL, it incurs a penalty of v_{20} or the nil penalty otherwise. The above clique potentials define the prior energy $U(f)$. The prior energy is then

$$U(f) = \sum_{i \in \mathcal{S}} V_1(f_i) + \sum_{i \in \mathcal{S}} \sum_{i' \in \mathcal{N}_i} V_2(f_i, f_{i'}) \quad (4.13)$$

The definitions of the above prior potentials are a generalization of that penalizing line process variables (Geman and Geman 1984; Marroquin 1985). The potentials may also be defined in terms of stochastic geometry (Baddeley and van Lieshout 1992).
 The conditional p.d.f., $p(d \mid f)$, of the observed data d, also called the *likelihood function* when viewed as a function of f given d fixed, has the following characteristics:

1. It is conditioned on pure nonNULL matches $f_i \neq 0$.

2. It is independent of the neighborhood system $\mathcal{N}$.

3. It depends on how the model object is observed in the scene, which in
   turn depends on the underlying transformations and noise.

Assume (1) that D and d are composed of types of features which are invariant
under the class of transformations considered[2]; and (2) that they are related
via the observation model

$$d_1(i) = D_1(f_i) + e_1(i), \qquad d_1(i, i') = D_2(f_i, f_{i'}) + e_2(i, i') \qquad (4.14)$$

where e is additive, independent, zero-mean Gaussian noise. The assumptions
of the independent and Gaussian noise may not be accurate but offer an
approximation when an accurate observation model is not available.

Then the likelihood function is a Gibbs distribution with the energy

$$U(d \mid f) = \sum_{i \in S, f_i \neq 0} V_1(d_1(i) \mid f_i) + \sum_{i \in S, f_i \neq 0} \sum_{i' \in S - \{i\}, f_{i'} \neq 0} V_2(d_2(i, i') \mid f_i, f_{i'})$$

$$(4.15)$$

where the constraints, $f_i \neq 0$ and $f_{i'} \neq 0$, restrict the summations to take
over the nonNULL matches. The likelihood potentials are

$$V_1(d_1(i) \mid f_i) = \begin{cases} \sum_{k=1}^{K_1} [d_1^{(k)}(i) - D_1^{(k)}(f_i)]^2 / \{2[\sigma_1^{(k)}]^2\} & \text{if } f_i \neq 0 \\ 0 & \text{otherwise} \end{cases}$$

$$(4.16)$$

and

$$V_2(d_2(i, i') \mid f_i, f_{i'}) = \begin{cases} \sum_{k=1}^{K_2} [d_2^{(k)}(i, i') - D_2^{(k)}(f_i, f_{i'})]^2 / \{2[\sigma_2^{(k)}]^2\} \\ \qquad\qquad \text{if } i' \neq i \text{ and } f_i \neq 0 \text{ and } f_{i'} \neq 0 \\ \\ 0 \qquad\qquad\qquad \text{otherwise} \end{cases}$$

$$(4.17)$$

where $[\sigma_n^{(k)}]^2$ $(k = 1, \ldots, K_n$ and $n = 1, 2)$ are the variances of the corresponding noise components. The vectors $D_1(f_i)$ and $D_2(f_i, f_{i'})$ are the "mean
vectors", conditioned on f_i and $f_{i'}$, for the random vectors $d_1(i)$ and $d_2(i, i')$,
respectively.

Using $U(f \mid d) = U(f) + U(d \mid f)$, we obtain the posterior energy

$$U(f \mid d) = \begin{aligned}[t] &\sum_{i \in S} V_1(f_i) + \\ &\sum_{i \in S} \sum_{i' \in \mathcal{N}_i} V_2(f_i, f_{i'}) + \\ &\sum_{i \in S: f_i \neq 0} V_1(d_1(i) \mid f_i) + \\ &\sum_{i \in S: f_i \neq 0} \sum_{i' \in S - \{i\}: f_{i'} \neq 0} V_2(d_2(i, i') \mid f_i, f_{i'}) \end{aligned}$$

$$(4.18)$$

There are several parameters involved in the posterior energy: the noise variances $[\sigma_n^{(k)}]^2$ and the prior penalties v_{n0}. Only the relative, not absolute,
values of $[\sigma_n^{(k)}]^2$ and v_{n0} are important because the solution f^* remains the

---

[2]The discovery and computation of visual invariants is an active area of research; see
(Mundy and Zisserman 1992).

same after the energy E is multiplied by a factor. The v_{n0} in the MRF prior potential functions can be specified to achieve the desired system behavior. The higher the prior penalties v_{n0}, the fewer features in the scene will be matched to the NULL for the minimal energy solution.

Normally, symbolic relations are represented internally by a number. The variances $[\sigma_n^{(k)}]^2$ for those relations are zero. One may set corresponding $[\sigma_n^{(k)}]^2$ to 0^+ (a very small positive number), which is consistent with the concept of discrete distributions. Setting $[\sigma_n^{(k)}]^2 = 0^+$ causes the corresponding distance to be infinitely large when the symbolic relations compared are not the same. This inhibits symbolically incompatible matches, if an optimal solution is sought and thus imposes the desired symbolic constraint. A method for learning $[\sigma_n^{(k)}]^2$ parameters from examples will be presented in Chapter 8.

4.2.2   Matching to Multiple Objects

The MAP configuration f^* derived in the above is the optimal mapping from the scene to the model object under consideration. In other words, it is the optimal labeling of the scene in terms of the model object.

Suppose there are L potential model objects. Then L MAP solutions, $f^{(1)}, \ldots, f^{(L)}$, can be obtained after matching the scene to each of the models in turn.[3] However, any feature in the scene can have only one match of model feature. To resolve this, we use the following method of cost minimization.

Rewrite $E(f) = U(f \mid d)$ in (4.18) in the form

$$E(f) = \sum_{i \in \mathcal{S}} E_1(f_i) + \sum_{i \in \mathcal{S}} \sum_{i' \in \mathcal{S}, i' \neq i} E_2(f_i, f_{i'}) \triangleq \sum_{i \in \mathcal{S}} E(f_i \mid f_{\mathcal{N}_i}) \qquad (4.19)$$

where

$$E_1(f_i) = \begin{cases} v_{10} & \text{if } f_i = 0 \\ V_1(d_1(i) \mid f_i) & \text{otherwise} \end{cases} \qquad (4.20)$$

and

$$E_2(f_i, f_{i'}) = \begin{cases} v_{20} & \text{if } i' \in \mathcal{N}_i \ \& \ (f_i = 0 \text{ or } f_{i'} = 0) \\ V_2(d_2(i, i') \mid f_i, f_{i'}) & \text{if } i' \neq i \text{ and } f_i \neq 0 \text{ and } f_{i'} \neq 0 \\ 0 & \text{otherwise} \end{cases}$$
$$(4.21)$$

are local posterior energies of orders one and two, respectively. $E(f_i \mid f_{\mathcal{N}_i}) = E_1(f_i) + \sum_{i' \in \mathcal{S}} E_2(f_i, f_{i'})$ is the cost incurred by the local match $i \rightarrow f_i$ given the rest of the matches. It will be used as the basis for selecting the best-matched objects for i in matching to multiple model objects. The image

---

[3]Fast indexing of model objects (Lamdan and Wolfson 1988) is a topic not studied in this work.

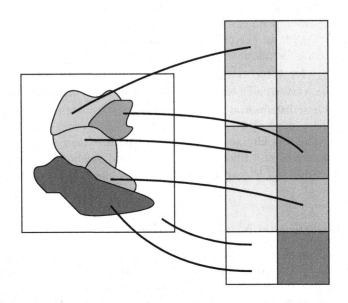

Figure 4.5: Mapping from the scene to multiple model objects. Different textures represent different structures. Bold lines represent submappings. Note that the background and nonmodel structure are mapped to the NULL structure (the blank square). From (Li 1992b) with permission; ©1992 Elsevier.

feature i is considered to come from object ℓ_i if the decision incurs the least cost

$$\ell_i = \arg \min_{\ell \in \{1,...,L\}} E(f_i^{(\ell)} \mid f_{\mathcal{N}_i}^{(\ell)}) \qquad (4.22)$$

The final label for i is feature number $f_i^{(\ell_i)}$ of object ℓ_i. Note, however, that the MAP principle is applied to matching to a single model object, not to multiple objects; the simple rule (4.22) does not maximize the posterior since at least the partition functions are different for matching to different objects.

Applying (4.22) to every i yields an overall mapping from the scene to the models, composed of several submappings, as illustrated in Fig. 4.5. On the right, the squares with different textures represent the candidate model structures to which the scene is to be matched. Among them, the blank square represents the NULL model. On the left is the scene structure. The regions, each corresponding to a subpart of a model structure, are overlapping (not separated). The recognition of the overlapping scene is (1) to partition the scene structure into parts such that each part is due to a single object and (2) to find correspondences between features in each part and those in the corresponding model object. The parts of the scene corresponding to the background and unmodeled objects should be mapped to the NULL, or in other words assigned the NULL label.

4.2.3 Extensions

The model above may be extended in a number of ways. With the assumption that different constraints are independent of each other, embedding a higher constraint can be achieved by adding a new energy term. Matching with different types of object features (e.g., points and lines) can be treated as coupled MRF's.

Incorporating Higher Constraints

In matching schemes based on invariants, the features chosen to represent the object modeled and the scene must be invariant to the expected transformations from the object to the observation for the process of cognition to be accomplished. The more complicated the transformations are, the higher the order of features needed for the invariant object representation (Li 1992a); the order needed may be higher than two.

In previous subsections, only constraints of up to second-order were considered. Incorporation of higher-order constraints can be achieved by adding higher-order energy terms. A clique of order $n > 2$ is an n-tuple $\{i_1, \ldots, i_n\}$ in which i_r and i_s $(r \neq s)$ are neighbors to each other. The incorporation is done as follows. First, the following n-th order a priori clique potentials are added to the prior energy $U(f)$:

$$V_{n0}(f_{i_1}, \ldots, f_{i_n}) = \begin{cases} v_{n0} & \text{if } f_{i_k} = 0 \quad \exists i_k \in \{i_1, \ldots, i_n\} \\ 0 & \text{otherwise} \end{cases} \quad (4.23)$$

where v_{n0} is a constant of prior penalty. Second, the likelihood energy for the nth order observation has the likelihood potentials

$$V_n(d_n(i_1, \ldots, i_n) \mid f_{i_1}, \ldots, f_{i_n}) = \\ \sum_{k=1}^{K_n} [d_n^{(k)}(i_1, \ldots, i_n) - D_n^{(k)}(f_{i_1}, \ldots, f_{i_n})]^2 / \{2[\sigma_n^{(k)}]^2\} \quad (4.24)$$

The corresponding posterior can be obtained using the Bayes rule, resulting in the nth order energy

$$E_n(f_{i_1}, \ldots, f_{i_n}) = \begin{cases} v_{n0}, & \text{if } f_{i_k} = 0 \; \exists i_k \in \{i_1, \ldots, i_n\} \\ V_n(d_n(i_1, \ldots, i_n) \mid f_{i_1}, \ldots, f_{i_n}), & \text{otherwise} \end{cases} \quad (4.25)$$

Adding together all the energy terms yields

$$E(f) = \sum_{i \in \mathcal{S}} E_1(f_i) + \sum_{i,i' \in \mathcal{S}} E_2(f_i, f_{i'}) + \cdots \\ + \sum_{i_1, \ldots, i_H \in \mathcal{S}} E_H(f_{i_1}, \ldots, f_{i_H}) \quad (4.26)$$

where H is the highest-order.

Coupled MRF's for Matching with Different Features

Let us consider the situation where an object consists of different types of features, such as points and lines. Obviously, a point in the scene should not be matched to a line in an object model. This is a symbolic constraint. In this case, the positivity condition of MRF in (2.8) does not hold any more if the configuration space $\mathbb{F}$ is still defined as the simple product as in (4.4) for a single MRF.

To overcome this limitation, we partition the whole set $\mathcal{L}$ of labels to a few admissible sets for different types of sites. This results in a few coupled MRF's. These MRF's are coupled to each other via inter-relations d_n $(n \geq 2)$. For example, the distance between a point and a line can constrain the two different types of features. Furthermore, they are also coupled via the label NULL which is a "wildcard" compatible with all types of features.

If there are two different types of features, then $\mathcal{L}$ can be partitioned into two admissible sets, with each set consisting of indices to features of the same type. In the most general case, each of the m sites has its own set of labels $\mathcal{L}_i \subseteq \mathcal{L}$ $(i = 1, \ldots, m)$, each $\mathcal{L}_i$ being determined using the symbolic unary constraints; and the label for site i assumes a value $f_i \in \mathcal{L}_i^+$, where $\mathcal{L}_i^+ = \{0\} \cup \mathcal{L}_i$. Then, the configuration space is defined as

$$\mathbb{F} = \mathcal{L}_1^+ \times \mathcal{L}_2^+ \times \cdots \times \mathcal{L}_m^+ \qquad (4.27)$$

In this situation, the energy $E(f)$ has the same form as usual and the solution is still found by $f^* = \arg\min_{f \in \mathbb{F}} E(f)$. The only difference is in the definition of the configuration space $\mathbb{F}$ in which the solution is searched for.

Relationships with Low Level MRF Models

Let us compare the present model with low-Level vision MRF models prototyped by Geman and Geman (1984). The present model is similar to the MRF models for piecewise constant image restoration, edge detection, and texture segmentation in that the labels are discrete. Of course, their prior distributions must be different to cope with different tasks.

In surface reconstruction involving discontinuities (Marroquin 1985; Blake and Zisserman 1987; Chou and Brown 1990; Szeliski 1989; Geiger and Girosi 1991), there are commonly two coupled MRF's: a surface field and a line process field. The former field is defined on $\mathcal{S}_1$, the domain of an image grid. It assumes configurations in the space $\mathcal{L}_1^{\mathcal{S}_1}$ where $\mathcal{L}_1$ is a real interval. The latter is defined on $\mathcal{S}_2$, the dual of $\mathcal{S}_1$. It assumes configurations in the space $\mathcal{L}_2^{\mathcal{S}_2}$, where $\mathcal{L}_2$ is the set of labels such as {edge, nonedge}. These fields are coupled to each other by the interaction between the line process variable and the neighboring pixels.

The concept of discontinuity in the high-level is the relational bond in the scene RS. For example, when no f_i or $f_{i'}$ assumes the NULL value, i and i' are relationally constrained; otherwise, when $f_i = 0$ or $f_{i'} = 0$, the relational bond between i and i' is broken. This corresponds to the line process.

The main difference between this high-level model and those low-Level models is in the encoding of higher-order relational constraints. Low-level models use unary observations only, such as pixel intensity; although intensity difference between neighboring pixels is also used, it is derived directly from the intensity. The present model uses relational measurements of any order. This is important for high-level problems in which contextual constraints play a more important role. Moreover, in the present model, the neighborhood system is nonhomogeneous and anisotropic, which also differs from the image case.

The matching method we have presented is based on a prerequisite that invariants are available for object representation under the group of transformations concerned. If geometric variants are also used as sources of constraints, object poses have to be resolved during the computation of matching; see the next section.

4.2.4 Experiments

The following presents some experiments. Given a number of model objects, a scene is generated. Properties and relations in model objects and the scene are measured using the same program. Only the second-order energy E_2 is taken into consideration. The energy is minimized by using a relaxation labeling algorithm (see Section 9.3.2). In the computation of the minimal solution, interactions or compatibilities are represented by integers of only 8 bits and good results are achieved; this demonstrates the error-tolerant aspect of the model. The optimal matching result is displayed by aligning the matched object features to the scene while the unmatched are not displayed. The alignment is performed between the matched pairs by using the least squares fitting method (Umeyama 1991). The parameter $v_{20} = 0.7$ is fixed for all the experiments. Parameters $[\sigma_2^{(k)}]^2$ vary for different applications.

Matching Objects of Point Patterns

There are three model objects as shown in Fig. 4.6(a)–Fig. 4.6(c). Each of the objects consists of three types of point features, shown in different sizes. The scene in (d) is generated from the model objects as follows: (1) Take a subset of features from each of the three objects, (2) do a transformation (rotation and translation) on each of the subsets, (3) mix the subsets together after that, (4) randomly deviate the locations of the points using either Gaussian or uniform noise, and (5) add spurious point features.

In this case of point matching, there is only one unary property (i.e., the point size) denoted $d_1(i)$. Each $d_1(i)$ takes a value $\{1, 2, 3\}$. There is only a

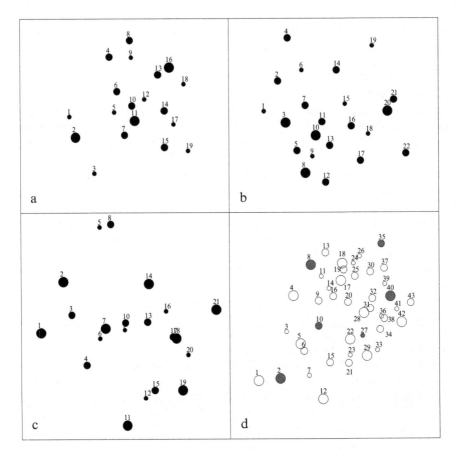

Figure 4.6: Matching objects of points. (a–c) The model objects. (d) The scene.

binary relation; i.e., the Euclidean distance between two points $d_2(i, i') = \text{dist}(i, i')$. The points, their sizes, and their distances constitute an RS. The unary property is symbolic, and this restricts the set of admissible labels for each $i \in \mathcal{S}$ as $\mathcal{L}_i = \{I \mid D_1(I) = d_1(i), \forall I \in \mathcal{L}\}$. The parameter is chosen as $[\sigma_2^{(1)}]^2 = 0.1$.

Figure 4.7 shows the matching results in which the matched object points are aligned with the scene in Fig. 4.7(d). The black points in Fig. 4.7(a) correspond to points 5, 6, 7, 11, 12, 14, 15, 17, and 19 of the object in Fig. 4.6(a). Those in Fig. 4.7(b) correspond to points 3, 5, 8, 9, 10, 11, 12, 13, 16, 17, and 18 of the object in Fig. 4.6(b). Those in Fig. 4.7(c) are points 4, 6, 7, 9, 10, 11, 12, 13, 14, 15, 16, 17, 18, 19, 20, and 21 of the object in Fig. 4.6(c). In (d) is shown the union of all the individual results. The spurious points 2, 8, 10, 27, 35, and 40 in Fig. 4.6(d) have found no counterparts in the three objects. They are correctly classified as the NULL. There is one mismatch: Point 19 in Fig. 4.6(d) is matched to the NULL while its correct home should be point 10 of Fig. 4.6(a).

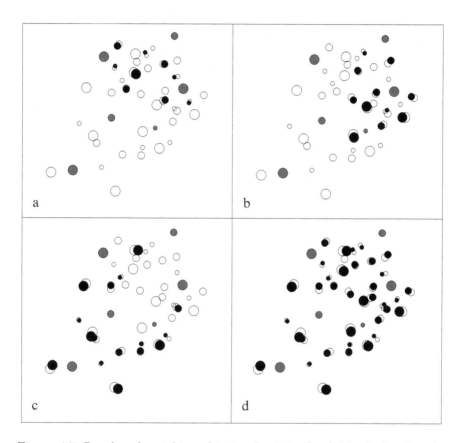

Figure 4.7: Results of matching objects of points. (a–c) Matched points (in black) from the respective models aligned with the scene. (d) All matched points aligned with the scene.

Matching Objects of Line Patterns

There are five objects made of lines, as shown in Fig. 4.8(a)–Fig. 4.8(e). The scene in Fig. 4.8(f) consists of a subset of deviated lines taken from the first three objects Fig. 4.8(a)–Fig. 4.8(c), shown as dotted lines, and spurious line features shown as dashed lines. Four types of binary relations are measured:

(1) $d_2^{(1)}(i, i')$: the angle between lines i and i';

(2) $d_2^{(2)}(i, i')$: the distance between the midpoints of the lines;

(3) $d_2^{(3)}(i, i')$: the minimum distance between the endpoints of the lines; and

(4) $d_2^{(4)}(i, i')$: the maximum distance between the endpoints of the lines.

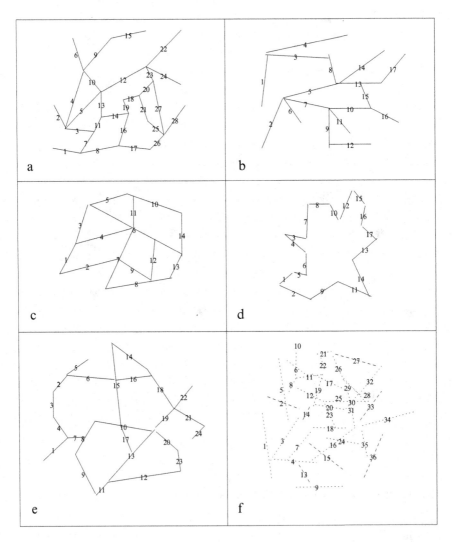

Figure 4.8: Matching objects of lines. (a–e) The five model objects. (f) The scene.

No unary relations are used. The value for the prior clique potential is fixed at $v_{20} = 0.70000$. The values for weighting the binary measurements are $1/[\sigma_2^{(1)}]^2 = 0.36422$, $1/[\sigma_2^{(2)}]^2 = 0.20910$, $1/[\sigma_2^{(3)}]^2 = 0.44354$, and $1/[\sigma_2^{(4)}]^2 = 0.74789$, which are estimated using a supervised learning procedure to be presented in Chapter 8.

Figure 4.9 shows the matching results in which the matched object lines from the first three objects are aligned with the scene in Fig. 4.9(d). The solid lines in Fig. 4.9(a) correspond to lines 11, 13, 14, 16, 17, 18, 19, 21, 25, 26, and 28 of the object in Fig. 4.8(a). Those in Fig. 4.9(b) correspond to lines 5, 8, 10, 11, 13, 14, 15, 16, and 17 of the object in Fig. 4.8(b). Those in Fig. 4.9(c) correspond to points 1, 2, 3, 5, 8, 9, 11, 12, and 14 of the object in Fig. 4.8(c). Objects in Fig. 4.9(d) and Fig. 4.9(e) do not have matches. In Fig. 4.9(d) is shown the union of all individual results. The spurious lines 2, 13, 14, 15, 27, 33, and 36 in Fig. 4.8(f) have found no counterparts in the object models. They are correctly classified as the NULL.

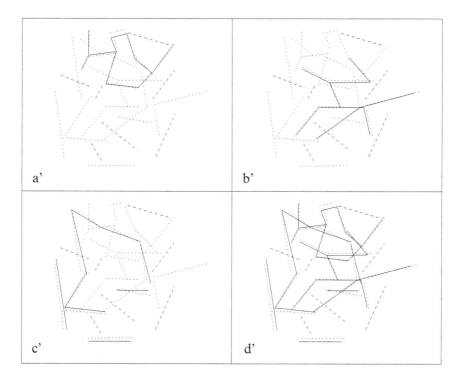

Figure 4.9: Results of matching objects of lines. (a–c) Matched lines (solid) from the respective models aligned with the scene. (d) All matched lines aligned with the scene.

Matching Curved Objects under Similarity Transformations

In the experiment shown in Fig. 4.10. There are eight model jigsaw objects. The scene contains rotated, translated, and scaled parts of the model objects, some of which are considerably occluded. Boundaries are computed from the image using the Canny detector followed by hysteresis and edge linking. After that, corners of the boundaries are located as $p_1, \ldots, p_m$.

The sites correspond to the corners on the scene curve and the labels correspond to the feature points on the model curve considered. The neighbors of a site are defined as the five forward points and the five backward points. Invariant relations are derived from the boundaries as well as the corners based on a similarity-invariant representation of curves (Li 1993). No unary properties are used ($K_1 = 0$). Only binary relations are used, which are of the following five types ($K_2 = 5$):

(1) $d_2^{(1)}(i, i')$: ratio of curve arc length $\widehat{p_i p_{i'}}$ and chord length $\overline{p_i p_{i'}}$,

(2) $d_2^{(2)}(i, i')$: ratio of curvature at p_i and $p_{i'}$,

(3) $d_2^{(3)}(i, i')$: invariant coordinates vector,

(4) $d_2^{(4)}(i, i')$: invariant radius vector, and

(5) $d_2^{(5)}(i, i')$: invariant angle vector

which are derived from the boundaries and the corners using a similarity-invariant representation of curves (Li 1993).

The parameters involved are supplied by an automated optimal estimation procedure (see Chapter 8) with the values $v_{20} = 0.7$, $1/\sigma_2^{(1)} = 0.00025$, $1/\sigma_2^{(2)} = 0$, $1/\sigma_2^{(3)} = 0.04429$, $1/\sigma_2^{(4)} = 0.02240$, and $1/\sigma_2^{(5)} = 0.21060$. The minimal labeling f^* is found by using a deterministic relaxation labeling algorithm (Hummel and Zucker 1983). The final result of recognition is shown in Fig. 4.11, in which model objects are aligned with the corresponding (sub)parts in the scene. Note that most of the model objects share common structures of round extrusion and intrusion. This means extensive ambiguities exist, which has to be resolved by using context.

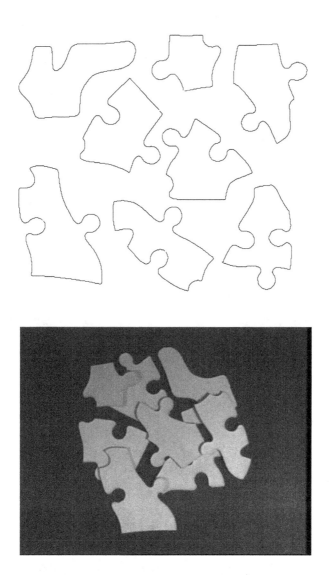

Figure 4.10: (Top) The eight model jigsaw objects. From (Li 1997b) with permission; ©1997 Kluwer. (Bottom) An overlapping scene.

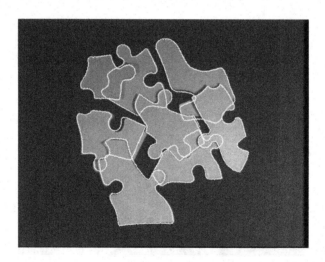

Figure 4.11: (Top) Boundaries detected from the scene. (Bottom) Matched objects are aligned with the scene. From (Li 1994a) with permission; ©1994 IEEE.

4.3 Optimal Matching to Multiple Overlapping Objects

Previously, matching between the scene and model objects was performed by considering one model object at a time. Once a part of the scene has been matched to a model object, it is excluded from subsequent matching; or better still, multiple matching results are obtained, each being optimal w.r.t. a single model object, as was done in Section 4.2 (see also (Li 1998a)). Inconsistencies may exist among the individual results because the matching to one model is done independently of the other models. Post-processing may be applied to obtain an overall consistent solution.

When a scene contains multiple mutually occluded objects, two sources of contextual constraints are required: *between-object constraints* (BOCs) and *within-object constraints* (WOCs). WOCs of an object, which describe the particular structure of the object itself, are used to identify instances of that object. BOCs, which are the constraints on features belonging to different objects, are used to discriminate between objects and unmodeled features.

Here, a statistically optimal, MAP-MRF-based formulation for recognition of multiple, partially occluded objects is presented. The MAP solution is defined w.r.t. all model objects, not just individual ones. Such a solution is optimal overall, and consistent by itself. A two-stage MAP estimation approach is proposed to reduce the computational cost. The first stage finds feature correspondence between the scene and *each* model object. The second stage solves the original, target MAP-MRF estimation problem in a much reduced space constructed from the stage 1 solutions. The energy functions for the two MAP estimation problems are formulated. BOCs are encoded in the prior distribution modeled as a Markov random field (MRF). WOCs are encoded in the likelihood distribution modeled as a Gaussian. This way, both BOCs and WOCs are incorporated into the posterior distribution. Experimental results are incorporated into the presentation to illustrate the theoretical formulation.

4.3.1 Formulation of MAP-MRF Estimation

Figure 4.12 illustrates an example of object and scene representation. Let $\mathcal{O}^{(all)} = \{\mathcal{O}^{(1)}, \ldots, \mathcal{O}^{(L)}\}$ be a number of L model objects ($L = 8$ on the left in Fig. 4.12). The objects in the scene are rotated, translated, scaled, and partially occluded versions of the model objects (Fig. 4.12, middle). The objects and the scene are represented by features (e.g., corners on boundary curves on the right in Fig. 4.12) and constraints on the feature such as unary properties and binary relations between features (the interested reader is referred to (Li 1997a) for an invariant representation of curved objects). The task is to recognize (separate and identify) the objects in the scene, optimally in the MAP sense, w.r.t. the L given objects. This can be done via feature

matching, which is aimed at establishing feature correspondence between the
scene and the model objects based on partial observation of the objects.

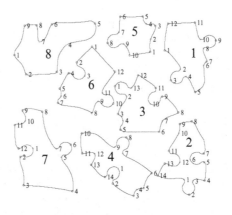

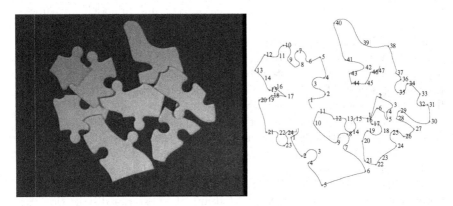

Figure 4.12: Top: Eight model jigsaw objects. Lower left: A scene. Lower right:
Three boundary curves and the corner features extracted from the scene.

In this problem, we have $\mathcal{S} = \{1, \ldots, m\}$ corresponds to the set of m
image features. On the right in Fig. 4.12, there are three sequences of corner
features in the scene, represented by three $\mathcal{S}$ sets with $m = 24$, $m = 47$, and
$m = 6$.

Let $\mathcal{L}^{(\alpha)} = \{1, \ldots, M^{(\alpha)}\}$ be a set of $M^{(\alpha)}$ labels corresponding to the set
of $M^{(\alpha)}$ features for model object α ($M^{(1)} = 12$ for model object 1 on the
left in Fig. 4.12). A virtual label, called the NULL and numbered 0, is added
to represent everything not belonging to $\mathcal{L}^{(\alpha)}$ (such as features due to other
model objects and those due to background and noise). This augments $\mathcal{L}^{(\alpha)}$
into $\mathcal{L}^{(\alpha)^+} = \{0, 1, \ldots, M^{(\alpha)}\}$. Without confusion, the notation $\mathcal{L}$ is still used
to denote the augmented set $\mathcal{L}^+$ unless there is a need to elaborate. The set

of all model features plus the NULL is $\mathcal{L}^{(all)} = \mathcal{L}^{(1)} \cup \mathcal{L}^{(2)} \cdots \mathcal{L}^{(L)}$. It consists of $\#\mathcal{L}^{(all)}$ elements where $\#\mathcal{L}^{(all)} = 1 + \sum_{\alpha=1}^{L} M^{(\alpha)}$.

The overall matching from $\mathcal{S}$ to the L model objects is represented by a label configuration $f = \{f_1^{(\alpha_1)}, \ldots, f_m^{(\alpha_m)}\}$. It is a mapping from the set of the sites to the set of the labels, $f : \mathcal{S} \to \mathcal{L}^{(all)}$. Three things are told by a label $f_i^{(\alpha_i)} \in \mathcal{L}^{(\alpha_i)}$. (i) It separates image features belonging to a model object from those not belonging to any in the following way: If $f_i^{(\alpha_i)} \neq 0$ (a nonNULL label), then image feature i belongs to an object; otherwise, if $f_i^{(\alpha_i)} = 0$, it belongs to the background, noise, or an unmodeled object. (ii) If $f_i^{(\alpha_i)} \neq 0$, α_i indicates that image feature i belongs to model object α_i. (iii) If $f_i^{(\alpha_i)} \neq 0$, $f_i^{(\alpha_i)}$ indexes the corresponding feature of object α_i, to which image feature i is matched.

The MAP solution for matching the scene to all the objects is defined by

$$f^* = \arg \max_{f \in \mathbb{F}^{(all)}} P(f \mid d, \mathcal{O}^{(all)}) \tag{4.28}$$

where $P(f \mid d, \mathcal{O}^{(all)})$ is the posterior probability of the labeling f given the observation d and the L object models, and $\mathbb{F}^{(all)}$ is the space of all admissible configurations (solutions). When all the labels are admissible for all the sites, $\mathbb{F}^{(all)}$ is the Cartesian product of the m $\mathcal{L}^{(all)}$'s; that is, $\mathbb{F}^{(all)} = \prod_{i=1}^{m} \mathcal{L}^{(all)}$.

Assuming that f, which is a realization of a family of m random variables, is a Markov random field (MRF), then its posterior is a Gibbs distribution $P(f \mid d, \mathcal{O}^{(all)}) \propto e^{-E^{(all)}(f)}$ where

$$E^{(all)}(f) \triangleq U(f | \mathcal{O}^{(all)}) + U(d \mid f, \mathcal{O}^{(all)}) \tag{4.29}$$

is the posterior energy consisting of the prior energy $U(f \mid \mathcal{O}^{(all)})$ and the likelihood energy $U(d \mid f, \mathcal{O}^{(all)})$. The solution to problem (4.28) equivalently minimizes the posterior energy: $f^* = \arg \min_{f \in \mathbb{F}^{(all)}} E^{(all)}(f)$.

The objective of (4.28), finding the minimum f^* in $\mathbb{F}^{(all)}$, is a formidable job since the configuration space $\mathbb{F}^{(all)}$ consists of a huge number of $\#\mathbb{F}^{(all)} = (1 + \sum_{\alpha=1}^{L} M^{(\alpha)})^m$ elements when all the model features (labels) are admissible. In the following, a two-stage MAP-MRF estimation approach is proposed to tackle this problem.

Formulation of Energy Functions

The prior distribution of $f^{(\alpha)} = \{f_1^{(\alpha)}, \ldots, f_m^{(\alpha)}\}$ for matching w.r.t. object α is assumed to be an MRF and hence is a Gibbs distribution $P(f^{(\alpha)} \mid \mathcal{O}^{(\alpha)}) \propto e^{-U(f^{(\alpha)} \mid \mathcal{O}^{(\alpha)})}$. The prior energy takes the form

$$U(f^{(\alpha)} \mid \mathcal{O}^{(\alpha)}) = \sum_{i \in \mathcal{S}} V_1(f_i^{(\alpha)}) + \sum_{i \in \mathcal{S}} \sum_{i' \in \mathcal{N}_i} V_2(f_i^{(\alpha)}, f_{i'}^{(\alpha)}) \tag{4.30}$$

where $\mathcal{N}_i$ is the set of neighbors for i, and $V_1(f_i^{(\alpha)})$ and $V_2(f_i^{(\alpha)}, f_{i'}^{(\alpha)})$ are single- and pair-site clique prior potential functions, respectively for $f^{(\alpha)}$. The clique potentials are defined based on (4.11(and (4.12) as

$$V_1(f_i^{(\alpha)}) = \begin{cases} 0 & \text{if } f_i \neq 0 \\ v_{10} & \text{if } f_i = 0 \end{cases} \tag{4.31}$$

$$V_2(f_i^{(\alpha)}, f_{i'}^{(\alpha)}) = \begin{cases} 0 & \text{if } f_i^{(\alpha)} \neq 0 \text{ and } f_{i'}^{(\alpha)} \neq 0 \\ v_{20} & \text{if } f_i^{(\alpha)} = 0 \text{ or } f_{i'}^{(\alpha)} = 0 \end{cases} \tag{4.32}$$

where $v_{10} > 0$ and $v_{20} > 0$ are penalty constants for NULL labels. These definitions encode BOCs, that is, constraints between different objects and between an object and the background. In a way, this is similar to the line process model (Geman and Geman 1984) for differentiating edge and nonedge elements.

The likelihood distribution $p(d \mid f^{(\alpha)}, \mathcal{O}^{(\alpha)})$ describes the statistical properties of the features seen in the scene and is therefore conditioned on pure nonNULL matches $(f_i^{(\alpha)} \neq 0)$ only. The likelihood is a Gibbs distribution with the energy function, which is based on (4.15)

$$U(d \mid f^{(\alpha)}, \mathcal{O}^{(\alpha)}) = \sum_{i \in \mathcal{S}, f_i^{(\alpha)} \neq 0} V_1(d_1(i) \mid f_i^{(\alpha)}) \tag{4.33}$$

$$\sum_{i \in \mathcal{S}, f_i^{(\alpha)} \neq 0} \sum_{i' \in \mathcal{S} \setminus i, f_{i'}^{(\alpha)} \neq 0} V_2(d_2(i, i') \mid f_i^{(\alpha)}, f_{i'}^{(\alpha)})$$

where $d_1(i)$ is the set of unary properties of image feature i, $d_2(i, i')$ is the set of binary relations between i and i', and $V_1(d_1(i) \mid f_i^{(\alpha)})$ and $V_2(d_2(i, i') \mid f_i^{(\alpha)}, f_{i'}^{(\alpha)})$ are the potentials in the likelihood distributions (where the distributions may be assumed to be Gaussian). The likelihood potentials encode WOCs, that is, constraints on image features belonging to object α only. Both BOCs and WOCs are incorporated into the posterior distribution with the posterior energy $E^{(\alpha)}(f^{(\alpha)}) = U(f^{(\alpha)} \mid \mathcal{O}^{(\alpha)}) + U(d \mid f^{(\alpha)}, \mathcal{O}^{(\alpha)})$.

The posterior in stage 2 is the target posterior distribution $P(f \mid d, \mathcal{O}^{(all)})$ of the original problem (1), with the posterior energy $U(f \mid d, \mathcal{O}^{(all)}) = U(f \mid \mathcal{O}^{(all)}) + U(d \mid f, \mathcal{O}^{(all)})$. In this stage, if f_i is non-NULL, then it is associated with a model α, so it should be read as $f_i^{(\alpha)}$ (and $f_{i'}$ as $f_{i'}^{(\alpha')}$). In the following derivation, it is assumed that model objects α and α' are independent of each other when $\alpha \neq \alpha'$.

The prior energy is

$$U(f \mid \mathcal{O}^{(all)}) = \sum_{i \in \mathcal{S}} V_1(f_i \mid \mathcal{O}^{(all)}) + \sum_{i \in \mathcal{S}} \sum_{i' \in \mathcal{N}_i} V_2(f_i, f_{i'} \mid \mathcal{O}^{(all)}) \tag{4.34}$$

The single-site prior potentials are defined as $V_1(f_i \mid \mathcal{O}^{(all)}) = V_1(f_i^{(\alpha)})$, which is the same as that in (4.11) for matching to a single model object α. The pair-site potential $V_2(f_i, f_{i'} \mid \mathcal{O}^{(all)}) = V_2(f_i^{(\alpha)}, f_{i'}^{(\alpha')} \mid \mathcal{O}^{(\alpha)}, \mathcal{O}^{(\alpha')})$, where

$$V_2(f_i^{(\alpha)}, f_{i'}^{(\alpha')} \mid \mathcal{O}^{(\alpha)}, \mathcal{O}^{(\alpha')}) = \begin{cases} V_2(f_i^{(\alpha)}, f_{i'}^{(\alpha)}) & \text{if } \alpha = \alpha' \\ v_{20} & \text{otherwise} \end{cases} \quad (4.35)$$

where $V_2(f_i^{(\alpha)}, f_{i'}^{(\alpha)})$ are as defined in (4.12) for matching to object α. Substituting (4.12) into (4.35) yields

$$V_2(f_i, f_{i'} \mid \mathcal{O}^{(all)}) = \begin{cases} 0 & \text{if } (\alpha = \alpha') \text{ and } (f_i \neq 0) \text{ and } (f_{i'} \neq 0) \\ v_{20} & \text{otherwise} \end{cases}$$

$$(4.36)$$

where f_i is associated with object α and $f_{i'}$ with object α'. The definitions above are an extension of (4.12) for dealing with multiple objects. In (4.12), features due to other objects (not belonging to object α) are all labeled as NULL ; (4.36) simply takes this into consideration.

The likelihood energy is

$$U(d \mid f, \mathcal{O}^{(all)}) = \sum_{i \in \mathcal{S}, f_i \neq 0} V_1(d_1(i) \mid f_i, \mathcal{O}^{(all)}) + \quad (4.37)$$

$$\sum_{i \in \mathcal{S}, f_i \neq 0} \sum_{i' \in \mathcal{S} \setminus i, f_{i'} \neq 0} V_2(d_2(i, i') \mid f_i, f_{i'}, \mathcal{O}^{(all)})$$

The single-site likelihood potentials are $V_1(d_1(i) \mid f_i, \mathcal{O}^{(all)}) = V_1(d_1(i) \mid f_i^{(\alpha)})$, defined in the same way as for (4.15). The pair-site likelihood potentials are $V_2(d_2(i, i') \mid f_i, f_{i'}, \mathcal{O}^{(all)}) = V_2(d_2(i, i') \mid f_i^{(\alpha)}, f_{i'}^{(\alpha')}, \mathcal{O}^{(\alpha)}, \mathcal{O}^{(\alpha')})$, where

$$V_2(d_2(i, i') \mid f_i^{(\alpha)}, f_{i'}^{(\alpha')}, \mathcal{O}^{(\alpha)}, \mathcal{O}^{(\alpha')}) = \begin{cases} V_2(d_2(i, i') \mid f_i^{(\alpha)}, f_{i'}^{(\alpha)}) & \text{if } \alpha = \alpha' \\ 0 & \text{otherwise} \end{cases}$$

$$(4.38)$$

where $V_2(d_2(i, i') \mid f_i^{(\alpha)}, f_{i'}^{(\alpha)})$ are defined in the same way as for (4.15) when matching to object α.

Some parameters have to be determined, such as v_{10} and v_{20} in the MRF prior, in order to define the MAP solutions completely. They may be estimated by using a supervised learning algorithm (see Chapter 8).

4.3.2 Computational Issues

Finding Solution in Two Stages

Stage 1 solves L subproblems, $f^{(\alpha)} = \arg\max_{f \in \mathbb{F}^{(\alpha)}} P(f \mid d, \mathcal{O}^{(\alpha)})$ for $\alpha = 1, \ldots, L$, resulting in L MAP solutions $\{f^{(1)}, \ldots, f^{(L)}\}$. Then, a reduced configuration space is constructed from the L solutions. In stage 2, the solution of (4.28) w.r.t. all the L objects is found in the reduced space.

Stage 1 matches the scene to each of the L objects individually (which can be done in parallel for all model objects). Let $P(f \mid d, \mathcal{O}^{(\alpha)}) \propto e^{-E^{(\alpha)}(f)}$ be the posterior distribution of f for matching the scene to object α $(1 \leq \alpha \leq L)$. The MAP-MRF solution for this is $f^{(\alpha)} = \{f_1^{(\alpha)}, \ldots, f_m^{(\alpha)}\} = \arg\min_{f \in \mathbb{F}^{(\alpha)}} E^{(\alpha)}(f)$ where $f_i^{(\alpha)}$ denotes the corresponding model feature in object α. The configuration space $\mathbb{F}^{(\alpha)}$ for object α consists of only $\#\mathbb{F}^{(\alpha)} = (1 + M^{(\alpha)})^m$ elements. For the L objects, the total size is $\sum_1^L (1 + M^{(\alpha)})^m$, much smaller than $\#\mathbb{F}^{(all)}$ which is $(1 + \sum_{\alpha=1}^L M^{(\alpha)})^m$.

Two things are told in $f^{(\alpha)}$: First, it separates image features belonging to object α from the other image features in the following way: If image feature i belongs to object α, then $f_i^{(\alpha)} \neq 0$ (a nonNULL label); otherwise, $f_i^{(\alpha)} = 0$. Second, if $f_i^{(\alpha)} \neq 0$, $f_i^{(\alpha)}$ is the model feature to which image feature i is matched.

$f^{(\alpha)}$ is optimal w.r.t. model object α but not to another one, so inconsistencies may exist among the L MAP solutions. A feature $i \in \mathcal{S}$ in the scene may have been matched to more than one model feature belonging to different objects; that is, there may exist more than one $\alpha \in \{1, \ldots, L\}$ for which $f_i^{(\alpha)} \neq 0$. For example, in Fig. 4.13, model instances found by $f^{(2)}$, $f^{(3)}$, and $f^{(7)}$ compete for a common part of the scene, which is mostly due to the common structures, such as the round extrusions and intrusions, of the matched objects. (Figure 4.13 also shows that a MAP solution allows multiple instances of a model object: e.g., $f^{(7)}$ contains two instances of model object 7)

Stage 2 solves the original MAP problem of (4.28) w.r.t. all the L objects in a reduced solution space constructed from the stage 1 solutions $\{f^{(1)}, \ldots, f^{(L)}\}$ (see the next subsection for the construction of the reduced space). This also resolves possible inconsistencies among the L MAP solutions because only one label f_i assigned to i in the overall solution f. Figure 4.14 shows the overall optimal result, which is the output of stage 2, for the MAP recognition of the scene w.r.t. all the models. It is consistent by itself.

Reduced Solution Space

Consider an illustration in Table 4.1, where a scene with $m = 12$ features is matched to $L = 5$ model objects, resulting in five MAP solutions $f^{(\alpha)}$ $(\alpha = 1, \ldots, 5)$. Let $\mathcal{S}' \subset \mathcal{S}$ be the set of sites that according to the stage 1 solution have been matched to more than one nonNULL label, $\mathcal{S}' = \{i \in \mathcal{S} \mid f_i^{(\alpha)} \neq 0$ for more than one $\alpha\}$. For example, $\mathcal{S}' = \{8, 9, 10, 11\}$ for the $f^{(\alpha)}$'s in Table 4.1. We can derive the reduced set of admissible labels for i, denoted by $\mathcal{L}_i^{(all)}$, from the $f^{(\alpha)}$'s as follows.

- For $i \in \mathcal{S}'$, $\mathcal{L}_i^{(all)}$ consists of all the nonNULL labels assigned to i by the $f^{(\alpha)}$'s, plus the NULL label; that is, $\mathcal{L}_i^{(all)} = 0 \cup \{f_i^{(\alpha)} \neq$

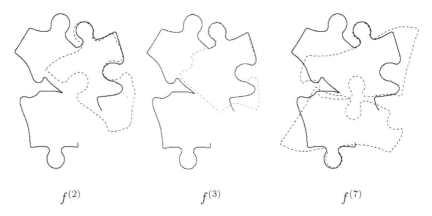

$$f^{(2)} \qquad\qquad f^{(3)} \qquad\qquad f^{(7)}$$

Figure 4.13: $f^{(2)}$, $f^{(3)}$, and $f^{(7)}$ compete for a common part of the scene.

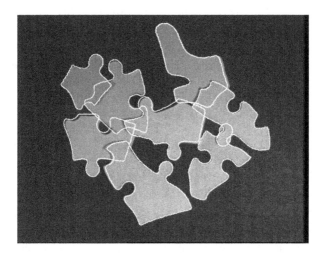

Figure 4.14: The overall matching and recognition result.

$0 \mid \alpha = 1,\ldots,L\}$. For the case in Table 4.1, for example, $\mathcal{L}_8^{(all)} = \{0, 10^{(2)}, 1^{(4)}, 7^{(5)}\}$, $\mathcal{L}_9^{(all)} = \{0, 9^{(2)}, 6^{(5)}\}$, $\mathcal{L}_{10}^{(all)} = \{0, 7^{(2)}, 3^{(3)}, 5^{(5)}\}$, and $\mathcal{L}_9^{(all)} = \{0, 4^{(3)}, 4^{(5)}\}$.

- For $i \notin \mathcal{S}'$, $\mathcal{L}_i^{(all)}$ consists in a unique label; e.g., $\mathcal{L}_6^{(all)} = \{3^{(4)}\}$, and $\mathcal{L}_3^{(all)} = \{0\}$.

Object α contributes one or no label to $\mathcal{L}_i^{(all)}$, as opposed to $M^{(\alpha)}$ labels before the reduction. For $i \in \mathcal{S}'$, the size of $\mathcal{L}_i^{(all)}$ is at most $L + 1$; for

Table 4.1: Matching and recognition of an image containing $m = 12$ features to $L = 5$ objects.

$i =$	1	2	3	4	5	6	7	8	9	10	11	12
$f^{(1)}$	0	0	0	0	0	0	0	0	0	0	0	0
$f^{(2)}$	0	0	0	0	0	0	0	$10^{(2)}$	$9^{(2)}$	$7^{(2)}$	0	0
$f^{(3)}$	0	0	0	0	0	0	0	0	0	$3^{(3)}$	$4^{(3)}$	0
$f^{(4)}$	0	0	0	$5^{(4)}$	$4^{(4)}$	$3^{(4)}$	$2^{(4)}$	$1^{(4)}$	0	0	0	0
$f^{(5)}$	0	0	0	0	0	0	0	$7^{(5)}$	$6^{(5)}$	$5^{(5)}$	$4^{(5)}$	$3^{(5)}$
f^*	0	0	0	$5^{(4)}$	$4^{(4)}$	$3^{(4)}$	$2^{(4)}$	$7^{(5)}$	$6^{(5)}$	$5^{(5)}$	$4^{(5)}$	$3^{(5)}$

$i \notin \mathcal{S}'$, the size of $\mathcal{L}_i^{(all)}$ is one, whereas before the reduction, the size was $1 + \sum_{\alpha=1}^{L} M^{(\alpha)}$ for every i.

The reduced space is constructed as $\mathbb{F}_{reduced}^{(all)} = \mathcal{L}_1^{(all)} \times \mathcal{L}_2^{(all)} \times \cdots \times \mathcal{L}_m^{(all)}$, where $\times$ is the Cartesian product of sets. Its size is much reduced. For the case in Table 4.1, the previous size of the raw solution space $\#\mathbb{F}^{(all)} = (\sum_{\alpha=1}^{5} M^{(\alpha)} + 1)^{12}$ configurations (e.g., 31384283770 for $M^{(\alpha)} = 10$) is reduced to $4 \times 3 \times 4 \times 3 = 144$. It is so small that an exhaustive search is affordable.

Stage 2 performs the target minimization in $\mathbb{F}_{reduced}^{(all)}$. In an iterative search algorithm, only those labels on the sites $i \in \mathcal{S}'$ are subject to changes, whereas those not in $\mathcal{S}'$ are fixed. This is equivalent to maximizing the conditional posterior $f_{\mathcal{S}'}^* = \arg \max_{f_{\mathcal{S}'} \in \mathbb{F}_{\mathcal{S}'}^{(all)}} P(f_{\mathcal{S}'} \mid d, f_{\mathcal{S}-\mathcal{S}'}, \mathcal{O}^{(all)})$, where $f_{\mathcal{S}'} = \{f_i \mid i \in \mathcal{S}'\}$ is the set of labels to be updated, $f_{\mathcal{S}-\mathcal{S}'} = \{f_i \mid i \in \mathcal{S} - \mathcal{S}'\}$ is the set of labels that are fixed during the maximization, and $\mathbb{F}_{\mathcal{S}'}^{(all)} = \prod_{i \in \mathcal{S}'} \mathcal{L}_i^{(all)}$.

A crucial question for the validity of the two-stage approach is whether the solution of (4.28) is contained in the reduced solution space $\mathbb{F}_{reduced}^{(all)}$. The necessary and sufficient condition is that $f^{(\alpha)}$ contains correct matches for object α (it is also allowed to contain some spurious matches). Now that the global solution can be found in $\mathbb{F}_{reduced}^{(all)}$ (e.g., by an exhaustive search), this means that the two-stage strategy can find the global solution of (4.28) if $\mathcal{L}_i^{(\alpha)}$ derived from $f^{(\alpha)}$ contains the correct matching components for the original problem.

The optimization in MAP matching and recognition is combinatorial. While an optimum is sought in a global sense, many optimization algorithms are based on local optimization. The Hummel-Zucker relaxation labeling algorithm (Hummel and Zucker 1983) is preferable in terms of the minimized energy value and computational costs and is used in the implementation. It converges after dozens of iterations. The computational time is dominated by relaxation labeling in the first stage and is roughly the complexity of the whole system.

4.4 Pose Computation

Pose computation aims to estimate the transformation needed to map an object model from the model coordinate system into the sensory data (Ayache and Faugeras 1986; Faugeras and Hebert 1986; Bolles and Horaud 1986; Stockman 1987; Haralick et al. 1989; Grimson 1990; Umeyama 1991). In this section, we derive two MRF models for pose estimation. The first is a model for pose clustering from corresponding point data containing multiple poses and outliers. The second model attempts to solve 3D matching and pose simultaneously from a 2D image without using view invariants.

4.4.1 Pose Clustering and Estimation

The problem of pose clustering is stated as follows. Let a set of corresponding points be given as the data, $d = \{(p_i, P_i) \mid i \in \mathcal{S}\}$, where p_i's are the *model* features, P_i are the *scene* features[4] and $\mathcal{S} = \{1, \ldots, m\}$ indexes the set of the matched pairs. Let f_i be the geometric transformation from p_i to P_i, and consider the set $f = \{f_1, \ldots, f_m\}$ as a configuration of the "pose field". In the case of noiseless, perfect correspondences, the following m equations, each transforming a model feature to a scene feature, should hold simultaneously:

$$P_i = f_i(p_i) \qquad i \in \mathcal{S} \tag{4.39}$$

We want to find the optimal pose configuration in the MAP sense; i.e., $f^* = \arg\min_f U(f \mid d)$.

Assume that each f_i is confined to a certain class $\mathcal{L}$ of pose transformations such that the admissible pose space is $\mathbb{F} = \mathcal{L}^m$. This imposes constraints on the parameters governing f_i. The number of transformation parameters (degree of freedom) needed depends on the class of transformation and the representation adopted for the pose transformation. In the case of the 3D–3D Euclidean transformation, for example, it can consists of an orthogonal rotation O_i followed by a translation T_i (i.e., $f_i = (O_i, T_i)$); the relation between the corresponding points is $P_i = f_i(p_i) = O_i p_i + T_i$. The simple matrix representation needs 12 parameters: nine elements in the rotation matrix O_i plus three elements in the translation vector T_i. The rotation angle representation needs six parameters: three for the three rotation angles and three for the translation. Quaternions provide still another choice. A single pair (p_i, P_i) alone is usually insufficient to determine a pose transformation f_i; more are needed for the pose to be fully determined.

---

[4]Note that, in this section, the uppercase notations are for models and the lowercase notations for the scene.

If all the pairs in the data, d, are inliers and are due to a single transformation, then all f_i, $i \in \mathcal{S}$, which are points in the pose space, must be close
to each other; and the errors $\|P_i - f_i(p_i)\|$, where $\| \cdot \|$ is the Euclidean dis-
tance, must all be small. Complications increase when there are multiple pose
clusters and outlier pairs. When there are multiple poses, f_i's should form
distinct clusters. In this case, the set f is divided into subsets, each giving a
consistent pose transformation from a partition of $\{p_i\}$ to a partition of $\{P_i\}$.
Figure 4.15 illustrates a case in which there are two pose clusters and some
outliers. Outlier pairs, if contained in the data, should be excluded from the
pose estimation because they can cause large errors. Multiple pose identification with outlier detection has a close affinity to the prototypical problem of
image restoration involving discontinuities (Geman and Geman 1984) and to
that of matching overlapping objects using data containing spurious features
(Li 1994a).

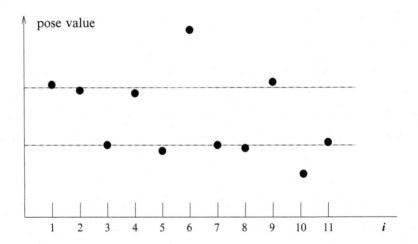

Figure 4.15: Pose clusters in one-dimensional parameter space. Poses f_1, f_2,
f_4, and f_9, due to pairs 1, 2, 4, and 9, agree to one transformation and poses
f_3, f_5, f_7, f_8, and f_{11} agree to another. Poses f_6 and f_{10} form isolated points
so that pair 6 and pair 10 are outliers.

Now we derive the MAP-MRF formulation. The neighborhood system is
defined by

$$\mathcal{N}_i = \{i' \in \mathcal{S} \mid [\text{dist}(p_i, p_{i'})]^2 \le r, \ i' \ne i\} \qquad (4.40)$$

where $\text{dist}(A, B)$ is some suitably defined measure of distance between model
features and the scope r may be reasonably related to the size of the largest
model object. We consider cliques of up to order two, and so the clique set
$\mathcal{C} = \mathcal{C}_1 \cup \mathcal{C}_2$, where $\mathcal{C}_1 = \{\{i\} \mid i \in \mathcal{S}\}$ is the set of single-site (first-order)
cliques and $\mathcal{C}_2 = \{\{i, i'\} \mid i' \in \mathcal{N}_i, i \in \mathcal{S}\}$ the pair-site (second-order) cliques.

Under a single pose transformation, nearby model features are likely to appear together in the scene, whereas model features distantly apart tend to be less likely. This is the coherence of spatial features. We characterize this using the Markovianity condition $p(f_i \mid f_{\mathcal{S}-\{i\}}) = P(f_i \mid f_{\mathcal{N}_i})$. The positivity condition $P(f) > 0$ also holds for all $f \in \mathbb{F}$, where $\mathbb{F}$ is the set of admissible transformations.

The MRF configuration f follows a Gibbs distribution. The two-site potentials determine interactions between the individual f_i's. They may be defined as

$$V_2(f_i, f_{i'}) = g(\|f_i - f_{i'}\|) \tag{4.41}$$

where $\|\cdot\|$ is a norm in the pose space and $g(\cdot)$ is some function. To be able to separate different pose clusters, the function $g(\cdot)$ should stop increasing as $\|f_i - f_{i'}\|$ becomes very large. A choice is

$$g_{\alpha,T}(\eta) = \min\{\eta^2, \alpha\} \tag{4.42}$$

where $\alpha > 0$ is a threshold parameter; this is the same as that used in the line process model for image restoration with discontinuities (Geman and Geman 1984; Marroquin et al. 1987; Blake and Zisserman 1987). It may be any APF defined in (5.28). Its value reflects the cost associated with the pair of pose labels f_i and $f_{i'}$ and will be large when f_i and $f_{i'}$ belong to different clusters. But it cannot be arbitrarily large since a large value such as might be given by a quadratic g tends to force the f_i and f_i' to stay in one cluster as the result of energy minimization, even when they should not. Using an APF imposes piecewise smoothness.

The single-site potentials $V_1(f_i)$ may be used to force f_i to stay in the admissible set $\mathcal{L}$ if such a force is needed. For example, assume $f_i = (O_i, T_i)$ is a 2D-2D Euclidean transformation. Then, the rotation matrix $O_i = [o_{i,r,s} \mid r, s = 1, 2]$ must be orthogonal. The unary potential for the orthogonality constraint can be expressed by $(o_{i,1,1}o_{i,2,1} + o_{i,1,2}o_{i,2,2})^2 + (o_{i,1,1}o_{i,1,2} + o_{i,2,1}o_{i,2,2})^2$. It has the value of zero only when O_i is orthogonal. If no scale change is allowed, then the scaling factor should be exactly one, and an additional term $[\det(O_i) - 1]^2$ can be added, where $\det(O_i)$ is the determinant. Adding these two gives the single-site potential as

$$\begin{aligned} V_1(f_i) &= a\left[(o_{i,1,1}o_{i,2,1} + o_{i,1,2}o_{i,2,2})^2 + (o_{i,1,1}o_{i,1,2} + o_{i,2,1}o_{i,2,2})^2\right] \\ &\quad + b[\det(O_i) - 1]^2 \end{aligned} \tag{4.43}$$

where a and b are the weighting factors. In this case, V_1 imposes the orthogonality. It is also possible to define $V_1(f_i)$ for other classes of transformations. Summing all prior clique potentials yields the following prior energy

$$U(f) = \sum_{i \in \mathcal{S}} V_1(f_i) + \sum_{i \in \mathcal{S}} \sum_{i' \in \mathcal{N}_i} V_2(f_i, f_{i'}) \tag{4.44}$$

which defines the prior distribution $P(f)$.

The likelihood function is derived below. Assume that the features are point locations and that they are subject to the additive noise model, $P_i = f_i(p_i) + e_i$, where $e_i \sim N(0, \sigma^2)$ is a vector of i.i.d. Gaussian noise. Then the distribution of the data d conditional on the configuration f is

$$P(d \mid f) \propto e^{-U(d \mid f)} \qquad\qquad (4.45)$$

where the likelihood energy is

$$U(d \mid f) = \sum_{i \in \mathcal{S}} \|f_i(p_i) - P_i\|^2 / [2\sigma^2] \qquad\qquad (4.46)$$

The location $f_i(p_i)$ is the conditional "mean" of the random variable P_i. The quantity $f_i(p_i) - P_i$ reflects the error between the location $f_i(p_i)$ predicted by f_i and the actual location P_i.

After that, the posterior energy follows immediately as $U(f \mid d) = U(f) + U(d \mid f)$. The optimal solution is $f^* = \arg\min_f U(f \mid d)$. As the result of energy minimization, inlier pairs undergoing the same pose transformation will form a cluster, whereas outlier pairs will form isolated points in the pose space, as illustrated in Fig. 4.15.

4.4.2   Simultaneous Matching and Pose Estimation

In the previous pose estimation formulation, a set of matched pairs is assumed to be available. Here we assume the situation in which the matching has not been done and pose estimation has to be performed during the matching. Pose estimation during matching is practiced when invariants are unavailable or difficult to compute; e.g., because the class of transformations is not linear or involves projections. In the following, an MRF model for simultaneous 3D-from-2D matching and pose estimation is derived without using view invariants. Matching and pose estimation are jointly sought as in (Wells 1991). The formulation is an extension of that given in Section 4.2.

Let $\mathcal{S} = \{1, \ldots, m\}$ index a set of m points on a 3D *model* object, $\{p_i \mid i \in \mathcal{S}\}$. Let $\mathcal{L} = \{1, \ldots, M\}$ be the label set indexing a set of M *scene* points in 2D, $\{P_I \mid I \in \mathcal{L}\}$, and $\mathcal{L}^+ = \{0\} \bigcup \mathcal{L}$ be the augmented set with 0 representing the NULL label. Let $f = \{f_1, \ldots, f_m\}$, $f_i \in \mathcal{L}^+$, denote the matching from the $\{p_i\}$ to $\{P_I\} \bigcup$ NULL. When i is assigned the virtual point 0, $f_i = 0$, it means that there is no corresponding point found in the physically existing point set $\mathcal{L}$. Let $\mathcal{T}$ be the *projective* pose transformation from the 3D model points p_i to the matched 2D image points P_{f_i} ($f_i \neq 0$). We have $P_{f_i} = \mathcal{T}(p_i)$, for all i for which $f_i \neq 0$, under an exact pose.

Now we derive the MAP-MRF formulation. The neighborhood system is defined by

$$\mathcal{N}_i = \{i' \in \mathcal{S} \mid \|p_i - p_{i'}\|^2 \le r, \ i' \neq i\} \qquad\qquad (4.47)$$

The single-site potential is an extension of (4.11) as

$$V_1(f_i, \mathcal{T}) = \begin{cases} v_{10} & \text{if } f_i = 0 \\ v(\mathcal{T}(p_i), P_{f_i}) & \text{otherwise} \end{cases} \qquad (4.48)$$

where v_{10} is a constant. The function $v(\mathcal{T}(p_i), P_{f_i})$ encodes the prior knowledge about $\mathcal{T}$. It may include prior terms, such as V_1 in the previous subsection for the admissibility of pose transformations. If the p.d.f. of the pose is known (e.g., to be a normal distribution centered at a known mean pose, which is assumed in (Wells 1991)), then $v(\mathcal{T}(p_i), P_{f_i})$ is a multivariate Gaussian function. The pair-site potential is defined as in (4.12)

$$V_2(f_i, f_{i'}, \mathcal{T}) = \begin{cases} v_{20} & \text{if } f_i = 0 \text{ or } f_{i'} = 0 \\ 0 & \text{otherwise} \end{cases} \qquad (4.49)$$

where v_{20} is a constant.

The likelihood function characterizes the distribution of the errors and relates to the observation model and the noise in it. Given $f_i \neq 0$, the model point $p_i = (x_i, y_i, z_i)$ is projected to a point $\mathcal{T}(p_i) \triangleq \hat{P}_{f_i} = (\hat{X}_i, \hat{Y}_i)$ by the projective transformation $\mathcal{T}$. In the inexact situation, $\hat{P}_{f_i} \neq P_{f_i}$, where $P_{f_i} = (X_i, Y_i) \triangleq d_1(i)$ is the corresponding image point actually observed.

Assume the additive noise model

$$P_{f_i} = \mathcal{T}(p_i) + e_i = \hat{P}_{f_i} + e_i \qquad (4.50)$$

where $e_i \sim N(0, \sigma^2)$ is a vector of i.i.d. Gaussian noise. Then the likelihood function is

$$p(d_1(i) \mid f_i, \mathcal{T}) = \left(\frac{1}{\sqrt{2\pi\sigma^2}}\right)^2 e^{-V_1(d_1(i) \mid f_i, \mathcal{T})} \qquad (4.51)$$

where

$$V_1(d_1(i) \mid f_i, \mathcal{T}) = [(X_i - \hat{X}_i)^2 + (Y_i - \hat{Y}_i)^2]/[2\sigma^2] \qquad (4.52)$$

is the unary likelihood potential. The joint likelihood is then $p(d_1 \mid f, \mathcal{T}) = \prod_i p(d_1(i) \mid f_i, \mathcal{T})$, where d_1 denotes the set of unary properties.

We also make use of the distances as an additional binary constraint. The distance, $\|p_i - p_{i'}\|$, between the two model points in 3D is projected to the distance

$$d_2(i, i') = \|\hat{P}_{f_i} - \hat{P}_{f_{i'}}\| = \sqrt{(X_i - X_{i'})^2 + (Y_i - Y_{i'})^2} \qquad (4.53)$$

in 2D. Its p.d.f. can be derived, based on the distribution of the projected points given in (4.50), in the following way. Let $X = (X_i, Y_i, X_{i'}, Y_{i'})$. These random variables are assumed independent, so their joint conditional p.d.f. is

$$p(X_i, Y_i, X_{i'}, Y_{i'} \mid f_i, f_{i'}, \mathcal{T}) =$$
$$\left(\frac{1}{\sqrt{2\pi\sigma^2}}\right)^4 e^{-[(X_i - \hat{X}_i)^2 + (Y_i - \hat{Y}_i)^2 + (X_{i'} - \hat{X}_{i'})^2 + (Y_{i'} - \hat{Y}_{i'})^2]/[2\sigma^2]}$$

$$(4.54)$$

Introduce new random variables, $Z(X) = (Z_1, Z_2, Z_3, Z_4)$, as

$$\begin{aligned}
Z_1 &= \sqrt{(X_i - X_{i'})^2 + (Y_i - Y_{i'})^2} \\
Z_2 &= Y_i \\
Z_3 &= X_{i'} \\
Z_4 &= Y_{i'}
\end{aligned} \tag{4.55}$$

each of which is a function of the X variables. Note that we are deriving the p.d.f. of Z_1. The inverse of $Z(X)$, denoted by $X = X(Z)$, is determined by

$$\begin{aligned}
X_i &= \sqrt{Z_1^2 - (Z_2 - Z_4)^2} + Z_3 \\
Y_i &= Z_2 \\
X_{i'} &= Z_3 \\
Y_{i'} &= Z_4
\end{aligned} \tag{4.56}$$

The Jacobian of the inverse is defined to be the determinant

$$J = \det\left[\nabla Z(X)\right] = \frac{Z_1}{\sqrt{Z_1^2 - (Z_2 - Z_4)^2}} \tag{4.57}$$

which is a function of the Z variables. The joint conditional p.d.f. $p_Z(Z)$ for Z can be derived from the joint p.d.f. (4.54) using the relation (Grimmett 1982)

$$p_Z(Z \mid f_i, f_{i'}, T) = p_X(X(Z) \mid f_i, f_{i'}, T) \times |J| \tag{4.58}$$

The conditional distribution of $Z_1 = d_2(i, i')$ is then the conditional marginal

$$p(d_2(i, i') \mid f_i, f_{i'}, T) = p_{Z_1}(Z_1 \mid f_i, f_{i'}, T) = \tag{4.59}$$

$$\int_{-\infty}^{+\infty} \int_{-\infty}^{+\infty} \int_{-\infty}^{+\infty} p_Z(Z_1, Z_2, Z_3, Z_4 \mid f_i, f_{i'}, T) \, dZ_2 \, dZ_3 \, dZ_4$$

which is a function of $X_i, Y_i, X_{i'}, Y_{i'}$. This gives the binary likelihood potential $V(d_2(i, i') \mid f_i, f_{i'}, T)$. The joint p.d.f. of the set of binary features, d_2, is approximated by the "pseudo-likelihood"

$$p(d_2 \mid f, T) = \prod_{i \in \mathcal{S}} \prod_{i' \in \mathcal{N}_i} p(d_2(i, i') \mid f_i, f_{i'}, T) \tag{4.60}$$

The joint p.d.f. of $d = \{d_1, d_2\}$ is approximated by

$$p(d \mid f, T) = p(d_1 \mid f, T) \, p(d_2 \mid f, T) \tag{4.61}$$

Now the posterior energy can be obtained as

$$\begin{aligned}
U(f, T \mid d) = \ & \textstyle\sum_{i \in \mathcal{S}} V_1(f_i, T) + \sum_{i \in \mathcal{S}} \sum_{i' \in \mathcal{N}_i} V_2(f_i, f_{i'}, T) + \\
& \textstyle\sum_{i \in \mathcal{S}: f_i \neq 0} V_1(d_1(i) \mid f_i, T) + \\
& \textstyle\sum_{i \in \mathcal{S}: f_i \neq 0} \sum_{i' \in \mathcal{S}: f_{i'} \neq 0} V_2(d_2(i, i') \mid f_i, f_{i'}, T)
\end{aligned} \tag{4.62}$$

The optimal solution is $(f^*, \mathcal{T}^*) = \arg\min_{(f,\mathcal{T})} U(f, \mathcal{T} \mid d)$. The nonNULL labels in f^* represents the matching from the model object considered to the scene and $\mathcal{T}^*$ determines the pose transformation therein. The model points that are assigned the NULL label are either spurious or due to other model objects. Another round of matching-pose operations may be formed on these remaining points in terms of another model object.

4.4.3 Discussion

Minimizing the energies derived in this section is difficult. The dimensionality of the search space is high. As a guide to the search, the energies are inefficient unless some strong prior constraints are available, such as the normal prior distribution of poses assumed in (Wells 1991). However, the derived models may be useful for verifying the matching and pose estimation results. Assume that pose candidates are found by using techniques such as the Hough transform or geometric indexing. The energies may be used as global cost measures for the matching and pose.

4.5 Face Detection and Recognition

Face detection finds the face areas (usually rectangles) in an image, giving the locations and sizes of the faces detected. Consider a subwindow (data d) of an image at each location in the image plane and each scale. The basic detection problem is to classify the subwindow as face or nonface. This dichotomy could be done based on computing the ratio of likelihood densities of face and nonface followed by comparing the ratio with a confidence threshold (Dass et al. 2002).

Let $\mathcal{S}$ be the set of m pixel locations in the subwindow to be classified and let the label set $\mathcal{L}$ consist of admissible pixel intensities. Dass, Jain, and Lu (2002) proposed the following auto-model to model the likelihood

$$p(d \mid f) = \frac{\exp\left\{\sum_i \alpha_i f_i + \sum_i \sum_{i' \in \mathcal{N}_i} \beta_{i,i'} f_i f_{i'}\right\}}{\sum_{f_1} \sum_{f_2} \cdots \sum_{f_N} \exp\left\{\sum_i \alpha_i f_i + \sum_i \sum_{i' \in \mathcal{N}_i} \beta_{i,i'} f_i f_{i'}\right\}} \quad (4.63)$$

where α_i and $\beta_{i,i'}$ are the auto-model parameters. For computational tractability, the pseudo-likelihood (see Section 7.1.2) approximation is used instead of (4.63)

$$PL = \prod_{i=1}^{m} \frac{\exp\left\{\alpha_i f_i + \sum_{i' \in \mathcal{N}_i} \beta_{i,i'} f_i f_{i'}\right\}}{\sum_{f_i} \exp\left\{\alpha_i f_i + \sum_{i' \in \mathcal{N}_i} \beta_{i,i'} f_i f_{i'}\right\}} \quad (4.64)$$

Two such auto-models could be used. The first assumes homogeneous correlations for all the sites. This is described by two pairwise parameters: $\beta_{i,i'} = \beta_h$ when i' is a horizontal neighbor of i and $\beta_{i,i'} = \beta_v$ when i' is a vertical neighbor with constants β_h and β_v.

The second model assumes inhomogeneous parameters $\beta_{i,i'}$ across sites i but isometric for different directions; that is, $\beta_{i,i'} = \beta_i$ is dependent on i only. Thus, the pseudo-likelihood is

$$\text{PL} = \prod_{i=1}^{m} \frac{\exp\left\{\alpha_i f_i + \beta_i \sum_{i' \in \mathcal{N}_i} f_i f_{i'}\right\}}{\sum_{f_i} \exp\left\{\alpha_i f_i + \beta_i \sum_{i' \in \mathcal{N}_i} f_i f_{i'}\right\}} \tag{4.65}$$

where m pairwise parameters are needed.

The parameters in the PL could be estimated using the maximum pseudo-likelihood (MPL) on a training set. The detection decision of classifying a subwindow as face or nonface is based on the pseudo-likelihoods of faces and nonfaces. Hence, two sets of parameters need to be estimated, one from a training set of faces and the other from a training set of nonface subwindows (Dass et al. 2002).

In the detection stage, a subwindow d is classified into a face or nonface based on the log pseudo-likelihood ratio (LPR)

$$\text{LPR} = \log \frac{PL_{face}}{PL_{nonface}} \tag{4.66}$$

The LPR is compared with a confidence value, and thereby a decision is made.

Works on MRF modeling for face recognition in the MAP-MRF framework have been reported in several papers, e.g., (Huang et al. 2004; Park et al. 2005). In (Huang et al. 2004), a face image is divided into m blocks represented by sites $d = \{d_1, \ldots, d_m\}$. The label set $L = \{1, \ldots, M\}$ corresponds to the M ID's. Assuming added Gaussian noise, the data term $p(d \mid f)$ is a Gaussian function. A pairwise "smoothness" term is imposed on pairs of labels as $P(f_i, f_{i'}) = \delta(f_i, f_{i'})$. In (Park et al. 2005), straight lines, corresponding to sites, are extracted from a face image. By attaching properties and binary relations to the straight lines, a face is then represented as an ARG. A partial matching is used to match two ARGs and select the best match.

Chapter 5

Discontinuities in MRF's

Smoothness is a generic assumption underlying a wide range of physical phenomena. It characterizes the coherence and homogeneity of matter within a scope of space (or an interval of time). It is one of the most common assumptions in computer vision models, in particular, those formulated in terms of Markov random fields (MRF's) (Geman and Geman 1984; Elliott et al. 1984; Marroquin 1985) and regularization (Poggio et al. 85a). Its applications are seen widely in image restoration, surface reconstruction, optical flow and motion, shape from X, texture, edge detection, region segmentation, visual integration, and so on.

The assumption of uniform smoothness implies smoothness *everywhere*. However, improper imposition of it can lead to undesirable oversmoothed solutions. This occurs when the uniform smoothness is violated, for example at discontinuities, where abrupt changes occur. It is necessary to take care of discontinuities when using smoothness priors. Therefore, how to apply the smoothness constraint while preserving discontinuities has been one of the most active research areas in image processing and low-Level vision (see, e.g., (Blake 1983; Terzopoulos 1983b; Geman and Geman 1984; Grimson and Pavlidis 1985; Marroquin 1985; Mumford and Shah 1985; Terzopoulos 1986b; Blake and Zisserman 1987; Lee and Pavlidis 1987; Koch 1988; Leclerc 1989; Shulman and Herve 1989; Li 1990b; Nordstrom 1990; Geiger and Girosi 1991; ?)).

This chapter[1] presents a systematic study on smoothness priors involving discontinuities. Through an analysis of the *Euler equation* associated with the energy minimization in MRF and regularization models, it is identified that the fundamental difference among different models for dealing with discontinuities lies in their ways of controlling the *interaction* between neighboring points. Thereby, an important necessary condition is derived for any regularizers or MRF prior potential functions to be able to deal with discontinuities.

---

[1]This chapter is based on Li (1995b) and Li (2000).

S.Z. Li, *Markov Random Field Modeling in Image Analysis*,
Advances in Pattern Recognition, DOI: 10.1007/978-1-84800-279-1_5,
© Springer-Verlag London Limited 2009

Based on these findings, a so-called *discontinuity adaptive* (DA) smoothness model is defined in terms of the Euler equation constrained by a class of
adaptive interaction functions (AIFs). The DA solution is C^1 continuous, allowing arbitrarily large but bounded slopes. Because of its continuous nature,
it is stable against changes in parameters and data. This is a good property
for regularizing ill-posed problems. The results provide principles for the selection of a priori clique potential functions in stochastic MRF models and
regularizers in deterministic regularization models. It is also shown that the
DA model includes as special instances most of the existing models, such as
the line process (LP) model (Geman and Geman 1984; Marroquin 1985), the
weak string and membrane models (Blake and Zisserman 1987), approximations of the LP model (Koch et al. 1986; Yuille 1987), minimal description
length (Leclerc 1989), biased anisotropic diffusion (Nordstrom 1990), and
mean field theory approximation (Geiger and Girosi 1991).

   The study of discontinuities is most sensibly carried out in terms of analytical properties, such as derivatives. For this reason, analytical regularization,
a special class of MRF model, is used as the platform for it. If we consider
that regularization contains three parts (Boult 1987) – the data, the class of
solution functions, and the regularizer – the following addresses mainly the
regularizer part. Regularization models are reviewed in connection with discontinuities. A necessary condition for the discontinuity adaptivity is made
explicit. Based on this, the DA model (for step discontinuities) is defined
and compared with other models. The model is then extended for surfaces
containing roof discontinuities. Some experimental results for step- and roof-
edge-preserving smoothing are presented.

5.1   Smoothness, Regularization,
and Discontinuities

In MRF vision modeling, the smoothness assumption can be encoded into an
energy via one of the two routes: analytic and probabilistic. In the analytic
route, the encoding is done in the *regularization* framework (Poggio et al. 85a;
Bertero et al. 1988). From the regularization viewpoint, a problem is said to
be "ill-posed" if it fails to satisfy one or more of the following criteria: the
solution exists, is unique and depends continuously on the data. Additional, a
priori, assumptions have to be imposed on the solution to convert an ill-posed
problem into a well-posed one. An important assumption of such assumptions
is the *smoothness* (Tikhonov and Arsenin 1977). It is incorporated into the
energy function whereby the cost of the solution is defined.

   From the probabilistic viewpoint, a regularized solution corresponds to
the *maximum a posteriori* (MAP) estimate of an MRF (Geman and Geman
1984; Marroquin et al. 1987). Here, the prior constraints are encoded into
the a priori MRF probability distribution. The MAP solution is obtained by

maximizing the posterior probability or equivalently minimizing the corresponding energy.

The MRF model is more general than the regularization model in (1) that it can encode prior constraints other than the smoothness and (2) that it allows arbitrary neighborhood systems other than the nearest ones. However, the analytic regularization model provides a convenient platform for the study of smoothness priors because of close relationships between the smoothness and the analytical continuity.

5.1.1 Regularization and Discontinuities

Consider the problem of restoring a signal f from the data $d = f + e$, where e denotes the noise. The regularization formulation defines the solution f^* to be the global minimum of an energy function $E(f)$, $f^* = \arg\min_f E(f)$. The energy is the sum of two terms

$$E(f) \doteq U(f \mid d) = U(d \mid f) + U(f) \tag{5.1}$$

The *closeness* term, $U(d \mid f)$, measures the cost caused by the discrepancy between the solution f and the data d

$$U(d \mid f) = \int_a^b \chi(x)[f(x) - d(x)]^2 dx \tag{5.2}$$

where $\chi(x)$ is a weighting function and a and b are the bounds of the integral. The *smoothness* term, $U(f)$, measures the cost caused by the irregularities of the solution f, the irregularities being measured by the derivative magnitudes $|f^{(n)}(x)|$. With identical independent additive Gaussian noise, the terms $U(f \mid d)$, $U(d \mid f)$ and $U(f)$ correspond to the energies in the posterior, the likelihood and the prior Gibbs distributions of an MRF, respectively (Marroquin et al. 1987).

The smoothness term $U(f)$, also called a *regularizer*, is the object of study in this work. It penalizes the irregularities according to the a priori smoothness constraint encoded in it. It is generally defined as

$$U(f) \doteq \sum_{n=1}^N U_n(f) = \sum_{n=1}^N \lambda_n \int_a^b g(f^{(n)}(x)) dx \tag{5.3}$$

where $U_n(f)$ is the nth order regularizer, N is the highest order to be considered, and $\lambda_n \geq 0$ is a weighting factor. A *potential function* $g(f^{(n)}(x))$ is the penalty against the irregularity in $f^{(n-1)}(x)$ and corresponds to prior clique potentials in MRF models. Regularizers differ in the definition of $U_n(f)$, more specifically in the selection of g.

Standard Regularization

In the standard regularization (Tikhonov and Arsenin 1977; Poggio et al. 85a), the potential function takes the pure quadratic form

$$g_q(\eta) = \eta^2 \tag{5.4}$$

With g_q, the more irregular $f^{(n-1)}(x)$ is at x, the larger $|f^{(n)}|$, and consequently the larger potential $g(f^{(n)})$ contributed to $U_n(f)$. The standard quadratic regularizer can have the more general form

$$U_n(f, w_n) = \lambda_n \int_a^b w_n(x)[f^{(n)}(x)]^2 dx \tag{5.5}$$

where $w_n(x)$ are the prespecified nonnegative continuous functions (Tikhonov and Arsenin 1977). It may also be generalized to multidimensional cases and to include cross derivative terms.

The quadratic regularizer imposes the smoothness constraint *everywhere*. It determines the constant *interaction* between neighboring points and leads to a *smoothing strength* proportional to $|f^{(n)}|$, as will be shown in the next section. The homogeneous or isotropic application of the smoothness constraint inevitably leads to oversmoothing at discontinuities at which the derivative is infinite.

If the function $w_n(x)$ can be prespecified in such a way that $w_n(x) = 0$ at x, where $f^{(n)}(x)$ is infinite, then the oversmoothing can be avoided. In this way, $w_n(x)$ acts as a *continuity controller* (Terzopoulos 1983b). It is further suggested that $w_n(x)$ may be discontinuous and not prespecified (Terzopoulos 1986b). For example, by regarding $w_n(x)$ as an unknown function, one could solve these unknowns using variational methods. But how well $w_n(x)$ can thus be derived remains unclear. The introduction of *line processes* (Geman and Geman 1984; Marroquin 1985) or *weak continuity constraints* (Blake and Zisserman 1987) provides a solution to this problem.

Line Process Model and Its Approximations

The LP model assumes *piecewise smoothness* whereby the smoothness constraint is *switched off* at points where the magnitude of the signal derivative exceeds a certain threshold. It is defined on a lattice rather than a continuous domain. Quantize the continuous interval $[a, b]$ into m uniformly spaced points $x_1, ..., x_m$ so that $f_i = f(x_i)$, $d_i = d(x_i)$, and $\chi_i = \chi(x_i)$. Introduce a set of binary line process variables $l_i \in \{0, 1\}$ into the smoothness term. If $w_i = w_1(x_i)$ in (5.5) also takes on a value in $\{0, 1\}$, then l_i is related to it by $l_i = 1 - w_i$. The on state, $l_i = 1$, of the line process variable indicates that a discontinuity is detected between neighboring points $i - 1$ and i; the off state, $l_i = 0$, indicates that the signal between the two points is continuous. Each turn-on of a line process variable is penalized by a quantity $\lambda\alpha$. These give the LP regularizer

$$U(f,l) = \lambda \sum_{i=2}^{m} [f_i - f_{i-1}]^2 [1 - l_i] + \lambda\alpha \sum_{i=2}^{m} l_i \qquad (5.6)$$

The energy for the LP model is

$$E(f,l) = U(f,l \mid d) = \sum_{i=1}^{m} \chi_i [f_i - d_i]^2 + \lambda \left(\sum_{i=2}^{m} [f_i - f_{i-1}]^2 [1 - l_i] + \alpha \sum_{i=2}^{m} l_i \right)$$
$$(5.7)$$

This is the *weak string* model whose 2D extension is the *weak membrane* model (Blake and Zisserman 1987). Equation (5.7) corresponds to the energy in the posterior distribution of the surface field f and the line process field l, the distribution being of the Gibbs form

$$P(f,l \mid d) = \frac{1}{Z} e^{-U(f,l \mid d)} \qquad (5.8)$$

where Z is a normalizing constant called the partition function.

The line process variables are determined as follows: If $[f_i - f_{i-1}]^2 < \alpha$, then it is cheaper to pay the price $[f_i - f_{i-1}]^2$ and set $l_i = 0$; otherwise it is more economical to turn on the variable $l_i = 1$ to insert a discontinuity with the cost of α. This is an interpretation of the LP model based on the concept of the weak continuity constraint introduced by Blake (1983) for edge labeling. An earlier idea of weak constraints is seen in Hinton's thesis work (Hinton 1978). In the LP model, the *interaction* is piecewise constant (1 or 0) and the *smoothing strength* at i is either proportional to $|f_i - f_{i-1}|$ or zero – see the next section. The concept of discontinuities can be extended to model-based recognition of overlapping objects (Li 1994a). There, the relational bond between any two features in the scene should be broken if the features are ascribed to two different objects.

Finding $f^* \in \mathbb{R}^m$ and $l^* \in \{0,1\}^m$ such that $U(f,l \mid d)$ is minimized is a mixture of real and combinatorial optimization. Algorithms for this can be classified as categories: stochastic (Geman and Geman 1984; Marroquin 1985) and deterministic (Koch et al. 1986; Yuille 1987; Blake and Zisserman 1987; Geiger and Girosi 1991). Some annealing techniques are often combined into them to obtain global solutions.

In stochastic approaches, f and l are updated according to some probability distribution parameterized by a temperature parameter. For example, Geman and Geman propose to use simulated annealing with the Gibbs sampler (Geman and Geman 1984) to find the global MAP solution. Marroquin (1985) minimizes the energy by a stochastic update in l together with a deterministic update in f using gradient-descent.

Deterministic approaches often use some classical gradient-based methods. Before these can be applied, the combinatorial minimization problem has to be converted into one of real minimization. By eliminating the line process, Blake and Zisserman (1987) convert the previous minimization problem into one that minimizes the following function containing only real variables:

$$U(f \mid d) = \sum_{i=1}^{m} \chi_i[f_i - d_i]^2 + \lambda \sum_{i=2}^{m} g_\alpha(f_i - f_{i-1}) \qquad (5.9)$$

where the truncated quadratic potential function (see (3.17))

$$g_\alpha(\eta) = \min\{\eta^2, \alpha\} \qquad (5.10)$$

shall be referred to as the *line process potential function*. Blake and Zisserman introduce a parameter p into $g_\alpha(\eta)$ to control the convexity of E, obtaining $g_\alpha^{(p)}(\eta)$. The parameter p varies from 1 to 0, which corresponds to the variation from a convex approximation of the function to its original form.

Koch, Marroquin, and Yuille (1986) and Yuille (1987) perform the conversion using the Hopfield approach (Hopfield 1984). Continuous variables $\bar{l}_i$ in the range $[0, 1]$ are introduced to replace the binary line process variables l_i in $\{0, 1\}$. Each $\bar{l}_i$ is related to an internal variable $v_i \in (-\infty, +\infty)$ by a sigmoid function

$$\bar{l}_i = \theta(v_i) = 1/(1 + e^{-v_i/\tau}) \qquad (5.11)$$

with $\tau > 0$ as the parameter whereby $\lim_{\tau \to 0} \bar{l}_i = l_i$. The energy with this treatment is

$$U(f, \bar{l} \mid d) = \sum_{i=1}^{m} \chi_i[f_i - d_i]^2 + \lambda \left(\sum_{i=2}^{m} [f_i - f_{i-1}]^2 [1 - \bar{l}_i] + \alpha \sum_{i=2}^{m} \bar{l}_i \right)$$
$$+ \sum_{i=2}^{m} \int_0^{\bar{l}_i} \theta^{-1}(l) dl \qquad (5.12)$$

It is shown that at stationary points where $dE/dt = 0$, there are $v_i = \lambda([f_i - f_{i-1}]^2 - \alpha)$ and hence the approximated line process variables (Yuille 1987)

$$\bar{l}_i = \frac{1}{1 + e^{-\lambda([f_i - f_{i-1}]^2 - \alpha)/\tau}} \qquad (5.13)$$

This gives the effective potential function

$$g_{\alpha,\tau}(\eta) = -\frac{\tau}{2} \ln[1 + e^{-\lambda(\eta^2 - \alpha)/\tau}] \qquad (5.14)$$

As the temperature decreases toward zero, $\bar{l}_i$ approaches l_i: $\lim_{\tau \to 0^+} \bar{l}_i = l_i$.

Geiger and Girosi (1991) approximate the line process using mean field theory. They introduce a parameter β into (5.8), giving an approximated posterior probability

$$P_\beta(f, l \mid d) = \frac{1}{Z_\beta} [P(f, l \mid d)]^\beta \qquad (5.15)$$

Using the saddle-point approximation method, they derive mean field equations which yield the approximated line process variables, which are identical to (5.13). The solution is in the limit $f^* = \arg\max_{f,l} \lim_{\beta \to \infty} P_\beta(f, l \mid d)$.

Li (1990b) proposes a continuous *adaptive regularizer* model. There, the smoothness constraint is applied without the switch-off as in the LP model. Its effect is decreased as the derivative magnitude becomes larger and is completely off only at the true discontinuities, where the derivative is infinite. This is an earlier form of the DA model in this work.

5.1.2 Other Regularization Models

Grimson and Pavlidis (1985) propose an approach in which the degree of interaction between pixels across edges is adjusted in order to detect discontinuities. Lee and Pavlidis (1987) investigate a class of smoothing splines that are piecewise polynomials. Errors of fit are measured after each successive regularization and used to determine whether discontinuities should be inserted. This process iterates until convergence is reached. Besl, Birch, and Watson (1988) propose a smoothing window operator to prevent smoothing across discontinuities based on robust statistics. Liu and Harris (1989) develop, based on a previous work (Harris 1987), a computational network in which surface reconstruction, discontinuity detection and estimation of first and second derivatives are performed cooperatively.

Mumford and Shah (1985) define an energy on a continuous domain

$$U(f, \{a_i\}, k \mid d) = \int_a^b [f(x) - d(x)]^2 dx + \lambda \sum_{i=0}^k \int_{a_i}^{a_{i+1}} [f'(x)]^2 dx + \alpha k \quad (5.16)$$

where λ and α are constants, k is the number of discontinuities, and the sequence $a = a_0 < a_1 < ... < a_k < a_{k+1} = b$ indicates the locations of discontinuities. The minimal solution $(f^*, \{a_i\}^*, k^*)$ minimizes $U(f, \{a_i\}, k \mid d)$ over each value of the integer k, every sequence $\{a_k\}$, and every function $f(x)$ continuously differentiable on each interval $a_i \le x \le a_{i+1}$. The minimization over k is a hard problem.

Using the minimal description length principle, Leclerc (1989) presents the following function for restoration of piecewise constant image f from noisy data d

$$U(f \mid d) = \sum_{i=1}^m [f_i - d_i]^2 + \lambda \sum_{i=2}^m [1 - \delta(f_i - f_{i-1})] \quad (5.17)$$

where $\delta(\cdot) \in \{0, 1\}$ is the Kronecker delta function. To minimize the function, he approximates the delta function with the exponential function parameterized by τ

$$U(f \mid d) = \sum_{i=1}^m [f_i - d_i]^2 + \lambda \sum_{i=2}^m [1 - e^{(f_i - f_{i-1})^2/\tau}] \quad (5.18)$$

and approaches the solution by a continuation in τ toward 0.

5.2   The Discontinuity Adaptive MRF Model

By analyzing the smoothing mechanism in terms of the Euler equation, it
will be clear that a major difference between different models lies in their
way of controlling the interaction between neighboring points and adjusting
the smoothing strength. The DA model is defined based on the principle that
wherever a discontinuity occurs, the interaction should diminish.

5.2.1   Defining the DA Model

We focus on the models that involve only the first-order derivative $f' = f^{(1)}$
and consider the general string model

$$E(f) = U(f \mid d) = \int_a^b u(f \mid d)dx \qquad (5.19)$$

where

$$u(f \mid d) = \chi(x)[f(x) - d(x)]^2 + \lambda g(f'(x)) \qquad (5.20)$$

Solutions f^* minimizing $U(f \mid d)$ must satisfy the associated Euler-Lagrange
differential equation or simply the *Euler equation* (Courant and Hilbert 1953)

$$u_f(f, f') - \frac{d}{dx} u_{f'}(f, f') = 0 \qquad (5.21)$$

with the boundary conditions

$$f(a) = f_a, \quad f(b) = f_b \qquad (5.22)$$

where f_a and f_b are prescribed constants. In the following discussion of solutions to the differential equation, the following assumptions are made: Both
$\chi(x)$ and $d(x)$ are continuous and $f(x)$ is continuously differentiable.[2] Writing
out the Euler equation yields

$$2\chi(x)[f(x) - d(x)] - \lambda \frac{d}{dx} g'(f'(x)) = 0 \qquad (5.23)$$

A potential function g is usually chosen to be (a) even such that $g(\eta) = g(|\eta|)$
and (b) the derivative of g can be expressed as

$$g'(\eta) = 2\eta h(\eta) \qquad (5.24)$$

where h is called an *interaction function*. Obviously, h thus defined is also
even. With these assumptions, the Euler equation can be expressed as

---

[2]In this work, the continuity of $\chi(x)$ and $d(x)$ and differentiability of $f(x)$ are assumed
for the variational problems defined on continuous domains (Courant and Hilbert 1953;
Cesari 1983). However, they are not necessary for discrete problems where $[a, b]$ is quantized
into discrete points. For example, in the discrete case, $\chi(x_i)$ is allowed to take a value in
{1,0}, indicating whether datum $d(x_i)$ is available or not.

$$\chi(x)[f(x) - d(x)] - \lambda\frac{d}{dx}[f'(x)h(f'(x))] = 0 \qquad (5.25)$$

The magnitude $|g'(f')| = |2f'h(f')|$ relates to the *strength* with which a regularizer performs smoothing, and $h(f'(x))$ determines the *interaction* between neighboring pixels.

Let $\eta \doteq f'(x)$. A *necessary condition* for any regularization model to be adaptive to discontinuities is (?; Li 1995b)

$$\lim_{\eta\to\infty} |g'(\eta)| = \lim_{\eta\to\infty} |2\eta h(\eta)| = C \qquad (5.26)$$

where $C \in [0, \infty)$ is a constant. The condition above with $C = 0$ entirely prohibits smoothing at discontinuities where $\eta \to \infty$, whereas with $C > 0$ it allows limited (bounded) smoothing. In any case, however, the interaction $h(\eta)$ must be small for large $|\eta|$ and approaches 0 as $|\eta|$ goes to ∞. This is an important guideline for selecting g and h for the purpose of the adaptation.

Definition 5.1. An *adaptive interaction function* (AIF) h_γ parameterized by γ (> 0) is a function that satisfies

$$\begin{array}{ll} \text{(i)} & h_\gamma \in C^1 \\ \text{(ii)} & h_\gamma(\eta) = h_\gamma(-\eta) \\ \text{(iii)} & h_\gamma(\eta) > 0 \\ \text{(iv)} & h'_\gamma(\eta) < 0 \quad (\forall \eta > 0) \\ \text{(v)} & \lim_{\eta\to\infty} |2\eta h_\gamma(\eta)| = C < \infty \end{array} \qquad (5.27)$$

The class of AIFs, denoted by $\mathbb{H}_\gamma$, is defined as the collection of all such h_γ. □

The continuity requirement (i) guarantees the twice differentiability of the integrand $u(f \mid d)$ in (5.20) w.r.t. f', a condition for the solution f to exist (Courant and Hilbert 1953). However, this can be relaxed to $h_\gamma \in C^0$ for discrete problems. The evenness of (ii) is usually assumed for spatially unbiased smoothing. The positive definiteness of (iii) keeps the interaction positive such that the sign of $\eta h_\gamma(\eta)$ will not be altered by $h_\gamma(\eta)$. The monotony of (iv) leads to decreasing interaction as the magnitude of the derivative increases. The bounded asymptote property of (v) provides the adaptive discontinuity control, as stated earlier. Other properties follow: $h_\gamma(0) = \sup_\eta h_\gamma(\eta)$, $h'_\gamma(0) = 0$, and $\lim_{\eta\to\infty} h_\gamma(\eta) = 0$, the last meaning zero interaction at discontinuities. The definition above characterizes the properties AIFs should possess rather than instantiating some particular functions. Therefore, the following definition of the DA model is rather broad.

Definition 5.2. The DA solution f is defined by the Euler equation (5.25) constrained by $h = h_\gamma \in \mathbb{H}_\gamma$. □

The DA solution is C^1 continuous.[3] Therefore the DA solution and its derivative never have discontinuities in them. The DA overcomes oversmoothing by allowing its solution to be steep, but C^1 continuous, at point x, where data $d(x)$ is steep. With every possible data configuration d, every $f \in C^1$ is possible.

There are two reasons to define the DA in terms of the constrained Euler equation: First, it captures the essence of the DA problem; the DA model should be defined based on the h_γ therein. Second, some h_γ that meets the requirements on discontinuities may not have its corresponding g_γ (and hence the energy) in closed form, where g_γ is defined below.

Definition 5.3. The *adaptive potential function* (APF) corresponding to an $h_\gamma \in \mathbb{H}_\gamma$ is defined by

$$g_\gamma(\eta) = 2 \int_0^\eta \eta' h_\gamma(\eta') d\eta' \qquad (5.28)$$

$\square$

The energy function (5.19) with $g = g_\gamma$ is called an *adaptive string*. The following are some properties of g_γ. Basically, g_γ is one order higher than h_γ in continuity; it is even, $g_\gamma(\eta) = g_\gamma(-\eta)$; and its derivative function is odd, $g'_\gamma(\eta) = -g'_\gamma(-\eta)$; however, it is not necessary for $g_\gamma(\infty)$ to be bounded. Furthermore, g_γ is strictly monotonically increasing as $|\eta|$ increases because $g_\gamma(\eta) = g_\gamma(|\eta|)$ and $g'_\gamma(\eta) = 2\eta h_\gamma(\eta) > 0$ for $\eta > 0$. This means a larger $|\eta|$ leads to a larger penalty $g_\gamma(\eta)$. It conforms to the original spirit of the standard quadratic regularizers determined by g_q. The line process potential function g_α does not have such a property: Its penalty is fixed and does not increase as $|\eta|$ increases beyond $\sqrt{\alpha}$. The idea that large values of η are equally penalized is questionable (Shulman and Herve 1989).

In practice, it is not always necessary to know the explicit definition of g_γ. The most important factors are the Euler equation and the constraining function h_γ. Nonetheless, knowing g_γ is helpful for analyzing the convexity of $E(f)$.

For a given $g_\gamma(\eta)$, there exists a region of η within which the smoothing strength $|g'_\gamma(\eta)| = |\eta h_\gamma(\eta)|$ increases monotonically as $|\eta|$ increases and the function g_γ is convex:

$$B_\gamma = \{\eta \mid g''_\gamma(\eta) > 0\} \stackrel{\triangle}{=} (b_L, b_H) \qquad (5.29)$$

The region B_γ is referred to as the *band of convexity* because $g(\eta)$ is convex in this region. The lower and upper bounds b_L, b_H correspond to the two extrema of $g'_\gamma(\eta)$, which can be obtained by solving $g''_\gamma(\eta) = 0$, and we have

---

[3]See (Cesari 1983) for a comprehensive discussion about the continuity of solutions of the class of problems to which the DA belongs.

Table 5.1: Four choices of AIFs, the corresponding APFs and bands.

AIF	APF	Band
$h_{1\gamma}(\eta) = e^{-\frac{\eta^2}{\gamma}}$	$g_{1\gamma}(\eta) = -\gamma e^{-\frac{\eta^2}{\gamma}}$	$B_{1\gamma} = (-\sqrt{\frac{\gamma}{2}}, \sqrt{\frac{\gamma}{2}})$
$h_{2\gamma}(\eta) = \frac{1}{[1+\frac{\eta^2}{\gamma}]^2}$	$g_{2\gamma}(\eta) = -\frac{\gamma}{1+\frac{\eta^2}{\gamma}}$	$B_{2\gamma} = (-\sqrt{\frac{\gamma}{3}}, \sqrt{\frac{\gamma}{3}})$
$h_{3\gamma}(\eta) = \frac{1}{1+\frac{\eta^2}{\gamma}}$	$g_{3\gamma}(\eta) = \gamma \ln(1 + \frac{\eta^2}{\gamma})$	$B_{3\gamma} = (-\sqrt{\gamma}, \sqrt{\gamma})$
$h_{4\gamma}(\eta) = \frac{1}{1+\frac{|\eta|}{\gamma}}$	$g_{4\gamma}(\eta) = \gamma|\eta| - \gamma^2 \ln(1 + \frac{|\eta|}{\gamma})$	$B_{4\gamma} = (-\infty, +\infty)$

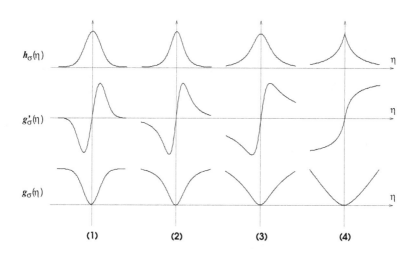

Figure 5.1: The qualitative shapes of the four DA functions. From (Li 1995b) with permission; ©1995 IEEE.

$b_L = -b_H$ when g is even. When $b_L < \eta < b_H$, $g''_\gamma(\eta) > 0$ and thus $g_\gamma(\eta)$ is strictly convex. For h_γ defined with $C > 0$, the bounds are $b_L \to -\infty$ and $b_H \to +\infty$.

Table 5.1 instantiates four possible choices of AIFs, the corresponding APFs and the bands. Figure 5.1 shows their qualitative shapes (a trivial constant may be added to $g_\gamma(\eta)$). Figure 5.2 gives a graphical comparison

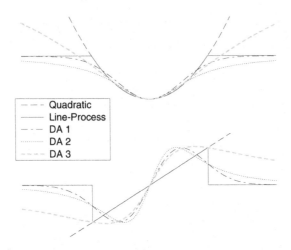

Figure 5.2: Comparison of different potential functions $g(\eta)$ (top) and their first derivatives $g'(\eta) = 2\eta h(\eta)$ (bottom). From (Li 1995b) with permission; ©1995 IEEE.

of the $g(\eta)$s for the quadratic, the LP model, and the first three DA models listed in Table 5.1 and their derivatives, $g'(\eta) = 2\eta h(\eta)$.

The fourth AIF, $h_{4\gamma} = 1/[1 + |\eta|/\gamma]$, allows bounded but nonzero smoothing at discontinuities: $\lim_{\eta \to \infty} \eta h_{4\gamma}(\eta) = \gamma$. It is interesting because $g''_{4\gamma}(\eta) = [2\eta h_{4\gamma}(\eta)]' > 0$ for all η (except at $\eta = 0$) and leads to strictly convex minimization. In fact, a positive number C in (5.27) leads to a convex AIF and hence energy function.[4] The convex subset of models for discontinuity adaptivity and M-estimation is appealing because they have some inherent advantages over nonconvex models, both in stability and computational efficiency. A discussion on convex models will be given in Section 6.1.5; see also (Li et al. 1995).

Figure 5.2 also helps us visualize how the DA performs smoothing. Like the quadratic $g_q(\eta) = \eta^2$, the DA allows the smoothing strength $|g'_\gamma(\eta)|$ to increase monotonically as η increases within the band B_γ. Outside the band, the smoothing decreases as η increases and becomes zero as $\eta \to \infty$. This differs from the quadratic regularizer, which allows boundless smoothing when $\eta \to \infty$. Also unlike the LP model, which shuts down smoothing abruptly just beyond its band $B_\alpha = (-\sqrt{\alpha}, \sqrt{\alpha})$, the DA decreases smoothing continuously toward zero.

---

[4]This is due to the following theorem: If $g(\cdot)$ is convex on $\mathbb{R}$, a real-valued energy function $E(f) = \int \chi(x)[f(x) - d(x)]^2 + \lambda g(f'(x))dx$ is convex w.r.t. f for all $f \in C^1$ and fixed $\chi, d \in C^1$.

5.2.2 Relations with Previous Models

The first three instantiated models behave in a way similar to the quadratic prior model when $\eta^2/\gamma \ll 1$ or $\eta^2 \ll \gamma$, as noticed by (Li 1990b). This can be understood by looking at the power series expansion of $g_\gamma(\eta)$, $g_\gamma(\eta) = \sum_{k=1}^{\infty} c_k (\eta^2/\gamma)^k$, where c_k are constants with $c_1 > 0$ (the expansion can also involve a trivial additive constant c_0). When $\eta^2/\gamma \ll 1$,

$$g_\gamma(\eta) = c_1(\eta^2/\gamma) + O(\eta^4/\gamma^2) \tag{5.30}$$

Thus, in this situation with sufficiently small η^2/γ, the adaptive model inherits the convexity of the quadratic model.

The interaction function h also well explains the differences between various regularizers. For the quadratic regularizer, the interaction is constant everywhere

$$h_q(\eta) = 1 \tag{5.31}$$

and the smoothing strength is proportional to $|\eta|$. This is why the quadratic regularizer leads to oversmoothing at discontinuities where η is infinite. In the LP model, the interaction is piecewise constant

$$h_\alpha(\eta) = \begin{cases} 1 & |\eta| < \sqrt{\alpha} \\ 0 & \text{otherwise} \end{cases} \tag{5.32}$$

Obviously, it inhibits oversmoothing by switching off smoothing when $|\eta|$ exceeds $\sqrt{\alpha}$ in a binary manner.

In the LP approximations using the Hopfield approach (Koch et al. 1986; Yuille 1987) and mean field theory (Geiger and Girosi 1991), the line process variables are approximated by (5.13). This approximation effectively results in the interaction function

$$h_{\alpha,T}(\eta) = 1 - \bar{l}(\eta) = 1 - \frac{1}{1 + e^{-\lambda(\eta^2 - \alpha)/T}} = \frac{e^{-\lambda(\eta^2 - \alpha)/T}}{1 + e^{-\lambda(\eta^2 - \alpha)/T}} \tag{5.33}$$

As the temperature τ decreases toward zero, (5.33) approaches $h_\alpha/2$; that is, $\lim_{\tau \to 0^+} h_{\alpha,\tau} = h_\alpha/2$. Obviously, $h_{\alpha,\tau}(\eta)$ with nonzero τ is a member of the AIF family (i.e., $h_{\alpha,\tau}(\eta) \in \mathbb{H}_\gamma$), and therefore the approximated LP models are instances of the DA model.

It is interesting to note an observation made by Geiger and Girosi: "sometimes a finite β solution may be more desirable or robust" (Geiger and Girosi (1991), pages 406–407) where $\beta = \frac{1}{T}$. They further suggest that there is "an optimal (finite) temperature (β)" for the solution. An algorithm is presented in (Hebert and Leahy 1992) for estimating an optimal β. The LP approximation with finite $\beta < +\infty$ or nonzero $T > 0$ is more an instance of the DA than the LP model, which they aimed to approximate. It will be shown in Section 5.2.4 that the DA model is indeed more stable than the LP model.

Anisotropic diffusion (Perona and Malik 1990) is a scale-space method for edge-preserving smoothing. Unlike fixed coefficients in the traditional isotropic scale-space filtering (Witkin 1983), anisotropic diffusion coefficients are spatially varying according to the gradient information. A so-called biased anisotropic diffusion (Nordstrom 1990) model is obtained if anisotropic diffusion is combined with a closeness term. Two choices of APFs are used in those anisotropic diffusion models: $g_{1\gamma}$ and $g_{2\gamma}$.

Shulman and Herve (1989) propose to use Huber's robust error penalty function (Huber 1981)

$$g_\beta(\eta) = \min\{\eta^2, \beta^2 + 2\beta|\eta - \beta|\} \qquad (5.34)$$

as the adaptive potential, where $\beta > 0$ plays a role similar to α in g_α. The above is a convex function and has the first derivative

$$g'_\beta(\eta) = 2\eta h_\beta(\eta) = 2\eta \qquad (5.35)$$

for $|\eta| \le \beta$ and

$$g'_\beta(\eta) = 2\eta h_\beta(\eta) = 2\beta\eta/|\eta| \qquad (5.36)$$

for other η. Comparing $g'_\beta(\eta)$ with (5.24), we find that the corresponding AIF is $h_\beta(\eta) = 1$ for $|\eta| \le \beta$ and $h_\beta(\eta) = \beta/|\eta|$ for other η. This function allows bounded but nonzero smoothing at discontinuities. The same function has also been applied by Stevenson and Delp (1990) to curve fitting. A comparative study of the DA model and robust statistics is presented in Chapter 6 (see also Li (1995a)).

The approximation (5.18) of Leclerc's minimal length model (Leclerc 1989) is in effect the same as the DA with APF 1. Equation (5.17) may be one of the best cost functions for the piecewise constant restoration; for more general piecewise continuous restoration, one needs to use (5.18) with a *nonzero* τ, which is a DA instance. Regarding the continuity property of domains, Mumford and Shah's model (Mumford and Shah 1985) and Terzopoulos's continuity-controlled regularization model (Terzopoulos 1986b) can also be defined on continuous domains like the DA model.

5.2.3   Discrete Data and 2D Cases

When the data d are available at discrete points, $d_i = d(x_i)$, $1 \le i \le m$, the Euler equation for the adaptive string is

$$\sum_i \chi(x)[f(x) - d_i]\delta(x - x_i) = \lambda\frac{d}{dx}[f'(x)h(f'(x))] \qquad (5.37)$$

where $\delta(\cdot)$ is the Dirac delta function. Integrating the equation above once yields

$$\sum_{i:x_i<x} \chi(x_i)[f(x_i) - d_i] = \lambda[f'(x)h(f'(x))] + c \qquad (5.38)$$

where c is a constant. Then, the solution f is determined by

$$f''(x) = 0 \qquad\qquad x_{i-1} < x < x_i, \ \ 1 < i \le m \ \ (5.39)$$

$$f(x_i^+) - f(x_i^-) = 0 \qquad\qquad 1 < i < m \ \ (5.40)$$

$$f'(x_i^+)h(f'(x_i^+)) - f'(x_i^-)h(f'(x_i^-)) = \frac{1}{\lambda}\chi(x_i)[f(x_i) - d_i] \qquad 1 < i < m$$
$$(5.41)$$

with appropriate boundary conditions at $x = x_1$ and $x = x_m$. Obviously, f in this case is piecewise C^1 continuous; more exactly, it is composed of consecutively joined line segments.

The adaptive string model can be extended to 2D and higher-order equivalents. The 2D adaptive membrane has

$$\chi[f - d] - \lambda\frac{\partial}{\partial x}[f_x h(f_x)] - \lambda\frac{\partial}{\partial y}[f_y h(f_y)] = 0 \qquad (5.42)$$

corresponding to the Euler equation (5.25), where data $d = d(x, y)$ and solution $f = f(x, y)$ represent images. Extending the DA to second-order, one obtains for the adaptive rod on 1D

$$\chi[f - d] + \lambda\frac{d}{dx^2}[f''h(f'')] = 0 \qquad (5.43)$$

and for the adaptive plate on 2D

$$\chi[f-d]+\lambda\frac{\partial}{\partial x^2}[f_{xx}h(f_{xx})]+2\lambda\frac{\partial}{\partial xy}[f_{xy}h(f_{xy})]+\lambda\frac{\partial}{\partial y^2}[f_{yy}h(f_{yy})] = 0 \ \ (5.44)$$

The discrete form of the adaptive membrane will be given in Section 5.2.5.

5.2.4 Solution Stability

The DA solution depends continuously on its parameters and the data, whereas the LP solution does not. An informal analysis follows. Consider an LP solution f_α obtained with α. The solution is a local equilibrium satisfying

$$\chi(f_\alpha - d) - \lambda f_\alpha' h_\alpha(f_\alpha') = 0 \qquad (5.45)$$

where $h_\alpha(f_\alpha') = 1$ or 0, depending on α and the configuration of f_α. When $|f_\alpha'(x)|^2$ is close to α for some x, a small change $\Delta\alpha$ may flip h_α over from one state to the other. This is due to the binary nonlinearity of h_α. The flipover leads to a significantly different solution. This can be expressed as

$$\lim_{\Delta\alpha \to 0}[f_{\alpha+\Delta\alpha} - f_\alpha] \ne 0(x) \qquad \exists \alpha \qquad (5.46)$$

where $0(x)$ is a function that is constantly zero in the domain $[a, b]$. The variation δf_α w.r.t. $\Delta\alpha$ may not be zero for some f_α and α, which causes instability. However, the DA solution, denoted f_γ, is stable

$$\lim_{\Delta\gamma\to 0}[f_{\gamma+\Delta\gamma} - f_\gamma] = 0(x) \qquad \forall\gamma \qquad\qquad (5.47)$$

where f_γ denotes the DA solution. Conclusions on the stability due to changes in parameter λ can be drawn similarly.

The same is also true w.r.t. the data. Given α and λ fixed, the solution depends on the data; i.e., $f = f[d]$. Assume a small variation δd in the data d. The solution $f[d]$ must change accordingly to reach a new equilibrium $f[d+\delta d]$ to satisfy the Euler equation. However, there always exists the possibility that $h_\alpha(f'_\alpha(x))$ may flip over for some x when $|f'_\alpha(x)|^2$ is near the α, resulting in an abrupt change in the LP solution f_α. This can be represented by

$$\lim_{\delta d\to 0(x)}[f_\alpha[d + \delta d] - f_\alpha[d]] \neq 0(x) \qquad \exists f_\alpha, d \qquad\qquad (5.48)$$

That is, the variation δf_α w.r.t. δd may not be zero for some f_α and d. However, the DA model is stable against such changes,

$$\lim_{\delta d\to 0(x)}[f_\gamma[d + \delta d] - f_\gamma[d]] = 0(x) \qquad \forall f_\gamma, d \qquad\qquad (5.49)$$

because of its continuous nature. From the analysis, it can be concluded that the DA better regularizes ill-posed problems than the LP.

To summarize, through an analysis of the associated Euler equation, a necessary condition is made explicit for MRF or regularization models to be adaptive to discontinuities. On this basis, the DA model is defined by the Euler equation constrained by the class of adaptive interaction functions (AIFs). The definition provides principles for choosing deterministic regularizers and MRF clique potential functions. It also includes many existing models as special instances.

The DA model has its solution in C^1 and adaptively overrides the smoothness assumption where the assumption is not valid, without the switching on and off of discontinuities in the LP model. The DA solution never contains true discontinuities. The DA model "preserves" discontinuities by allowing the solution to have arbitrarily large but *bounded* slopes. The LP model "preserves true discontinuities" by switching between small and *unbounded* (or large) slopes.

Owing to its continuous properties, the DA model possesses some theoretical advantages over the LP model. Unlike the LP model, it is stable against changes in parameters and in the data. Therefore it is better than the LP model in solving ill-posed problems. In addition, it is able to deal with problems on a continuous domain. Furthermore, it is better suited for analog VLSI implementation; see Section 10.3.

5.2.5   Computational Issues

The Euler equation (5.25) can be treated as a boundary value problem. It can also be solved by minimizing the corresponding energy (5.19) because a minimum of the energy is (sufficiently) a solution of the equation. (With $g'(\eta) = 2\eta h(\eta)$, the form of the energy need not be known in order to minimize it – see below.) The energy minimization approach is chosen here.

Because both $\chi(x)[f(x) - d(x)]^2$ and $g(f'(x))$ are bounded below, so is the energy. This means that a minimal solution, and hence a solution to the Euler equation, exists. If h_γ is chosen with $C = 0$, the corresponding energy $E(f)$ is nonconvex w.r.t. f, and the minimization is subject to local minima. See (Blake and Zisserman 1987) for an analysis of convexity in string and membrane models. The following presents a discrete method for finding a local minimum.

Sample the integral interval $[a, b]$ into m uniformly spaced points: $x_1 = a$, ..., $x_m = b$. For clarity and without loss of generality, let us assume that the bounds a and b are rescaled in such a way that the point spacing is one unit[5] (i.e., $x_i - x_{i-1} = 1$). Approximate the first derivative f'_i by the first-order backward difference $f'_i = \frac{f_i - f_{i-1}}{x_i - x_{i-1}} = f_i - f_{i-1}$. Then the energy (5.19) is approximated by

$$E(f) = \sum_{i=1}^{m} \chi_i [f_i - d_i]^2 + \lambda \sum_{i=1}^{m} \sum_{i' \in \mathcal{N}_i} g(f_i - f_{i'}) \qquad (5.50)$$

where $\mathcal{N}_i$ consists of the neighbors of i. Using the gradient-descent method, we can obtain the following updating equation: For $1 < i < m$,

$$\begin{aligned} f_i^{(t+1)} \leftarrow f_i^{(t)} \quad &- \quad 2\mu \{ \chi_i [f_i^{(t)} - d_i] \\ &- \quad \lambda \sum_{i' \in \mathcal{N}_i} (f_{i'}^{(t)} - f_i^{(t)}) \, h(f_{i'}^{(t)} - f_i^{(t)}) \} \end{aligned} \qquad (5.51)$$

The solution $f^{(t)}$ takes prescribed values at the boundary points at $i = 1$ and $i = m$ to meet the boundary condition (5.22), and the values may be estimated from data d_i near the boundaries. With initial $f^{(0)}$, the solution is in the limit $f^* = f^{(\infty)}$.

Equation (5.51) helps us see more of how the DA works. The smoothing at i is due to $\lambda \sum_{i' \in N_i} (f_{i'} - f_i) h(f_{i'} - f_i)$, where $N_i = \{i-1, i+1\}$. The contributions to smoothing are from the two neighboring points. That from site i' is proportional to the product of the two factors, $f_{i'} - f_i$ and $h_\gamma(f_{i'} - f_i)$. This is why we relate $|g'(\eta)| = |2\eta h(\eta)|$ to the strength of smoothing performed by regularizers. On the other hand, $h(\eta)$ acts as an adaptive weighting function to control the smoothing due to the difference $\eta = f_{i'} - f_i$. Therefore we regard it as the interaction. Two more remarks are made: First, the contributions from the two sides, $i - 1$ and $i + 1$, are treated separately or nonsymmetrically. Second, the sum of the contributions to smoothing is zero if the three points are aligned when $f_{i+1} - f_i = f_i - f_{i-1}$ and $h(\eta)$ is even since in this situation $\sum_{i' \in N_i} (f_{i'} - f_i) h(f_{i'} - f_i) = 0$.

---

[5] More advanced numerical methods using varying spacing, such as that of Weiss (1990), may be advantageous in obtaining more accurate solutions.

The updating rule for the adaptive membrane on 2D can be easily derived. In the 2D case (5.42), there is an additional term in the y dimension. This leads to the energy

$$E(f) = \sum_{i,j} \chi_{i,j}[f_{i,j} - d_{i,j}]^2 + \lambda g(f_{i,j} - f_{i-1,j}) + \lambda g(f_{i,j} - f_{i,j-1}) \quad (5.52)$$

Letting $\frac{\partial f_{i,j}}{\partial t} = -\mu \frac{\partial E}{\partial f_{i,j}}$ leads to the updating rule

$$f_{i,j}^{(t+1)} \leftarrow f_{i,j}^{(t)} - \quad (5.53)$$

$$2\mu \left\{ \chi_{i,j}[f_{i,j}^{(t)} - d_{i,j}] - \lambda \sum_{(x,y)\in\mathcal{N}_{i,j}} (f_{x,y}^{(t)} - f_{i,j}^{(t)})h(f_{x,y}^{(t)} - f_{i,j}^{(t)}) \right\}$$

where $\mathcal{N}_{i,j} = \{(i-1,j), (i+1,j), (i,j-1), (i,j+1)\}$ is the set of the four neighboring points[6] of (i,j). The update on the2D grid can be performed alternately on the white and black sites of a checkerboard to accelerate convergence.

There are three parameters in (5.51) that shall be determined for the DA model: λ, γ, and μ. Parameter μ is related to the convergence of the relaxation algorithm. There is an upper bound for μ for the system to be stable. There also exists an optimal value for μ (Blake and Zisserman 1987). Optimal choices of λ for quadratic regularization may be made using cross validation (Wahba 1980; Shahraray and Anderson 1989). In (Nadabar and Jain 1992), a least squares method (Derin and Elliott 1987) is presented for estimating MRF clique potentials for an LP model. Automated selection of the λ and γ parameters for the DA model is an unsolved problem. They are currently chosen in an ad hoc way.

The DA model with $C = 0$ leads to nonconvex dynamic systems, and direct minimization using gradient-descent only guarantees finding a local minimum. A GNC-like algorithm can be constructed for approximating the global solution; see Section 10.3.

5.3   Total Variation Models

When the L^2 norm is used in the formulation of an MRF prior, the solution lies in the so-called Sobolev space $W^{1,2}(\Omega)$ (Tikhonov and Arsenin 1977). The total variation (TV) models use the total variation norm (a seminorm[7])

---

[6]With the 4-neighborhood system, the model considers derivatives in the horizontal and vertical directions. With the 8-neighborhood system, the regularizer also includes the diagonal derivatives weighted by $1/\sqrt{2}$.

[7]A seminorm is a function defined on a vector space V, denoted by $\|v\|$, such that $\forall v, w \in V$ the following conditions hold: (1) $\|v\| \geq 0$; (2) $\|cv\| = |c|\|v\|$, where c is a scalar; and (3) $\|v + w\| \leq \|v\| + \|w\|$. Unlike in a norm, there may exist some $w \in V, w \neq 0$, but $\|w\| = 0$.

and provide another solution for discontinuity adaptive smoothing. There, a solution lies in a functional space, called bounded variation (BV) space, composed of functions with finite total variation (Rudin et al. 1992).

5.3.1 Total Variation Norm

Let f be a function in $\mathbb{R}^n$ defined on a bounded open set Ω. The total variation of f is defined as the TV norm

$$\|f\|_{\mathrm{TV}} = \int_\Omega |\nabla f|\, dx \qquad (5.54)$$

where $|\nabla f|$ is the L^1 norm of the gradient ∇f. In the case of images where $n = 2$, the gradient ∇f is approximated by the digital difference operator and the integration is replaced by summation. Consider all functions having finite TV norms on Ω

$$\|f\|_{\mathrm{TV}} = \int_\Omega |\nabla f|\, dx < \infty \qquad (5.55)$$

They form the bounded variation space $BV(\Omega)$. The Hamilton operator ∇ must be interpreted in the distributional sense in that the integration of a Dirac function is unity. This gives the key characteristic of $BV(\Omega)$ in which functions with discontinuities are included. Figure 5.3 shows four functions of one variable (one straight line, two smooth curves, and one step function) having the same total variation in the interval $[t_0, t_1]$. The mathematics to deal with functions in $BV(\Omega)$ has been well developed; see, e.g., (Chan and Shen 2005). This is the reason why solutions have so far been considered in $BV(\Omega)$ rather than directly in the larger space $L^1(\Omega)$.

5.3.2 TV Models

ROF Model

The ROF model of Rudin, Osher, and Fatemi (1992) provides the basic idea for TV models. Let d be the observed image, let the underlying true image be denoted by f, and suppose d is a function of f. In most situations in image restoration, this function can be modeled as a linear transformation plus additive noise n such that

$$d = Af + n \qquad (5.56)$$

where A is a linear operator. The object of image restoration is to reconstruct f from observation d. Generally, this is an ill-posed problem, and the solution is regarded as the optimizer of (1.27):

$$U(f \mid d) = U(d \mid f) + U(f) \qquad (5.57)$$

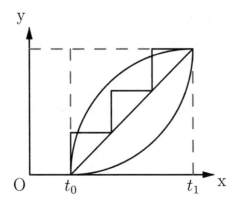

Figure 5.3: Four functions (the straight line, two curves, and the step) have the same total variation. This illustrates that the $BV(\Omega)$ space can include functions with discontinuities.

In the ROF model (Rudin et al. 1992), the data fidelity is the L^2 norm for $U(d \mid f) = \|Af - d\|^2$ when the noise is additive Gaussian. The solution is

$$f^* = \underset{f \in BV(\Omega)}{\arg \min} \ \|f\|_{\text{TV}} + \frac{1}{2}\lambda\|Af - d\|_{L^2} \tag{5.58}$$

where λ is the Lagrange multiplier, also a scale parameter. This TV model is equivalent to the following constrained optimization problem:

$$\begin{aligned} \min \quad & \|f\|_{\text{TV}} \\ \text{s.t.} \quad & \|Af - d\|_{L^2} \leq \sigma^2 \\ & f \in BV(\Omega) \end{aligned} \tag{5.59}$$

The solution can be obtained using the Euler-Lagrange equation of functional (5.58); that is,

$$\nabla \left[\frac{\nabla f}{|\nabla f|} \right] + \lambda(d - Af) = 0 \tag{5.60}$$

Several numerical algorithms have been used to find the optimal solution (Rudin et al. 1992; Chan et al. 1996; Carter 2001; Chambolle 2004; Darbon and Sigelle 2004; Chan and Shen 2005; Darbon and Sigelle 2005; Yin 2006).

While the ROF model could do edge-preserving smoothing, it may cause some problems, such as loss of contrast, distortion in geometry and texture, and causing staircases, especially at corners where the curvature of the level sets is high (Chan and Shen 2005).

TV+L^1 Model

The data fidelity in the ROF model could be replaced by an L^1 norm to achieve a better solution (Yin 2006) because the TV norm can be regarded as the L^1 norm of the gradient ∇f. (The L^1 estimator is more robust than the L^2 estimator (Chan and Shen 2005).) The optimum of the TV+L_1 model is

$$f^* = \underset{f \in BV(\Omega)}{\arg\min} \; \|f\|_{\mathrm{TV}} + \lambda \|Af - d\|_{L^1} \qquad (5.61)$$

which could be converted to the following constrained optimization:

$$\begin{aligned} \min \quad & \|f\|_{\mathrm{TV}} \\ \text{s.t.} \quad & \|Af - d\|_{L^1} \leq \sigma \\ & f \in BV(\Omega) \end{aligned} \qquad (5.62)$$

The corresponding Euler-Lagrange equation is deduced as

$$\nabla \left[\frac{\nabla f}{|\nabla f|} \right] + \lambda \frac{d - Af}{|d - Af|} = 0 \qquad (5.63)$$

In the TV+L^1 model, the Lagrange multiplier λ is related to the scale of $d - f^*$, so this model has the ability to extract texture by scale. Furthermore, this model is morphologically invariant (Chan and Esedoglu 2005; Yin 2006).

Meyer Model

The G-norm has been used to define the data fidelity for the extraction of oscillation patterns (Meyer 2001). G is a space dual to the subspace $\mathcal{BV}(\Omega)$ of $BV(\Omega)$, such that $\mathcal{BV}(\Omega) = \{ f \in BV(\Omega), |\nabla f| \in L^1 \}$. G is also a Banach space, which can be defined as

$$G = \{ v \mid v = \nabla \mathbf{g}, \; \mathbf{g} \in L^\infty(\mathbb{R}^n; \mathbb{R}^n) \} \qquad (5.64)$$

The G-norm of $v \in G$ is defined as

$$\|v\|_G = \inf_{v = \nabla \mathbf{g}} \| |\mathbf{g}|_{l^2} \|_{L^\infty} \qquad (5.65)$$

With the G-norm, the optimum is defined as

$$f^* = \underset{f \in BV(\Omega)}{\arg\min} \; \|f\|_{\mathrm{TV}} + \lambda \|Af - d\|_G \qquad (5.66)$$

It can be converted to the constrained optimization

$$\begin{aligned} \inf \quad & \|f\|_{\mathrm{TV}} \\ \text{s.t.} \quad & \|Af - d\|_G \leq \sigma \\ & f \in BV(\Omega) \end{aligned} \qquad (5.67)$$

However, it is impossible to derive the Euler-Lagrange equation for this problem. The problem (5.66) may be relaxed to some tractable form (Vese and Osher 2003).

5.3.3 Multichannel TV

In multichannel image applications, such as color imaging, multichannel re-mote sensing, and MRI images, an image pixel can consist of several values. The TV-norm can be extended to deal with multichannel image problems (Chan and Shen 2005). Let $\mathbf{f}(x) = [f^1(x), \ldots, f^K(x)]$ denote a K-channel image, in which a vectoral function is defined on a 2D domain Ω. The following describes three multichannel TV norms.

Channel-by-Channel TV

A simple treatment is the channel-by-channel method. It considers each chan-nel separately and models each channel using a scalar TV model; it then combines these $\|f^k\|_{\mathrm{TV}}, k = 1, 2, \ldots, K$ to get the final result. The channel-by-channel TV norm is defined as the sum of all the scalar TV norms:

$$\|\mathbf{f}\|_{\mathrm{TV}_C} = \sum_{k=1}^{K} \|f^k\|_{\mathrm{TV}} = \sum_{k=1}^{K} \int_{\Omega} |\nabla f^k| \, dx = \int_{\Omega} \left(\sum_{k=1}^{K} |\nabla f^k| \right) \, dx \quad (5.68)$$

The optimum can be computed by the system of Euler-Lagrange equations

$$\nabla \left[\frac{\nabla f^k}{|\nabla f^k| dx} \right] = 0, \quad k = 1, \cdots, K \quad (5.69)$$

While it is simple, the channel-by-channel TV norm ignores the connections between different channels, so it does not provide good solutions to applica-tions.

Blomgren-Chan TV

Blomgren and Chan (Blomgren 1998; Blomgren and Chan 1998) treat $\|f^k\|_{\mathrm{TV}}$, $k = 1, \ldots, K$ as a vector $(\|f^1\|_{\mathrm{TV}}, \ldots, \|f^k\|_{\mathrm{TV}})$ and form multichannel TV using the Euclidean norm of this vector. This gives the Blomgren-Chan TV norm

$$\|\mathbf{f}\|_{\mathrm{TV}_B} = \left[\sum_{k=1}^{K} \|f^k\|_{\mathrm{TV}}^2 \right]^{\frac{1}{2}} = \left[\sum_{k=1}^{K} \left(\int_{\Omega} |\nabla f^k| \, dx \right)^2 \right]^{\frac{1}{2}} \quad (5.70)$$

The corresponding Euler-Lagrange equations are given by

$$\alpha_k(\mathbf{f}) \cdot \nabla \left[\frac{\nabla f^k}{|\nabla f^k|} \right] = 0, \quad k = 1, \cdots, K \quad (5.71)$$

where the weight constant is calculated as the orientation cosine in the Eu-clidean space:

$$\alpha_k(\mathbf{f}) = \frac{\|f^k\|_{\mathrm{TV}}}{\|\mathbf{f}\|_{\mathrm{TV}_B}}, \quad k = 1, \cdots, K \quad (5.72)$$

Both $\|\mathbf{f}\|_{\mathrm{TV}_C}$ and $\|\mathbf{f}\|_{\mathrm{TV}_B}$ integrate the TV models of individual channels to form a multichannel TV norm. So the interactions between channels, if they exist, are accounted for in a global sense. However, unlike (5.69), the equations in (5.71) are coupled by the coefficients $\alpha_k(\mathbf{f})$.

Sapiro-Ringach TV

Sapiro and Ringach (Sapiro and Ringach 1996; Tang et al. 2001) considered the connections point by point in the image plane Ω, and integrated the pointwise norms (Zenzo 1986) over Ω to get the final multichannel TV. This defines the norm

$$\|\mathbf{f}\|_{\mathrm{TV}_S} = \int_\Omega \left(\sum_{k=1}^K |\nabla f^k|^2 \right)^{\frac{1}{2}} dx = \int_\Omega \|\nabla\mathbf{f}\|_F \, dx \qquad (5.73)$$

where $\nabla\mathbf{f} = (\nabla f^1, \nabla f^2, \cdots, \nabla f^K)$, and $\| \cdot \|_F$ is the Frobenius norm of a matrix. The Euler-Lagrange equations are then

$$\nabla \left[\frac{\nabla\mathbf{f}}{\|\nabla\mathbf{f}\|} \right] = 0 \qquad (5.74)$$

where

$$\|\nabla\mathbf{f}\|_F = \sqrt{\mathrm{trace}(\nabla\mathbf{f}(\nabla\mathbf{f})^T)} = \sqrt{\sigma_+^2 + \sigma_-^2} \qquad (5.75)$$

in which σ_+ and σ_- are the singular values of $\nabla\mathbf{f}$. Sapiro and Ringach also suggest other nonnegative functions $\phi(\sigma_+, \sigma_-)$ to replace $\|\nabla\mathbf{f}\|_F$ as alternative solutions.

5.4 Modeling Roof Discontinuities

The previous sections dealt with *step-edge* preserving smoothing, where the models assume that the underlying surface has zero first-order derivatives. For *roof-edges*, one might suggest using higher-order piecewise smoothness, such as the weak rod and plate models (Blake and Zisserman 1987). However, such models are usually not recommended because of their instability (Blake and Zisserman 1987).

In this section, roof-edge (as well as step-edge) preserving smoothing is performed by using first-order piecewise smoothness. The approach is the following. The image surface function is modeled using a parametric polynomial representation. Based on this, the problem of preserving C^1 discontinuities in the image surface function is converted to preserving C^0 discontinuities in the governing parameter functions; in other words, the piecewise smoothness is imposed on the parameter functions rather than directly on the image surface height function itself. In this way, only the first derivatives are required to preserve roof edges. This avoids the instability of using higher-order derivatives and is the advantage of such an MAP-MRF model. The model extends the ability of the existing first-order piecewise smoothness models (such as the line process and weak membrane models) in edge-preserving smoothing.

5.4.1 Roof-Edge Model

An *ideal roof edge* can be modeled as a joining point of two planar surfaces. Figure 5.4 shows a few examples of roof edges, noting that they are not necessarily local extrema in surface height. A planar surface takes the form $z(x) = a + bx$ if it is on a 1D domain or $z(x, y) = a + bx + cy$ if on 2D, where a, b, and c are some constant parameters. Let the parameters be denoted collectively by $f = \{a, b, c\}$ and a parameterized plane be denoted by $z = z_f(x, y) = a + bx + cy$. Each distinct set of constants $\{a, b, c\}$ represents a distinct plane. For a slanted plane, at least one of b and c is nonzero.

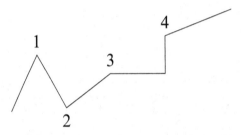

Figure 5.4: Examples of roof edges in 1D. Corner points 1, 2, and 3 are roof edge points; No. 4 is a roof to the right but a step to the left.

Let us look at Fig. 5.5 to see what happens when the line process model is used to smooth $z(x)$: the discontinuity in b (also in a) is smoothed out, and the information about the roof is lost. However, imposing the first-order piecewise smoothness on these parameters can solve this problem; see below.

A *general roof* can be modeled as a joining point of two nonlinear surfaces represented by $z = z_f(x, y) = a(x, y) + b(x, y)x + c(x, y)y$, where a, b, and c are now some functions of x and y. A step edge is a step (C^0) discontinuity in $z(x, y)$; a roof is signaled by a step (C^0) discontinuity in either $b(x, y)$ or $c(x, y)$. In addition, there is usually also a discontinuity in $a(x, y)$ at either a roof or a step discontinuity. If $a(x, y)$, $b(x, y)$, and $c(x, y)$ are (piecewise) constant, then $z_f(x, y)$ degenerates to be (piecewise) planar. If they are (piecewise) C^0 continuous, then $z_f(x, y)$ consists of (piecewise) nonlinear surface(s).

For notational convenience, let the points on the x-y grid be indexed by a single index $i \in \mathcal{S} = \{1, \ldots, m\}$, where $\mathcal{S}$ is called the set of *sites* in the MRF literature. When all the $z_i = z_f(x_i, y_i)$ belong to a single planar surface

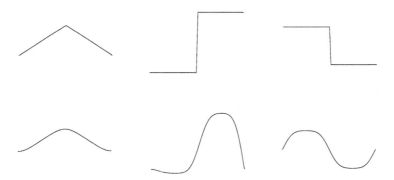

Figure 5.5: A clean ideal roof in 1D before (top) and after (bottom) being smoothed by the line process model. From left to right are the functions $z(x) = a(x) + b(x)x$, $a(x)$, and $b(x)$, respectively. The smoothing is performed on the $z(x)$. The $a(x)$ and $b(x)$ in the figure are computed from the respective $z(x)$ functions. The discontinuities in a and b, and hence the roof edge, are smoothed out.

of the form $z(x, y) = a + bx + cy$, the problem of finding the underlying a, b, and c parameter constants can be well solved; for example, by using the least squares method. However, when they are due to piecewise surfaces and the segmentation of the data into separate surfaces is unknown, finding the underlying parameters $f = \{a, b, c\}$ as piecewise functions is a nontrivial problem.

When the segmentation is unknown, we may consider the parameters to be some functions of (x, y), such as $a_i = a(x_i, y_i)$, $b_i = b(x_i, y_i)$, and $c_i = c(x_i, y_i)$, collectively denoted by $f_i = [a_i, b_i, c_i]^T$. As such, the surface heights is reconstructed as $z_i = z_{f_i}(x_i, y_i) = a_i + b_i x_i + c_i y_i$. Consider a_i, b_i, and c_i as random variables, f_i a random vector, and $f = \{f_1, \ldots, f_m\}$ a random field. We have the following remarks about the relationship between the a, b, c functions and the local configuration of f:

- In the case of a single planar surface, the parameter functions $a(x, y)$, $b(x, y)$, and $c(x, y)$ are constant over the x-y domain. Therefore, there must be $f_i = f_{i'}$ for all $i \neq i'$.

- In the case of a piecewise planar surface, the functions are piecewise constant over the x-y domain. Therefore, $f_i = f_{i'}$ holds only when both z_i and $z_{i'}$ belong to the same plane, and $f_i \neq f_{i'}$ otherwise. Discontinuities occur at boundary locations where two surfaces meet.

- In the case of a single nonlinear surface, the functions $a(x, y)$, $b(x, y)$, and $c(x, y)$ are C^0 continuous over the x-y domain. Therefore, there should be no discontinuity between f_i and $f_{i'}$ when i and i' are neighboring sites.

• In the case of a piecewise nonlinear surface containing roofs (i.e., C^1 discontinuities in z), the parameter functions are piecewise C^0 continuous over the x-y domain. Therefore, two neighboring parameter vectors f_i and $f_{i'}$ should be similar when z_i and $z_{i'}$ belong to a continuous part of the surface; otherwise, there should be a discontinuity between f_i and $f_{i'}$.

5.4.2 MAP-MRF Solution

First, let us define the prior for the parameter vector field $f = \{f_1, \ldots, f_m\}$, which is assumed to be Markovian. The $a(x,y)$, $b(x,y)$, and $c(x,y)$ are piecewise constant functions when $z(x,y)$ consists of piecewise planar surfaces, and therefore their partial derivatives w.r.t. x and y should be zero at nonedge locations. These constraints may be imposed by using the "prior energy"

$$U(f) = \sum_i \sum_{i' \in \mathcal{N}_i} g(\rho(f_i, f_{i'})) \tag{5.76}$$

where $\mathcal{N}_i$ is the set of neighbors of i (e.g., the four or eight directly adjacent points), g is an adaptive potential function satisfying certain conditions (Li 1995b), including the necessary condition mentioned earlier (e.g., $g(\eta) = 1/(1+\eta^2)$ used for experiments in (Li 1995b)), and

$$[\rho(f_i, f_{i'})]^2 = w_a[a_i - a_{i'}]^2 + w_b[b_i - b_{i'}]^2 + w_c[c_i - c_{i'}]^2 \tag{5.77}$$

is a weighted distance between f_i and $f_{i'}$ where $w > 0$ are the weights.

Now, let us define the observation model. Assume that the data $d = \{d_1, \ldots, d_m\}$, $d_i = z_{f_i}(x_i, y_i) + e_i$, is the true surface height z_i corrupted by independently and identically distributed (i.i.d.) Gaussian noise e_i. Then the conditional density of data $p(d \mid f)$ is a Gibbs distribution with the "likelihood energy"

$$U(d \mid f) = \sum_i [z_{f_i}(x_i, y_i) - d_i]^2 = \sum_i (a_i + b_i x_i + c_i y_i - d_i)^2 \tag{5.78}$$

This imposes the constraint from the data.

Now, the posterior $p(f \mid d)$ is a Gibbs distribution with the "posterior energy" function $U(f \mid d) = U(d \mid f) + U(f)$. For $z = a + bx + cy$, the energy can be written as

$$U(f \mid d) = \sum_i (a_i + b_i x_i + c_i y_i - d_i)^2 + \sum_i \sum_{i' \in \mathcal{N}_i} \tag{5.79}$$
$$g(\sqrt{w_a[a_i - a_{i'}]^2 + w_b[b_i - b_{i'}]^2 + w_c[c_i - c_{i'}]^2})$$

The MAP estimate is defined as $f^* = \arg\min_f U(f \mid d)$.

The proposed MAP-MRF model generalizes the line process model of Geman and Geman (1984). The latter is a special case of the former with $b_i = c_i = 0$. When $b_i = c_i = 0$, then the surface function is reduced to $z_i = a_i$ and the two models become equivalent.

5.4.3 Computational Issues

The posterior energy $E(f) = U(f \mid d)$ may be minimized by using a deterministic gradient-descent algorithm to achieve $\frac{\partial E(f)}{\partial a_i} = \frac{\partial E(f)}{\partial b_i} = 0$ for all i. The computation of the gradient is illustrated below with the surface function $z = a + bx + cy$ and the potential function $g(\eta) = \ln(1 + \eta^2)$. The gradient components are calculated as

$$\frac{1}{2}\frac{\partial E(f)}{\partial a_i} = (a_i + b_i x_i + b_i y_i - d_i) + \sum_{i' \in \mathcal{N}_i} w_a h(\rho(f_i, f_{i'}))(a_i - a_{i'}) \quad (5.80)$$

$$\frac{1}{2}\frac{\partial E(f)}{\partial b_i} = (a_i + b_i x_i + b_i y_i - d_i)x_i + \sum_{i' \in \mathcal{N}_i} w_b h(\rho(f_i, f_{i'}))(b_i - b_{i'}) \quad (5.81)$$

$$\frac{1}{2}\frac{\partial E(f)}{\partial c_i} = (a_i + b_i x_i + b_i y_i - d_i)y_i + \sum_{i' \in \mathcal{N}_i} w_c h(\rho(f_i, f_{i'}))(c_i - c_{i'}) \quad (5.82)$$

where

$$\rho(f_i, f_{i'}) = \sqrt{w_a[a_i - a_{i'}]^2 + w_b[b_i - b_{i'}]^2 + w_c[c_i - c_{i'}]^2} \quad (5.83)$$

and

$$h(\eta) = \frac{1}{1 + \eta^2} \quad (5.84)$$

determines the interactions between the two pixels i and i'. The updating is performed on (a_i, b_i, c_i) for all $i \in \mathcal{S}$. For example, the equation for updating b is $b_i^{(t+1)} - b_i^{(t)} \leftarrow \Delta_b$, where

$$\Delta_b = \mu \left\{ (a_i + b_i x_i + b_i y_i - d_i)x_i + \sum_{i' \in \mathcal{N}_i} w_b h(\rho(f_i, f_{i'}))(b_i - b_{i'}) \right\} \quad (5.85)$$

with $\mu > 0$ being a constant factor.

Annealing can be incorporated into the iterative gradient-descent process to help escape local minima. For example, introduce a "temperature" parameter T into the h function so that it becomes

$$h_T(\eta) = \frac{1}{1 + \eta^2/T} \quad (5.86)$$

At the beginning, T is set to a high value $T^{(0)}$. As the iteration continues, it is decreased toward the target value of 1 according to, for example, $T^{(t+1)} \leftarrow 0.9T^{(t)}$. This has been shown to be effective.

Randomization can also be incorporated into the gradient-descent. In a simple method called randomized gradient-descent, we introduce a random weight λ between the two terms in every update for every i and n. For example, for the b parameter, this gives

$$\Delta_b = \mu \left\{ (a_i + b_i x_i + b_i y_i - d_i)x_i + \lambda \sum_{i' \in \mathcal{N}_i} w_b h(\rho(f_i, f_{i'}))(b_i - b_{i'}) \right\} \quad (5.87)$$

When λ is constant all the time, it reduces to the deterministic gradient-descent. The randomized gradient-descent strategy helps avoid local minima and yields better results than the deterministic gradient-descent. It is more efficient than random sampling methods such as the Metropolis algorithm (Metropolis et al. 1953) and Gibbs sampler (Geman and Geman 1984) because the acceptance probability of any update is always one. A randomized neighborhood system (Li 1990a) also works well in avoiding local minima.

5.5   Experimental Results

5.5.1   Step-Edge-Preserving Smoothing

Two experimental results are presented in the following (more can be found in Chapter 3 of ?)). The first is the reconstruction of a real image of size 256×256 pixels (Fig. 5.6). Here, APF 1 ($g_{1\gamma}$) is used and the parameters are empirically chosen as $\lambda = 50$, $\gamma = 2$. The result shows that the reconstructed image is much cleaner with discontinuities well preserved.

The second experiment is the detection of step and roof edges from a simulated noisy pyramid image of size 128×128 pixels (Fig. 5.7). The detection process runs in three stages: (1) regularizing the input image and computing images of first derivatives in the two directions from the regularized image using finite difference, (2) regularizing the derivative images and (3) detecting steps and roofs by thresholding the regularized derivative images. APF 2 ($g_{2\gamma}$) is used and the parameters are empirically chosen as $\lambda = 10$, $\gamma_{target} = 10$ for the first stage of the regularization and $\lambda = 100$, $\gamma_{target} = 0.002$ for the second stage.

Edges in the horizontal and vertical directions are detected best, while those in the diagonal directions are not as well detected. This is because only derivatives in the two axis directions are considered in the DA discussed so far; changes in the diagonal directions are largely ignored. Regularizers using the 8-neighborhood system (see footnote 6 below (5.53) in this chapter) should help improve the detection of diagonal changes.

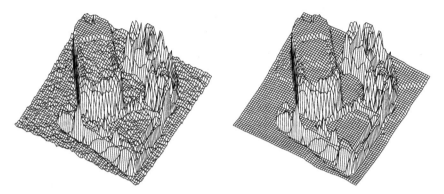

Figure 5.6: 3D plots of an image (left) and its reconstruction using DA (right). The plots are drawn after sampling the images at every 4 pixels in both directions into the size of 64 × 64 pixels. From (Li 1995b) with permission; ©1995 IEEE.

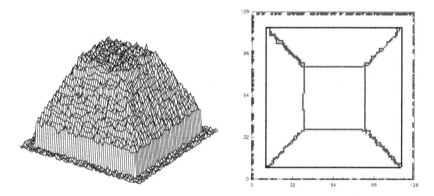

Figure 5.7: Step and roof edges (right) detected from a pyramid image (left). Step edges are shown as dots and roof edges as crosses. From (Li 1995b) with permission; ©1995 IEEE.

5.5.2 Roof-Edge-Preserving Smoothing

Range images are used because in such data there are geometrically well-defined roof and step edges. The 8-adjacency is used to define the MRF neighborhood system. The surfaces are assumed to be of the form $z(x_i, y_i) = a_i + b_i x_i + c_i y_i$. The a_i, b_i, and c_i parameters are initially estimated from the data by a local bilinear fit. The MAP estimates a^*, b^*, and c^* are computed by minimizing $U(f \mid d)$ (using an iterative gradient-descent combined with annealing). The smoothed surface is reconstructed as $z_i^* = a_i^* + b_i^* x_i + c_i^* y_i$. Roof edges are detected by thresholding the directional derivatives of b^* and

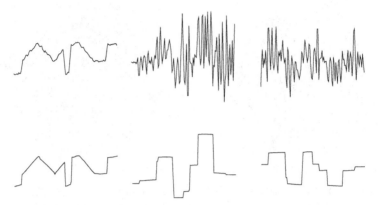

Figure 5.8: Smoothing 1D surface. Top (from left to right): surface data $d(x)$, and initial estimates for $a(x)$ and $b(x)$ obtained by using finite difference. Bottom (from left to right): The smoothed surface $z^*(x) = a^*(x) + b^*(x) \cdot x$ and the MAP estimates $a^*(x)$ and $b^*(x)$.

c^*, and step edges are detected by thresholding z^*; this is followed by a local maximum selection operation to get the final edge map.

Before presenting results for image surfaces on the 2D domain, let us look at a result for a 1D surface. For the clean ideal roof in Fig. 5.5, the proposed model produces results that are very close to the analytically calculated shapes shown in the top row of Fig. 5.5, with sharp steps in a^* and b^* and hence a sharp roof in z^*. For a noisy input (Fig. 5.8), the discontinuities in a^* and b^* are well preserved and there are no roundups over the discontinuities. This demonstrates significant improvement over the line process model result shown at the bottom of Fig. 5.5.

Now, let us look at two results for 2D surfaces. The first is obtained from a synthetic pyramid image composed of piecewise planar surfaces (Fig. 5.9) containing ideal roof edges. The smoothed image has clean surfaces and clear roof boundaries, as can be seen in the 3D plots. The a, b, c and edge maps before and after smoothing demonstrate the removal of noise and the preservation of discontinuities in the a, b, and c parameter functions. The detection of step edges (the outermost part of the edge map) and roof edges (the inner part) is nearly perfect even with the simple thresholding edge detector.

The second result is obtained on a Renault part image composed of free-form surfaces with roof edges (Fig. 5.10), noting that the original image is severely quantized in depth. In this case, a, b, and c are piecewise continuous nonlinear functions (of x and y) rather than piecewise constant. True roof edges emerge as a result of the edge-preserving smoothing. This result demonstrates that the proposed model also works well for free-form surfaces containing general roof edges.

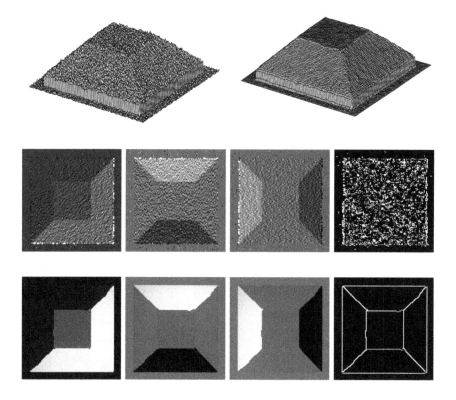

Figure 5.9: Row 1: 3D plots of a pyramid image before and after smoothing. Rows 2 and 3: The parameter and edge maps of the pyramid before (row 2) and after (row 3) smoothing; from left to right are parameter maps a, b, and c and the detected edges.

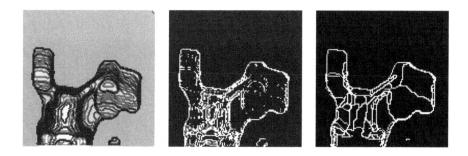

Figure 5.10: The Renault part image (left, shaded for visualization of the shape) and edges detected before (middle) and after (right) roof edge-preserving-smoothing.

Chapter 6

MRF Model with Robust Statistics

Robust statistical methods (Tukey 1977; Huber 1981; Rousseeuw 1984) are tools for statistics problems in which *outliers* are an issue. It is well known that the least squares (LS) error estimates can be arbitrarily wrong when outliers are present in the data. A robust procedure is aimed at making solutions insensitive to the influence of outliers. That is, its performance should be good with all-inlier data and should deteriorate gracefully with increasing number of outliers. The mechanism by which robust estimators deal with outliers is similar to that of the discontinuity adaptive MRF prior model studied in the previous chapter. This chapter provides a comparative study (Li 1995a) of the two kinds of models based on the results from the DA model and presents an algorithm (Li 1996b) to improve the stability of the robust M-estimator to the initialization.

The conceptual and mathematical comparison comes naturally from the parallelism of the two models: Outliers cause a violation of a distributional assumption, while discontinuities cause a violation of the smoothness assumption. Robustness to outliers is in parallel to adaptation to discontinuities. Detecting outliers corresponds to inserting discontinuities. The similarity of the two models suggests that results in either model could be used for the other.

Probably for this reason,there have seen considerable interests in applying robust techniques to solving image and vision problems. Kashyap and Eom (1988) developed a robust algorithm for estimating parameters in an autoregressive image model where the noise is assumed to be a mixture of a Gaussian and an outlier process. Shulman and Herve (1989) proposed to use Huber's robust M-estimator (Huber 1981) to compute optical flow involving discontinuities. Stevenson and Delp (1990) used the same estimator for curve fitting. Besl, Birch, and Watson (1988) proposed a robust M window

S.Z. Li, *Markov Random Field Modeling in Image Analysis*, 161
Advances in Pattern Recognition, DOI: 10.1007/978-1-84800-279-1_6,
© Springer-Verlag London Limited 2009

operator to prevent smoothing across discontinuities. Haralick, Joo, Lee, Zhuang, Vaidya, and Kim (1989), Kumar and Hanson (1989) and Zhuang, Wang, and Zhang (1992) used robust estimators to find pose parameters. Jolion, Meer, and Bataouche (1991) used the robust minimum volume ellipsoid estimator to identify clusters in feature space. Boyer, Mirza, and Ganguly (1994) present a procedure for surface parameterization based on a robust M-estimator. Black and Anandan (1993) and Black and Rangarajan (1994) applied a robust operator not only to the smoothness term but also to the data term. Li (1995a) presented a comparative study on robust models and discontinuity adaptive MRF models. He also devised a method for stabilizing robust M-estimation w.r.t. the initialization and convergence (Li 1996b).

A robust location estimator, which essentially seeks the mode of an outlier-contaminated distribution, can be extended to perform data clustering. In this connection, the mean shift algorithm (Fukunaga 1990) has been used successfully in vision problems such as segmentation (Cheng 1995; Comaniciu and Meer 1997; Comaniciu and Meer 1999).

As is well known, robust estimation procedures have a serious problem in that the estimates are dependent on the initial estimate value; this problem has been overcome by applying the principle of the graduated nonconvexity (GNC) method (Blake and Zisserman 1987) for visual reconstruction. The exchange of theoretical results and practical algorithms is useful to the image and vision communities because both MRF and robust models have applications in the4 areas.

6.1   The DA Prior and Robust Statistics

What do we mean by discontinuities and outliers? Unfortunately, their definitions are usually ambiguous. What we are certain of is that the likelihood of a discontinuity between a pair of neighboring pixels is related to the difference in pixel labels (such as pixel values), and an outlier is related to the distance between the location of the datum and the estimated value. Where the label difference is very large, there is likely to be a discontinuity between the pixels, and where the datum is very far from the cluster, it is likely an outlier.

A more concrete comparison can be made by analyzing the adaptation (to discontinuities) and the robustness (to outliers) in mathematical terms (Li 1995a). The adaptation is realized as follows. The interaction between related (e.g, neighboring) points must be decreased as the violation of the relational bond between them is increased and prohibited in the limit. This is true of both the MRF and the robust models. We give the necessary condition for such adaptation and then, based on this condition, a definition of a class of adaptive interaction functions for both models. The definition captures the essence of the adaptation ability and is general enough to offer in theory infinitely many suitable choices of such functions.

The problem of discontinuities and outliers also exists in other areas. In model-based object recognition, for example, there are two related sub-problems: first, separating the scene into different parts, each being due to a single object; and second, finding feature correspondences between each separate part of the scene and an object. The two subproblems have to be solved somewhat simultaneously. The process of matching while separating is similar to reconstruction with discontinuities and estimation with outliers. Indeed, matches to one object can be considered as outliers w.r.t. matches to a different object. Different groups of matches should not be constrained to each other. The separation can be done by inserting "discontinuities" between different groups. This view can be regarded as a generalization of the weak constraint (Hinton 1978; Blake and Zisserman 1987).

6.1.1 Robust M-Estimator

Robust statistical methods (Tukey 1977; Huber 1981) provide tools for statistics problems in which underlying assumptions are inexact. A robust procedure should be insensitive to departures from the underlying assumptions caused, for example, by *outliers*. That is, it should have good performance under the underlying assumptions, and the performance should deteriorate gracefully as the situation departs from the assumptions. Applications of robust methods in vision are seen in image restoration, smoothing and segmentation (Kashyap and Eom 1988; Jolion et al. 1991; Meer et al. 1991), surface and shape fitting (Besl et al. 1988; Stein and Werman 1992), and pose estimation (Haralick et al. 1989), where outliers are an issue.

There are several types of robust estimators. Among them are the M-estimator (maximum likelihood estimator), L-estimator (linear combinations of order statistics), R-estimator (estimator based on rank transformation) (Huber 1981), RM estimator (repeated median) (Siegel 1982) and LMS estimator (estimator using the least median of squares) (Rousseeuw 1984). We are concerned with the M-estimator.

The essential form of the M-estimation problem is the following. Given a set of m data samples $d = \{d_i \mid 1 \leq i \leq m\}$, where $d_i = f + \eta_i$, the problem is to estimate the location parameter f under noise η_i. The distribution of η_i is not assumed to be known exactly. The only underlying assumption is that $\eta_1, \ldots, \eta_m$ obey a symmetric, independent, identical distribution (symmetric i.i.d.). A robust estimator has to deal with departures from this assumption.

Let the residual errors be $\eta_i = d_i - f$ $(i = 1, \ldots, m)$ and the error penalty function be $g(\eta_i)$. The M-estimate f^* is defined as the minimum of a global error function

$$f^* = \arg\min_f E(f) \tag{6.1}$$

where

$$E(f) = \sum_i g(d_i - f) \tag{6.2}$$

Table 6.1: Robust functions.

Type	$h_\gamma(\xi)$	$g_\gamma(\xi)$	Range of ξ
Tukey	$=\begin{cases}(1-\xi^2)^2\\0\end{cases}$	$=\begin{cases}[1-(1-\xi^2)^3]/6\\1/6\end{cases}$	$\begin{array}{l}\lvert\xi\rvert\le 1\\\lvert\xi\rvert>1\end{array}$
Huber	$=\begin{cases}1\\\tau\frac{\text{sgn}(\xi)}{\xi}\end{cases}$	$=\begin{cases}\xi^2\\2\tau\lvert\xi\rvert-\tau^2\end{cases}$	$\begin{array}{l}\lvert\xi\rvert\le\tau\\\lvert\xi\rvert>\tau\end{array}$
Andrews	$=\begin{cases}\frac{\sin(\pi\xi)}{\pi u}\\0\end{cases}$	$=\begin{cases}[1-\cos(\pi\xi)]/\pi^2\\1/\pi^2\end{cases}$	$\begin{array}{l}\lvert\xi\rvert\le 1\\\lvert\xi\rvert>1\end{array}$
Hampel	$=\begin{cases}1\\a\frac{\text{sgn}(\xi)}{\xi}\\a\frac{c-\lvert\xi\rvert}{c-b}\frac{\text{sgn}(\xi)}{\xi}\\0\end{cases}$	$=\begin{cases}u^2/2\\a\lvert u\rvert-a^2/2\\ab-a^2/2+\\(c-b)a/2\left[1-\left(\frac{c-\lvert\xi\rvert}{c-b}\right)\right]\\ab-a^2/2+(c-b)a/2\end{cases}$	$\begin{array}{l}\lvert\xi\rvert\le a\\a<\lvert\xi\rvert\le b\\b<\lvert\xi\rvert\le c\\\\\lvert\xi\rvert>c\end{array}$

To minimize (6.2), it is necessary to solve the equation

$$\sum_i g'(d_i - f) = 0 \qquad (6.3)$$

This is based on *gradient descent*. When $g(\eta_i)$ can also be expressed as a function of η_i^2, its first derivative can take the form

$$g'(\eta_i) = 2\eta_i h(\eta_i) = 2(d_i - f)h(d_i - f) \qquad (6.4)$$

where $h(\eta)$ is an even function. In this case, the estimate f^* can be expressed as the weighted sum of the data samples

$$f^* = \frac{\sum_i h(\eta_i)\, d_i}{\sum_i h(\eta_i)} \qquad (6.5)$$

where h acts as the weighting function. This algorithm can be derived by using half-quadratic (HQ) optimization to be presented in Section 6.1.6.

In the LS regression, all data points are weighted the same with $h_\gamma(\eta) = 1$ and the estimate is $f^* = \frac{1}{m}\sum_{i=1}^m d_i$. When outliers are weighted equally as inliers, it will cause considerable bias and deterioration in the quality of the estimate. In robust M-estimation, the function h provides adaptive weighting. The influence from d_i is decreased when $\lvert\eta_i\rvert = \lvert d_i - f\rvert$ is very large and suppressed when it is infinitely large.

Table 6.1 lists some robust functions used in practice where $\xi = \eta/\gamma$. They are closely related to the adaptive interaction function and adaptive potential function defined in (5.27) and (5.28). Figure 6.1 shows their qualitative shapes in comparison with the quadratic and the line process models (note that a

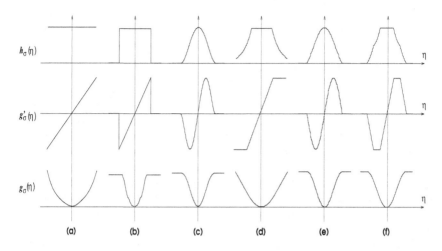

Figure 6.1: The qualitative shapes of potential functions in use. The quadratic prior (equivalent to LS) model in (a) is unable to deal with discontinuities (or outliers). The line process model (b), Tukey's (c), Huber's (d), Andrews' (e), and Hampel's (f) robust model are able to, owing to their property of $\lim_{\eta \to \infty} h_\gamma(\eta) = 0$. From (Li 1995a) with permission; ©1995 Elsevier.

trivial constant may be added to $g_\gamma(\eta)$). These robust functions are piecewise, as in the line process model. Moreover, the parameter γ in ξ is dependent on some scale estimate, such as the median of absolute deviation (MAD).

6.1.2   Problems with M-Estimator

Computationally, existing M-estimators have several problems affecting their performance. First, they are not robust to the initial estimate, a problem common to nonlinear regression procedures (Myers 1990), also encountered by vision researchers (Haralick et al. 1989; Meer et al. 1991; Zhuang et al. 1992). The convergence of the algorithm depends on the initialization. Even if the problem of convergence is avoided, the need for a good initial estimate cannot be ignored for convergence to the global estimate; this is because most M-estimators are defined as the global minimum of a generally *nonconvex* energy function and hence the commonly used gradient-based algorithms can get stuck at unfavorable local solutions. The M-estimator has the theoretical breakpoint of $\frac{1}{p+1}$, where p is the number of unknown parameters to be estimated, but, in practice, the breakpoint can be well below this value because of the problem of local minima.

Second, the definition of the M-estimator involves some scale estimate, such as the median of absolute deviation (MAD), and a parameter to be chosen. These are also sources of sensitivity and instability. For example, Tukey's biweight function (Tukey 1977) is defined as

$$h(\eta_i) = \begin{cases} \left(1 - \left(\frac{\eta_i}{cS}\right)^2\right)^2 & \text{if } |\eta_i| < cS \\ 0 & \text{otherwise} \end{cases} \qquad (6.6)$$

where S is an estimate of the spread, c is a constant parameter, and cS is the scale estimate. Possible choices include $S = \text{median}\{\eta_i\}$ with c set to 6 or 9, and $S = \text{median}\{|\eta_i - \text{median}\{\eta_i\}|\}$ (median of absolute deviation (MAD)) with $c = 1.4826$ chosen for the best consistency with the Gaussian distribution. Classical scale estimates such as the median and MAD are not very robust. The design of the scale estimates is crucial and needs devoted study.

Furthermore, the convergence of the M-estimator often is not guaranteed. Divergence can occur when initialization or parameters are not chosen properly. Owing to the problems above, the theoretical breakdown point can hardly be achieved.

In the following, an improved robust M-estimator, referred to as the *annealing M-estimator* (AM-estimator), is presented to overcome the above problems (Li 1996b). It has two main ingredients: a redefinition of the M-estimator and a GNC-like annealing algorithm.

6.1.3 Redefinition of M-Estimator

Resemblances between M-estimation with outliers and adaptive smoothing with discontinuities have been noted by several authors (Besl et al. 1988; Shulman and Herve 1989; Black and Anandan 1993; Black and Rangarajan 1994; Li 1995a). We can compare the M-estimator with the DA model studied in Chapter 5. The influence of the datum d_i on the estimate f is proportional to $\eta_i h(\eta_i)$. This compares with the smoothing strength $f'h(f')$ given after (5.25). A very large $|\eta_i|$ value, due to d_i being far from f, suggests an outlier. This is similar to saying that a very large $|f'(x)|$ value is likely due to a step (discontinuity) in the signal there. The resemblance suggests that the definition of the DA model can also be used to define M-estimators (Li 1996b).

We replace the scale estimate in the M-estimator by a parameter $\gamma > 0$ and choose to use the adaptive interaction function h_γ and the adaptive potential function g_γ for the M-estimation. However, h_γ need only be C^0 continuous for the location estimation from discrete data. Theoretically, the definitions give an infinite number of suitable choices of the M-estimators. Table 5.1 and Fig. 5.1 showed four such possibilities. With h_γ and g_γ, we can define the energy under γ as

$$E_\gamma(f) = \sum_i g_\gamma(d_i - f) \qquad (6.7)$$

and thereby the minimum energy estimate

$$f_\gamma^* = \arg\min_f E_\gamma(f) = \frac{\sum_i h_\gamma(\eta_i)\, d_i}{\sum_i h_\gamma(\eta_i)} \qquad (6.8)$$

AM-Estimator
Begin Algorithm
set $t = 0$, $f_\gamma^{(1)} = f^{LS}$; choose initial γ;
do {
    $t \leftarrow t + 1$;
    compute errors $\eta_i = d_i - f_\gamma^{(t-1)}$, $\forall i$;
    compute weighted sum $f_\gamma^{(t)} = \frac{\sum_i h_\gamma(\eta_i)\, d_i}{\sum_i h_\gamma(\eta_i)}$;
    if $(|f_\gamma^{(t)} - f_\gamma^{(t-1)}| < \epsilon)$ /* converged */
        $\gamma \leftarrow \text{lower}(\gamma)$;
} until $(\gamma < \delta)$ /* frozen */
$f^* \leftarrow f_\gamma^{(t)}$;
End Algorithm

Figure 6.2: The AM-estimation algorithm.

This defines a class of M-estimators that are able to deal with outliers as the traditional M-estimators do. Their performance in the solution quality is significantly enhanced by using an annealing procedure.

6.1.4 AM-Estimator

The annealing algorithm for the redefined M-estimator, called the AM-estimator (Li 1996b), is based on the idea of the GNC algorithm (Blake and Zisserman 1987). It aims to overcome the local minimum problem in the M-estimation (i.e., to obtain a good estimate regardless of the initialization). It also make the estimation free from parameters or at least insensitive to their choices.

The annealing is performed by continuation in γ, and the AM-estimator is defined in the limit

$$f^* = \lim_{\gamma \to 0^+} f_\gamma^* \qquad (6.9)$$

An algorithm that implements the AM-estimator algorithm is given in Fig. 6.2. Initially, γ is set to a value high enough to guarantee that the corresponding APF $g_\gamma(\eta)$ is convex in an interval. With such a γ, it is easy to find the unique minimum of the global error function $E_\gamma(f)$ using the gradient-descent method, regardless of the initial value for f. The minimum is then used as the initial value for the next phase of minimization under a lower γ to obtain the next minimum. As γ is lowered, $g_\gamma(\eta)$ may no longer be convex and local minima may appear. However, if we track the global minima for decreasing γ values, we may hopefully approximate the global minimum f^* under $\gamma \to 0^+$.

Obviously, whatever the initialization is, the first iteration always gives a value equal to the LS estimate

$$f^{LS} = \frac{1}{m} \sum_{i=1}^{m} d_i \qquad (6.10)$$

This is because for $\gamma \to +\infty$, which guarantees the strict convexity, all weights are the same as $h_\gamma(\eta_i) = 1$. The initial γ is chosen to satisfy

$$|\eta_i| = |d_i - f^{LS}| < b_H(\gamma) \qquad (6.11)$$

where $b_H(\gamma)$ $(= -b_L(\gamma))$ is the upper bound of the band in (5.29). This guarantees $g_\gamma''(\eta_i) > 0$ and hence the strict convexity of g_γ. The parameter γ is lowered according to the schedule specified by the function lower(γ). The parameters δ and ϵ in the convergence conditions are some small numbers. An alternative way is to decrease γ according to a fixed schedule regardless of whether or not $f^{(t)}$ converges at the current γ, which is equivalent to setting a big value for ϵ. In this case, the algorithm freezes after dozens of iterations. This quick annealing is used in our experiments.

The advantages of the AM-estimator are summarized below. First, the use of the annealing significantly improves the quality and stability of the estimate. The estimate is made independent of the initialization. Because the starting point for obtaining f_γ^* at current γ is the convergence point obtained with the previous γ value, the divergence problem with the traditional M-estimator is minimized. Second, the definition of the AM-estimator effectively eliminates scale parameters in the M-estimation because γ is finally set to zero (or a small number to whose value the final estimate is insensitive). This avoids the instability problem incurred by inappropriate selection of the scale parameters. Furthermore, it needs no order statistics, such as the median, and hence no sorting. This improves the computational efficiency.

6.1.5 Convex Priors for DA and M-Estimation

Encoding the edge-preserving ability into prior distributions may lead to *nonconvex* energy functions. This is the case in many models such as the well-known line-process model (Geman and Geman 1984; Blake and Zisserman 1987). Models with nonconvex energy functions have two disadvantages. The first is the instability of the solution (i.e., the energy minimum) w.r.t. the data (Bouman and Sauer 1993; Stevenson et al. 1994). A small change in the input might result in a drastic difference in the solutions. The phenomenon is also due to a hard decision-making property of nonconvex models (Blake and Zisserman 1987). As such, the solution often depends substantially on the method used to perform the minimization. The second disadvantage is the high computational cost associated with the solution-finding procedure. An annealing process, either deterministic or stochastic, is incorporated into a local search algorithm in order to locate the global minimum (Geman and Geman 1984; Blake and Zisserman 1987). This makes the minimization procedure inefficient.

There has been considerable interest in convex energy models with edge-preserving ability (Shulman and Herve 1989; Green 1990; Lange 1990; Bouman and Sauer 1993; Stevenson et al. 1994; Li et al. 1995). This class of models overcomes the problems mentioned above with nonconvex functions. First, in terms of defining the minimal solution, the convexity guarantees the stability w.r.t. the input and makes the solution less sensitive to changes in the parameters (Bouman and Sauer 1993). The second advantage is the computational efficiency in searching for the global solution. Because there is only one unique solution, gradient-based minimization techniques can be efficiently utilized. Time-consuming techniques for tackling the local minimum problem, such as continuation or annealing, are not necessary in convex minimization.

Shulman and Herve (1989) proposed to use

$$g(\eta) = \begin{cases} \eta^2 & |\eta| \leq \gamma \\ 2\gamma|\eta| - \gamma^2 & |\eta| > \gamma \end{cases} \tag{6.12}$$

for computing optical flows involving discontinuities. This is the error function used in Huber's robust M-estimator (Huber 1981). The function has also been applied to curve fitting (Stevenson and Delp 1990), surface reconstruction (Stevenson et al. 1994), and image expansion (Schultz and Stevenson 1994). It has been shown to be advantageous in terms of computational complexity and reconstruction quality (Stevenson et al. 1994).

Similar to the Huber function, the function of Green (1990)

$$g(\eta) = \ln(\cosh(\eta/\gamma)) \tag{6.13}$$

is approximately quadratic for small η and linear for large values. Lange (1990) suggested using

$$g(\eta) = \frac{1}{2} \left(|\eta/\gamma| + \frac{1}{1 + |\eta/\gamma|} - 1 \right) \tag{6.14}$$

which can be described by seven properties.

Bouman and Sauer (1993) construct a scale-invariant Gaussian MRF model by using

$$g(\eta) = |\eta|^p \tag{6.15}$$

where $1.0 \leq p \leq 2.0$. When $p = 2$, it becomes the standard quadratic function. When $p = 1$, the corresponding estimator is the sample median and allows discontinuities. The results show that edges are best preserved when $p = 1$ and deteriorated for $p > 1$. The reason will be given in the next section.

Stevenson, Schmitz, and Delp (1994) presentd a systematic study on both convex and nonconvex models and give the following four properties for a function g to have good behavior: (i) convex, (ii) symmetric (i.e., $g(\eta) = g(-\eta)$), (iii) $g(\eta) < \eta^2$ for $|\eta|$ large to allow discontinuities, and (iv) controlled continuously by a parameter γ. They define a class of convex potential functions

$$g(\eta) = \begin{cases} |\eta|^p & |\eta| \leq \gamma \\ (|\eta| + (\frac{p}{q}\gamma^{p-1})^{\frac{1}{q-1}} - \gamma)^q & \\ \quad + \gamma^p - (\frac{p}{q}\gamma^{p-1})^{\frac{q}{q-1}} & |\eta| > \gamma \end{cases} \tag{6.16}$$

with three parameters: γ, p, and q. When $1.0 \leq p \leq 2.0$ and $\gamma = \infty$, it is the same as that used by Bouman and Sauer; when $p = 2.0$ and $q = 1.0$, it is the Huber error function. The values $p = 1.8$ and $q = 1.2$ are suggested in that paper. We point out that (iii) is too loose and inadequate for preserving discontinuities. An example is $g(\eta) = \eta^2 - g_0$, where $g_0 > 0$ is a constant; it satisfies (iii) but is unable to preserve discontinuities. As pointed out in the next section, it is the derivative of $g(\eta)$ that determines how a model responds to discontinuities.

6.1.6   Half-Quadratic Minimization

The AM-estimator with convex priors can be explained by half-quadratic (HQ) minimization. HQ performs continuous optimization of a nonconvex function using the theory of convex conjugated functions (Rockafellar 1970). Since the introduction of HQ minimization to the field of computer vision (Geman and Reynolds 1992), HQ has now been used in M-estimation and mean-shift for solving image analysis problems.

   In HQ minimization, auxiliary variables are introduced into the original energy function. The resulting energy function then becomes quadratic w.r.t. the original variable when the auxiliary variables are fixed, and convex w.r.t. the auxiliary variable given the original variable (thus the name "half-quadratic"). An alternative minimization procedure is applied to minimize the new energy function. The convergence of HQ optimization is justified in (Nikolova and NG 2005; Allain et al. 2006). While HQ itself is a local minimization algorithm, global minimization via HQ can be achieved by applying the annealing M-estimator concept (Li 1996b).

Auxiliary Variables

Given a set of m data samples $d = \{d_i \mid 1 \leq i \leq m\}$, where $d_i = f + \eta_i$, the problem is to estimate the location parameter f under noise η_i. The distribution of η_i is not assumed to be known exactly. The only underlying assumption is that $\eta_1, \ldots, \eta_m$ obey a symmetric, independent, identical distribution (symmetric i.i.d.). A robust estimator has to deal with departures from this assumption.

Let the residual errors be $\eta_i = d_i - f$ and the error penalty function be $g(\eta_i)$, satisfying conditions of (5.27) and $g'(\eta) = 2\eta h(\eta)$. The M-estimate f^* is defined as the minimum of a global error function

$$f^* = \arg\min_f E(f) \tag{6.17}$$

where

$$E(f) = \sum_i g(\eta_i) \tag{6.18}$$

HQ minimization applies the theory of convex conjugated functions (Rockafellar 1970). For each term $g(\eta_i)$, introduce an auxiliary variable b_i and consider the dual function $G(b_i)$ of $g(\eta_i)$

$$G(b_i) = \sup_{\eta_i \in \mathbb{R}} \left\{ -\frac{1}{2} b_i \eta_i^2 + g(\eta_i) \right\} \tag{6.19}$$

$G(b_i)$ is convex w.r.t. b_i. We have reciprocally

$$g(\eta_i) = \inf_{b_i \in \mathbb{R}} \left\{ \frac{1}{2} b_i \eta_i^2 + G(b_i) \right\} \tag{6.20}$$

The infimum is reached at the explicit form (Charbonnier et al. 1997)

$$b_i = 2h(\eta_i) = \begin{cases} g''(0^+) & \text{if} \quad \eta_i = 0 \\ \frac{g'(\eta_i)}{\eta_i} & \text{if} \quad \eta_i \neq 0 \end{cases} \tag{6.21}$$

With the auxiliary variables $b = \{b_i\}$ in (6.18), we get

$$\tilde{E}(f, b) = \sum_i \left\{ \frac{1}{2} b_i (d_i - f)^2 + h(b_i) \right\} \tag{6.22}$$

The infimum of $\tilde{E}(f, b)$ with a fixed f is

$$E(f) = \min_b \{ \tilde{E}(f, b) \} \tag{6.23}$$

The optimal configuration can then be represented as

$$f^* = \arg\min_f E(f) = \arg\min_{f,b} \tilde{E}(f, b) \tag{6.24}$$

Alternate Minimization

The new energy function (6.22) is quadratic w.r.t. f when b is fixed and convex w.r.t. b given f. So, it can be efficiently optimized by using an alternate minimization algorithm.

Given (f^{t-1}, b^{t-1}) as the solution at the $(t-1)$th step, the tth step calculates

$$b^t = \arg\min_{b \in R} \tilde{E}(f^{t-1}, b) \tag{6.25}$$

$$f^t = \arg\min_{f \in \mathbb{F}} \tilde{E}(f, b^t) \tag{6.26}$$

The solutions to these two minimization problems can be found analytically as

$$b_i^t = 2h(d_i - f^{t-1}) \tag{6.27}$$

$$f^t = \frac{\sum_i b_i^t d_i}{\sum_i b_i^t} \tag{6.28}$$

If we choose the $g(\eta)$ to be one of the robust functions given in Table 6.1, the alternate minimization reduces to

$$f^t = \frac{\sum_i h(d_i - f^{t-1}) d_i}{\sum_i h(d_i - f^{t-1})} \tag{6.29}$$

which is (6.5). A convergence analysis of the alternate minimization can be found in (Nikolova and NG 2005).

The connection between the widely used mean-shift (MS) algorithm (Fukunaga 1990; Comaniciu and Meer 1997) and HQ optimization is explained in (Yuan and Li 2007). The MS algorithm maximizes the following kernel density estimation (KDE) w.r.t. the mean f:

$$p(d \mid f) = \sum_i w_i k((d_i - f)^2) \tag{6.30}$$

where k is the kernel function and the predetermined weights satisfy $\sum_i w_i = 1$. The MS algorithm is obtained immediately by solving the gradient equation of $p(d \mid f)$ via fixed-point iterations:

$$f^t = \frac{\sum_i w_i k' \left((d_i - f^{t-1})^2 \right) d_i}{\sum_i w_i k' \left((d_i - f^{t-1})^2 \right)} \tag{6.31}$$

By setting $g(t) = -k(t^2)$ in (6.18), we can see that maximizing (6.30) is equivalent to minimizing the error penalty function (6.18).

By applying HQ minimization to (6.30), we get the alternate maximization algorithm

$$b_i^t = -2k'((d_i - f^{t-1})^2) \tag{6.32}$$

$$f^t = \frac{\sum_i w_i b_i^t d_i}{\sum_i w_i b_i^t} \tag{6.33}$$

which is identical to the iteration in the MS algorithm and AM-estimator. This also explains the MS algorithm from the HQ optimization perspective.

Annealing HQ

The HQ algorithm can be used with an annealing schedule to approximate the global solution. Replacing the scale estimate in the HQ by a bandwidth parameter $\gamma > 0$ and using an adaptive potential function g_γ with the adaptive interaction function h_γ gives the energy function

$$E_\gamma(f) = \sum_i g_\gamma(d_i - f) \qquad (6.34)$$

The alternate minimization is then

$$b_i^t = 2h_\gamma(d_i - f^{t-1}) \qquad (6.35)$$

$$f^t = \frac{\sum_i b_i^t \, d_i}{\sum_i b_i^t} \qquad (6.36)$$

The local minimum problem can be overcome using the idea of AM-estimation (Li 1996b). Denote f_γ^* as the local minimum obtained under γ. The annealing is performed by continuation in γ as

$$f^* = \lim_{\gamma \to 0+} f_\gamma^* \qquad (6.37)$$

Initially, γ is set to a value large enough to guarantee that the corresponding $g_\gamma(d_i - f)$ is convex in an interval. With such γ, it is easy to find the unique minimum of the $E_\gamma(f)$ using HQ minimization. The minimum is then used as the initial value for the next phase of minimization under a lower γ to obtain the next minimum. This way, we track the global minimum for decreasing γ values and hopefully approximate the global minimum f^* under $\gamma \to 0^+$.

6.2 Experimental Comparison

Two experiments are presented. The first is a general comparison of two estimators, the AM-estimator and the M-estimator with Tukey's biweight function, with simulated data. The second deals with an application. Experimental results demonstrate that the AM-estimator is significantly better than the traditional M-estimator in estimation accuracy, stability, and breakdown point.

6.2.1 Location Estimation

Simulated data points in 2D locations are generated. The data set is a mixture of true data points and outliers. First, m true data points $\{(x_i, y_i) \mid i = 1, \ldots, m\}$ are randomly generated around $\bar{f} = (10, 10)$. The values of x_i and y_i obey an identical, independent Gaussian distribution with a fixed mean value of 10 and a variance value V. After that, a percentage λ of the m

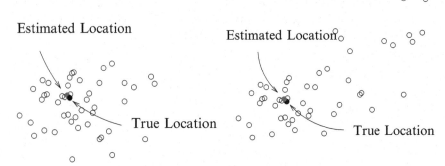

Figure 6.3: The AM-estimate of location. From (Li 1995a) with permission; ©1995 Elsevier.

data points are replaced by random outlier values. The outliers are uniformly distributed in a square of size 100×100 centered at $(b, b) \neq \bar{f}$. There are four parameters to control the data generation. Their values are:

1. the number of data points $m \in \{50, 200\}$,

2. the noise variance $V \in \{0, 2, 5, 8, 12, 17, 23, 30\}$,

3. the percentage of outliers λ from 0 to 70 with step 5, and

4. the outlier square centered parameter $b = 22.5$ or 50.

The experiments are done with different combinations of the parameter values. The AIF is chosen to be $h_{3\gamma}(\eta) = 1.0/(1 + \eta^2/\gamma)$. The schedule in lower($T$) is $\gamma \leftarrow \left(\frac{100}{t^2}\right)^{1.5} - 1$; when time $t \to \infty$, $\gamma \to 0^+$. It takes about 50 iterations for each of these data sets to converge.

Figure 6.3 shows two typical data distributions and estimated locations. Each of the two data sets contains 32 Gaussian-distributed true data points and 18 uniformly distributed outliers. The two sets differ only in the arrangement of outliers, while the true data points are common to both sets. The algorithm takes about 50 iterations for each of these data sets to converge. The estimated locations for the two data sets are marked in Fig. 6.3. The experiments show that the estimated locations are very stable regardless of the initial estimate, though the outlier arrangements are quite different in the two sets. Without the use of AM-estimation, the estimated location would have been much dependent on the initialization.

In a quantitative comparison, two quantities are used as the performance measures: (1) the mean error $\bar{e}$ versus the percentage of outliers (PO) λ and (2) the mean error $\bar{e}$ versus the noise variance (NV) V. Let the Euclidean

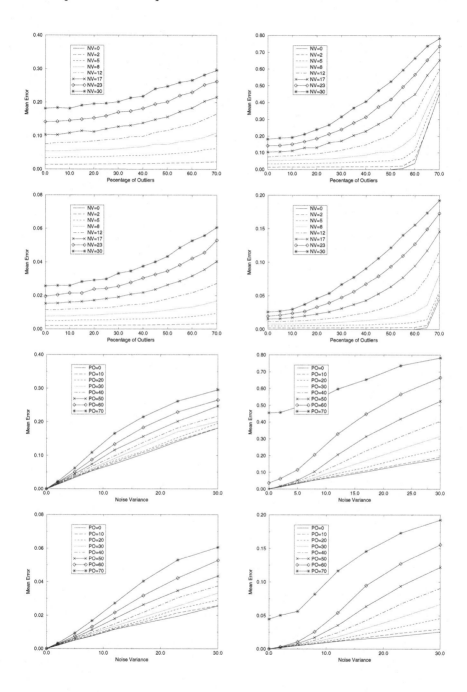

Figure 6.4: Mean error of the AM-estimate. From (Li 1996b) with permission; ©1996 Elsevier.

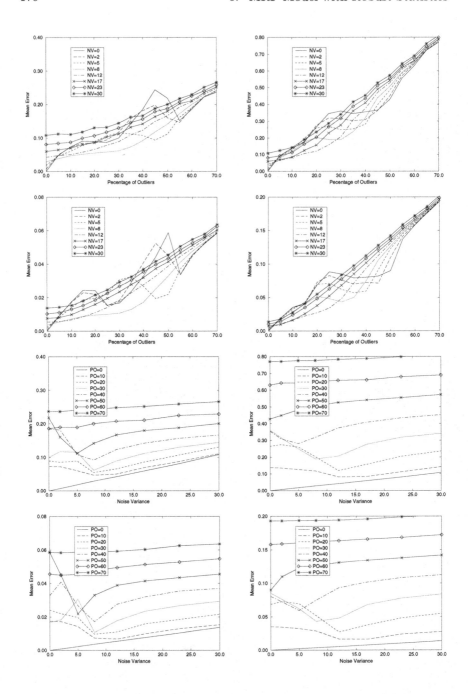

Figure 6.5: Mean error of the M-estimate. From (Li 1996b) with permission; ©1996 Elsevier.

error be $e = \|f^* - \bar{f}\| = \sqrt{(x^* - 10)^2 + (y^* - 10)^2}$, where f^* is the estimate and $\bar{f}$ is the true location.

Figures 6.4 and 6.5 show the mean errors of the AM-estimator and the M-estimator, respectively. Every statistic for the simulated experiment is made based on 1000 random tests, and the data sets are exactly the same for the two estimators compared. Outliers are uniformly distributed in a square centered at $b = 22.5$ (the left columns) or $b = 50$ (the right columns). The plots show the mean error versus percentage of outliers with $m = 50$ (row 1) and $m = 200$ (row 2) and the mean error vs. noise variance with $m = 50$ (row 3) and $m = 200$ (row 4). It can be seen that the AM-estimator has a very stable and elegant behavior as the percentage of outliers and the noise variance increase; in contrast, the M-estimator not only gives a higher error but also has an unstable behavior.

6.2.2 Rotation Angle Estimation

This experiment compares the AM-estimator with the M-estimator in computing the relative rotation of motion sequences. Consider the sequence of images in Fig. 6.6. Corners can be detected from these images as in Fig. 6.7

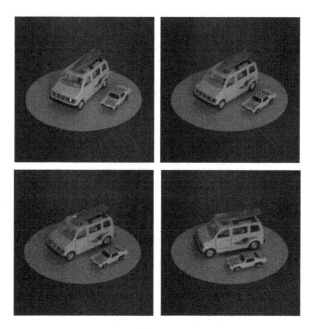

Figure 6.6: Part of a sequence of images rotating at 10 degrees between adjacent frames. The image size is 256×256.

Figure 6.7: Corners detected.

by using the Wang-Brady detector (Wang and Brady 1991). The data

$$d = \{(p_i, p'_i) \mid i = 1, \ldots, m\} \tag{6.38}$$

where $p_i = (x_i, y_i)$ and $p'_i = (x'_i, y'_i)$ represent a set of matched point pairs
between two images. A previous work (Wang and Li 1994) showed that when
the rotation axis $\ell = (\ell_x, \ell_y, \ell_z)^T$ is known, a unique solution can be computed using only one pair of corresponding points and the LS solution can
be obtained using m pairs by minimizing

$$E(f) = \sum_{i=1}^{m} \{A_i \tan(f/2) + B_i\}^2 \tag{6.39}$$

where

$$\begin{aligned} A_i &= \ell_z(\ell_y(x_i + x'_i) - \ell_x(y_i + y'_i)) \\ B_i &= \ell_x(x_i - x'_i) + \ell_y(y_i - y'_i). \end{aligned} \tag{6.40}$$

A unique solution exists for the LS problem (Wang and Li 1994). It is determined by the equation

$$\sum_{i=1}^{m} \left\{ [A_i \tan(f/2) + B_i] \cdot A_i/2 \cdot \sec^2(f/2) \right\} = 0 \tag{6.41}$$

where $f^* \neq 180°$; that is,

$$f^* = 2 \arctan \left(-\frac{\sum A_i B_i}{\sum A_i^2} \right) \tag{6.42}$$

The formulation above is based on an assumption that all the pairs
$\{(x_i, y_i), (x'_i, y'_i)\}$ are correct correspondences. This may not be true in practice. For example, due to acceleration and deceleration, turning and occlusion,

the measurements can change drastically and false matches (i.e., outliers) can occur. The LS estimate can get arbitrarily wrong when outliers are present in the data d. When outliers are present, the M-estimator can produce a more reliable estimate than the LS estimator. The AM-estimator further improves the M-estimator to a significant extent.

The AM-estimator minimizes, instead of (6.39),

$$E(f) = \sum_{i=1}^{m} g_\gamma(A_i \tan(f/2) + B_i) \tag{6.43}$$

where g_γ is an adaptive potential function. By setting $\frac{dE}{df} = 0$ and using $g'_\gamma(\eta) = 2\eta h_\gamma(\eta)$, one obtains

$$\sum_{i=1}^{m} \left\{ [A_i \tan(f/2) + B_i] \cdot h_\gamma(A_i \tan(f/2) + B_i) \cdot A_i/2 \cdot \sec^2(f/2) \right\} = 0 \tag{6.44}$$

Rearranging this equation gives the fixed-point equation

$$f = 2 \arctan \left(-\frac{\sum_{i=1}^{m} h_i A_i B_i}{\sum_{i=1}^{m} h_i A_i^2} \right) \tag{6.45}$$

where $h_i = h_\gamma(A_i \tan(f/2) + B_i)$. It is solved iteratively with decreasing γ values.

Figures 6.8 and 6.9 show the estimated rotation angles (in the vertical direction) between consecutive frames (the label on the horizontal axis is the frame number) and the corresponding standard deviations (in vertical bars) computed using the LS-, M-, and AM-estimators. From Fig. 6.8, we see that with 20% outliers, the M-estimator still works quite well while the LS-estimator has broken down. In fact, the breakdown point of the LS-estimator is less than 5%.

From Fig. 6.9, we see that the M- and AM-estimators are comparable when the data contains less than 20% outliers. Above this percentage, the

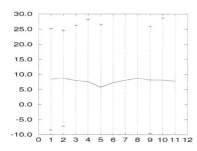 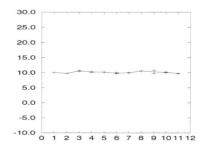

Figure 6.8: Rotation angles computed from the correspondence data containing 20% of outliers using the LS-estimator (left) and the M-estimator (right).

AM-estimator demonstrates its enhanced stability. The AM-estimator continues to work well when the M-estimate is broken down by outliers. The AM-estimator has a breakdown point of 60%. This illustrates that the AM-estimator has a considerably higher actual breakpoint than the M-estimator.

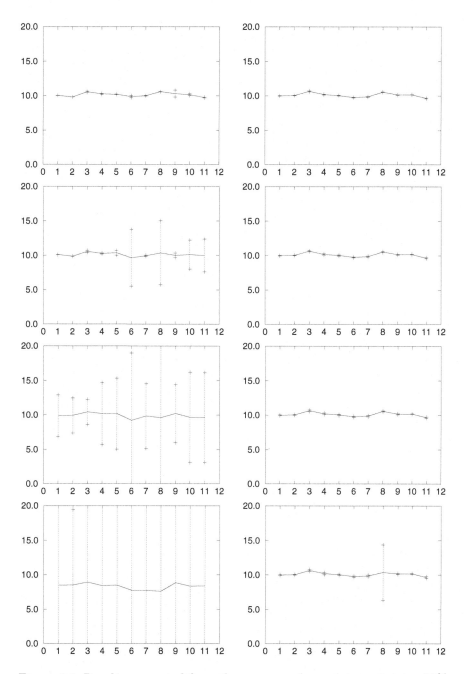

Figure 6.9: Results computed from the correspondence data containing 20% (row 1), 40% (row 2), 50% (row 3), and 60% (row 4) outliers using the M-estimator (left) and the AM-estimator (right).

Chapter 7

MRF Parameter Estimation

A probabilistic distribution function has two essential elements: the form of the function and the parameters involved. For example, the joint distribution of an MRF is characterized by a Gibbs function with a set of clique potential parameters; and the noise by a zero-mean Gaussian distribution parameterized by a variance. A probability model is incomplete if not all the parameters involved are specified, even if the functional form of the distribution is known. While formulating the forms of objective functions such as the posterior distribution has long been a subject of research in image and vision analysis, estimating the parameters involved has a much shorter history. Generally, it is performed by optimizing a statistical criterion, for example, using existing techniques such as maximum likelihood, coding, pseudo-likelihood, expectation-maximization,or Bayesian estimation.

The problem of parameter estimation can have several levels of complexity. The simplest is to estimate the parameters, denoted by θ, of a single MRF, F, from the data d, which are due to a clean realization, f, of that MRF. Treatments are needed if the data are noisy. When the noise parameters are unknown, they have to be estimated, too, along with the MRF parameters. The complexity increases when the given data are due to a realization of more than one MRF (e.g., when multiple textures are present in the image data, unsegmented). Since the parameters of an MRF have to be estimated from the data, partitioning the data into distinct MRF's becomes a part of the problem. The problem is even more complicated when the number of underlying MRF's is unknown and has to be determined. Furthermore, the order of the neighborhood system and the largest size of the cliques for a Gibbs distribution can also be part of the parameters to be estimated.

The main difficulty in ML estimation for MRF's is due to the partition function Z in the Gibbs distribution $P(f \mid \theta)$. Z is also a function of θ and has

S.Z. Li, *Markov Random Field Modeling in Image Analysis*,
Advances in Pattern Recognition, DOI: 10.1007/978-1-84800-279-1_7,
© Springer-Verlag London Limited 2009

to be taken into consideration during ML estimation. Since Z is calculated by summing over all possible configurations, maximizing $P(f \mid \theta)$ becomes intractable, in general, even for small problems.

The example in Fig. 7.1 illustrates the significance of getting correct model parameters for MRF labeling procedures to produce good results. The binary textures in row 1 are generated using the MLL model (2.42) with the β parameters being $(0.5, 0.5, 0.5, -0.5)$ (for the left) and $(-0.5, -0.5, -0.5, 0.5)$ (for the right). They have pixel values of 100 and 160, respectively. Identical independently distributed Gaussian noise $N(0, 60^2)$ is added to the pixel values, giving degraded images in row 2. Row 3 shows the result obtained using the true parameters of the texture and the noise. Since such accurate information usually is not available in practice, results cannot be as good as that. Rows 4 and 5 show results obtained by using incorrect parameters $(0.4, 0.4, 0.4, 0.4)$ and $(1.5, 1.5, 1.5, 1.5)$, respectively. These results are unacceptable.

An issue in parameter estimation is the evaluation of fit. After a set of parameters are estimated, one would like to evaluate the estimate. The evaluation is to test the goodness of fit: How well does the sample distribution fit the estimated MRF model with parameter θ. This is a statistical problem. Some validation schemes are discussed in (Cross and Jain 1983; Chen 1988) for testing the fit between the observed distribution and the assumed distribution with the estimated parameters. One available technique is the well-known χ^2 test for measuring the correspondence between two distributions.

7.1 Supervised Estimation with Labeled Data

Labeled data mean realization(s) of only a single MRF such as a certain type of texture. For a data set containing more than one MRF, the data should be partitioned into distinct subsets (e.g., texture regions) such that each subset is due to a distinct MRF. Such data are considered known-class data. From the viewpoint of pattern recognition, such estimation is *supervised*. The set of parameters, θ, for each MRF, F, are estimated using data that are a clean realization,[1] f, of that MRF. The algorithms described this section also provide the basis for more complicated unsupervised estimation with unlabeled data.

7.1.1 Maximum Likelihood

Given a realization f of a single MRF, the maximum likelihood (ML) estimate maximizes the conditional probability $P(f \mid \theta)$ (the likelihood of θ); that is,

---

[1]Some methods are needed to combine estimates obtained from more than one realization.

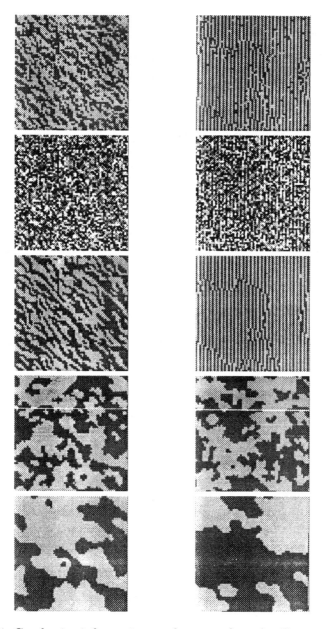

Figure 7.1: Good prior information produces good results. Row 1: True texture. Row 2: Images degraded by noise. Row 3: Restoration results with the exact parameters for generating the images in row 2. Rows 4 and 5: Images restored with incorrect parameters (see the text). From (Dubes and Jain 1989) with permission; ©1989 Taylor & Francis.

$$\theta^* = \arg \max_{\theta} P(f \mid \theta) \tag{7.1}$$

or its log-likelihood $\ln P(f \mid \theta)$. Note that in this case f is the data used for the estimation. When the prior p.d.f. of the parameters, $p(\theta)$, is known, the MAP estimation that maximizes the posterior density

$$P(\theta \mid f) \propto p(\theta) P(f \mid \theta) \tag{7.2}$$

can be sought. The ignorance of the prior p.d.f. leads the user to a rather diffuse choice. The prior p.d.f. is assumed to be flat when the prior information is totally unavailable. In this case, the MAP estimation reduces to the ML estimation.

Let us review the idea of the ML parameter estimation using a simple example. Let $f = \{f_1, \ldots, f_m\}$ be a realization of an identical independent Gaussian distribution with two parameters, the mean μ and the variance σ^2; that is, $f_i \sim N(\mu, \sigma^2)$. We want to estimate the two parameters $\theta = \{\mu, \sigma^2\}$ from f. The likelihood function of θ for fixed f is

$$p(f \mid \theta) = \frac{1}{(\sqrt{2\pi\sigma^2})^m} e^{-\sum_{i \in \mathcal{S}} (f_i - \mu)^2 / 2\sigma^2} \tag{7.3}$$

A necessary condition for maximizing $p(f \mid \theta)$, or equivalently maximizing $\ln p(f \mid \theta)$, is $\frac{\partial \ln p}{\partial \mu} = 0$ and $\frac{\partial \ln p}{\partial \sigma} = 0$. Solving this, we find as the ML estimate

$$\mu^* = \frac{1}{m} \sum_{i=1}^{m} f_i \tag{7.4}$$

and

$$(\sigma^2)^* = \frac{1}{m} \sum_{i=1}^{m} (f_i - \mu^*)^2 \tag{7.5}$$

A more general result can be obtained for the m-variate Gaussian distribution, which when restricted by the Markovianity is also known as the auto-normal (Besag 1974) MRF (see Section 2.2). Assuming $\mu = 0$ in (2.37) and thus θ consisting only of $B = \{\beta_{i,i'}\}$, the log-likelihood function for the auto-normal field is

$$\ln p(f \mid \theta) \propto -\frac{1}{m} \ln[\det(B)] + \ln(f^{\mathrm{T}} B f) \tag{7.6}$$

where f is considered a vector. To maximize this, one needs to evaluate the determinant $\det(B)$. The ML estimation for the Gaussian cases is usually less involved than for general MRF's.

The ML estimation for an MRF in general requires evaluation of the normalizing partition function in the corresponding Gibbs distribution. Consider a homogeneous and isotropic auto-logistic model in the 4-neighborhood system with the parameters $\theta = \{\alpha, \beta\}$ (this MRF will also be used in subsequent sections for illustration). According to (2.29) and (2.30), its energy function and conditional probability are, respectively,

$$U(f \mid \theta) = \sum_{\{i\} \in \mathcal{C}_1} \alpha f_i + \sum_{\{i,i'\} \in \mathcal{C}_2} \beta f_i f_{i'} \tag{7.7}$$

and

$$P(f_i \mid f_{\mathcal{N}_i}) = \frac{e^{\alpha f_i + \sum_{i' \in \mathcal{N}_i} \beta f_i f_{i'}}}{1 + e^{\alpha + \sum_{i' \in \mathcal{N}_i} \beta f_{i'}}} \tag{7.8}$$

The likelihood function is in the Gibbs form

$$P(f \mid \theta) = \frac{1}{Z(\theta)} \times e^{-U(f \mid \theta)} \tag{7.9}$$

where the partition function

$$Z(\theta) = \sum_{f \in \mathbb{F}} e^{-U(f \mid \theta)} \tag{7.10}$$

is also a function of θ. Maximizing $P(f \mid \theta)$ w.r.t. θ requires evaluation of the partition function $Z(\theta)$. However, the computation of $Z(\theta)$ is intractable even for moderately sized problems because there are a combinatorial number of elements in the configuration space $\mathbb{F}$. This is a main difficulty in parameter estimation for MRF's. Approximate formulae will be used for solving this problem.

The approximate formulae are based on the conditional probabilities $P(f_i \mid f_{\mathcal{N}_i})$, $i \in \mathcal{S}$. (The notation of parameters, θ, on which the probabilities are conditioned, is dropped temporarily for clarity.) Write the energy function in the form

$$U(f) = \sum_{i \in \mathcal{S}} U_i(f_i, f_{\mathcal{N}_i}) \tag{7.11}$$

Here, $U_i(f_i, f_{\mathcal{N}_i})$ depends on the configuration of the cliques involving i and $\mathcal{N}_i$, in which the labels f_i and $f_{\mathcal{N}_i}$ are mutually dependent. If only single- and pair-site cliques are considered, then

$$U_i(f_i, \mathcal{N}_i) = V_1(f_i) + \sum_{i':\{i,i'\} \in \mathcal{C}} V_2(f_i, f_{i'}) \tag{7.12}$$

The conditional probability (2.18) can be written as

$$P(f_i \mid f_{\mathcal{N}_i}) = \frac{e^{-U_i(f_i, f_{\mathcal{N}_i})}}{\sum_{f_i \in \mathcal{L}} e^{-\sum_i U_i(f_i, f_{\mathcal{N}_i})}} \tag{7.13}$$

Based on this, approximate formulae for the joint probability $P(f)$ are given in the following subsections.

7.1.2   Pseudo-likelihood

A simple approximate scheme is pseudo-likelihood (Besag 1975; Besag 1977). In this scheme, each $U_i(f_i, f_{\mathcal{N}_i})$ in (7.11) is treated as if $f_{\mathcal{N}_i}$ are given and the pseudo-likelihood is defined as the simple product of the conditional likelihood

$$
\mathrm{PL}(f) = \prod_{i \in S - \partial S} P(f_i \mid f_{\mathcal{N}_i}) = \prod_{i \in S - \partial S} \frac{e^{-U_i(f_i, f_{\mathcal{N}_i})}}{\sum_{f_i} e^{-U_i(f_i, f_{\mathcal{N}_i})}}   \tag{7.14}
$$

where ∂S denotes the set of points at the boundaries of S under the neighborhood system $\mathcal{N}^2$. Substituting the conditional probability (7.8) into (7.14), we obtain

$$
\mathrm{PL}(f) = \prod_i \frac{e^{\alpha_i f_i + \sum_{i' \in \mathcal{N}_i} \beta f_i f_{i'}}}{1 + e^{\alpha_i + \sum_{i' \in \mathcal{N}_i} \beta f_{i'}}}   \tag{7.15}
$$

as the pseudo-likelihood for the homogeneous and isotropic auto-logistic model. The pseudo-likelihood does not involve the partition function Z.

Because f_i and $f_{\mathcal{N}_i}$ are not independent, the pseudo-likelihood is not the true likelihood function, except in the trivial case of the nil neighborhood (contextual independence). In the large lattice limit, the consistency of the maximum pseudo-likelihood (MPL) estimate was proven by Geman and Graffigne (1987); that is, it converges to the truth with probability one.

The following example illustrates the MPL estimation. Consider the homogeneous and isotropic auto-logistic MRF model described by (7.7) and (7.8). Express its pseudo-likelihood (7.15) as $\mathrm{PL}(f \mid \theta)$. The logarithm is

$$
\ln \mathrm{PL}(f \mid \theta) = \sum_{i \in S - \partial S} \left\{ \alpha f_i + \beta f_i \sum_{i' \in \mathcal{N}_i} f_{i'} - \ln(1 + e^{\alpha + \beta \sum_{i' \in \mathcal{N}_i} f_{i'}}) \right\}   \tag{7.16}
$$

The MPL estimates $\{\alpha, \beta\}$ can be obtained by solving

$$
\frac{\partial \ln \mathrm{PL}(f \mid \alpha, \beta)}{\partial \alpha} = 0, \qquad \frac{\partial \ln \mathrm{PL}(f \mid \alpha, \beta)}{\partial \beta} = 0   \tag{7.17}
$$

7.1.3   Coding Method

Using the *coding method* (Besag 1974), the likelihood formula of the form (7.14) can be made a genuine one. The idea is to partition the set S into several disjoint sets $S^{(k)}$, called codings, such that no two sites in one $S^{(k)}$ are neighbors. Figure 7.2 illustrates the codings for the 4- and 8- neighborhood systems. For the 4-neighborhood system, two sets, S^1 and S^2, are needed, one composed of the black sites and the other of the white sites in a checkerboard. Four codings are needed for the 8-neighborhood system.

---

[2]The boundary sites are excluded from the product to reduce artificiality.

1	2	1	2	1
2	1	2	1	2
1	2	1	2	1
2	1	2	1	2
1	2	1	2	1

1	2	1	2	1
3	4	3	4	3
1	2	1	2	1
3	4	3	4	3
1	2	1	2	1

Figure 7.2: Codings for the 4-neighborhood (left) and 8-neighborhood (right) systems. A block is partitioned into different sets, marked by different numbers (say, pixels numbered k belong to set $\mathcal{S}^{(k)}$).

The fact that the sites within each $\mathcal{S}^{(k)}$ do not neighbor to each other provides computational advantages. Under the Markovian assumption, the variables associated with the sites in an $\mathcal{S}^{(k)}$, given the labels at all other sites, are mutually independent. This gives, for the likelihood, the simple product

$$P^{(k)}(f \mid \theta) = \prod_{i \in \mathcal{S}^{(k)}} P(f_i \mid f_{\mathcal{N}_i}, \theta) \qquad (7.18)$$

in which the need to evaluate Z is also dispensed with. Unlike the pseudo-likelihood, the definition above is a genuine expression for the likelihood function. Maximizing (7.18) gives coding estimates $\theta^{(k)}$.

A disadvantage is the inefficiency: Each of the ML estimates $\theta^{(k)}$ is estimated based on information on only one coding. It is not clear how to combine the results optimally. The simplest scheme of the arithmetic average

$$\theta^* = \frac{1}{K} \sum_k \arg\max_\theta P^{(k)}(f \mid \theta) \qquad (7.19)$$

is usually used.

Consider the homogeneous and isotropic auto-logistic MRF model described by (7.7) and (7.8). Its log coding likelihood takes a form similar to that for the pseudo-likelihood

$$\ln P^{(k)}(f \mid \theta) = \sum_{i \in \mathcal{S}^{(k)}} \left\{ \alpha f_i + \beta f_i \sum_{i' \in \mathcal{N}_i} f_{i'} - \ln(1 + e^{\alpha + \beta \sum_{i' \in \mathcal{N}_i} f_{i'}}) \right\} \qquad (7.20)$$

where $k = 1, 2$. The ML estimates $\{\alpha^{(k)}, \beta^{(k)}\}$ $(k = 1, 2)$ can be obtained by solving

$$\frac{\partial \ln P^{(k)}(f \mid \alpha, \beta)}{\partial \alpha} = 0, \qquad \frac{\partial \ln P^{(k)}(f \mid \alpha, \beta)}{\partial \beta} = 0 \qquad (7.21)$$

Taking the arithmetic averages, we obtain the final estimate as

$$\hat{\theta} = \left\{ \frac{\alpha^1 + \alpha^2}{2}, \frac{\beta^1 + \beta^2}{2} \right\} \tag{7.22}$$

7.1.4   Mean Field Approximations

Mean field approximation from statistical physics (Chandler 1987; Parisi 1988) provides other possibilities. It originally aims to approximate the behavior of interacting spin systems in thermal equilibrium. We use it to approximate the behavior of MRF's in equilibrium. Generally, the mean of a random variable X is given by $\langle X \rangle = \sum_X XP(X)$, so we define the mean field $\langle f \rangle$ by the mean values

$$\langle f_i \rangle = \sum_{f \in \mathbb{F}} f_i P(f) = Z^{-1} \sum_f f_i e^{-\frac{1}{T}U(f)} \tag{7.23}$$

The mean field approximation makes the following assumption for calculating $\langle f_i \rangle$: The actual influence of $f_{i'}$ ($i' \neq i$) can be approximated by the influence of $\langle f_{i'} \rangle$. This is reasonable when the field is in equilibrium. It leads to a number of consequences. Equation (7.12) is approximated by the mean field local energy expressed as

$$U_i(f_i \mid \langle f_{\mathcal{N}_i} \rangle) = V_1(f_i) + \sum_{i':\{i,i'\} \in \mathcal{C}} V_2(f_i, \langle f_{i'} \rangle) \tag{7.24}$$

and the conditional probability is approximated by the mean field local probability

$$P(f_i \mid \langle f_{\mathcal{N}_i} \rangle) = {Z'}_i^{-1} e^{-\frac{1}{T}U_i(f_i \mid \langle f_{\mathcal{N}_i} \rangle)} \tag{7.25}$$

where

$$Z'_i = \sum_{f_i \in \mathcal{L}} e^{-\frac{1}{T}U_i(f_i \mid \langle f_{\mathcal{N}_i} \rangle)} \tag{7.26}$$

is called the mean field local partition function. Note that in the mean field approximation above, the values $\langle f_{\mathcal{N}_i} \rangle$ are assumed given. The mean field values are approximated by

$$\langle f_i \rangle \approx \sum_{f_i \in \mathcal{L}} f_i P(f_i \mid \langle f_{\mathcal{N}_i} \rangle) \tag{7.27}$$

The mean field local probability (7.25) is an approximation to the marginal distribution

$$P(f_i) = \sum_{f_{S-\{i\}}} Z^{-1} e^{-\frac{1}{T}U(f)} \tag{7.28}$$

Because the mean field approximation "decouples" the interactions in equilibrium, the mean field approximation of the joint probability is the product of the mean field local probabilities

$$P(f) \approx \prod_{i \in \mathcal{S}} P(f_i \mid \langle f_{\mathcal{N}_i} \rangle) = \prod_{i \in \mathcal{S}} Z'^{-1}_i e^{-\frac{1}{T} U_i(f_i \mid \langle f_{\mathcal{N}_i} \rangle)} \tag{7.29}$$

and the mean field partition function is the product of the mean field local partition functions

$$Z \approx Z' = \sum_f e^{-\frac{1}{T} \sum_{i \in \mathcal{S}} U_i(f_i \mid \langle f_{\mathcal{N}_i} \rangle)} = \prod_{i \in \mathcal{S}} \sum_{f_i} e^{-\frac{1}{T} U_i(f_i \mid \langle f_{\mathcal{N}_i} \rangle)} \tag{7.30}$$

When $\mathcal{L}$ is continuous, such as for the real line, all the probability functions above are replaced by the corresponding p.d.f.s and the sums by integrals.

The mean field approximation (7.29) bears the form of the pseudo-likelihood function (Besag 1975; Titterington and Anderson 1994). Indeed, if $\langle f_{i'} \rangle$ ($i' \in \mathcal{N}_i$) in the mean field computation is replaced by the current $f_{i'}$, the two approximations become the same. However, unlike the pseudo-likelihood, the mean values $\langle f_i \rangle$ and the mean field conditional probabilities $P(f_i \mid \langle f_{\mathcal{N}_i} \rangle)$ are computed iteratively using (7.25), (7.26), and (7.27) given an initial $\langle f^{(0)} \rangle$.

The above is one of various approaches for mean field approximation. In Section 10.2, we will present an approximation for labeling with a discrete label set using saddle-point approximation (Peterson and Soderberg 1989). The reader is also referred to (Bilbro and Snyder 1989; Yuille 1990; Geiger and Girosi 1991) for other approaches. A study of mean field approximations of various GRFs and their network implementations is presented in the Ph.D thesis of Elfadel (1993). Its application in MRF parameter estimation is proposed by (Zhang 1992; Zhang 1993).

7.1.5 Least Squares Fit

MRF parameter estimation can also be done by using a least squares (LS) fit procedure (Derin and Elliott 1987). The procedure consists of the following steps: (1) Find the relationship between the joint probabilities $P(f_i, f_{\mathcal{N}_i})$ and the parameters θ; (2) use histogram techniques to estimate all the $P(f_i, f_{\mathcal{N}_i})$; (3) construct an overdetermined system of equations in terms of the probabilities and the parameters; and (4) solve it using the LS method.

Consider the MLL model with the label set $\mathcal{L} = \{1, \ldots, M\}$. For the 8-neighborhood system shown in Fig. 7.3, there are ten types of clique potential parameters, shown in Fig. 2.3. Denote all the parameters as a $K \times 1$ vector

$$\theta = [\alpha_1, \ldots, \alpha_M, \beta_1, \ldots, \beta_4, \gamma_1, \ldots, \gamma_4, \xi]^T = [\theta_1, \ldots, \theta_K]^T \tag{7.31}$$

Let

$$U_i(f_i, f_{\mathcal{N}_i}, \theta) = \sum_{c:i \in c} V_c(f) \tag{7.32}$$

be the sum of the potential functions over all the cliques containing i. Define the indicator functions

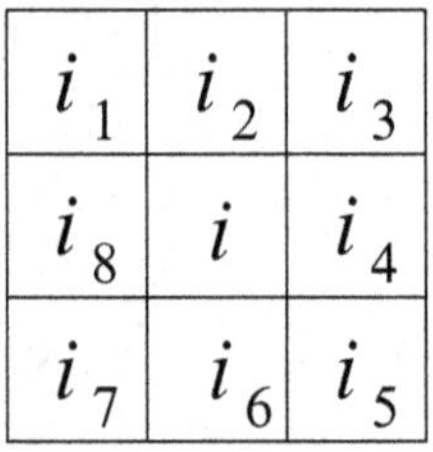

i_1	i_2	i_3
i_8	i	i_4
i_7	i_6	i_5

Figure 7.3: Pixel i and its eight neighbors in the second-order neighborhood system.

$$\chi(f_1,\ldots,f_k) = \begin{cases} 1 & \text{if } f_1 = f_2 \cdots = f_k \\ -1 & \text{otherwise} \end{cases} \tag{7.33}$$

for expressing (2.42) and

$$\chi_I(f_i) = \begin{cases} 1 & \text{if } f_i = I \\ -1 & \text{otherwise} \end{cases} \tag{7.34}$$

for (2.43). In terms of θ and χs, (7.32) can be written as

$$U_i(f_i, f_{\mathcal{N}_i}, \theta) = \theta^T N_i(f_i, f_{\mathcal{N}_i}) = \sum_{k=1}^{K} \theta_k n_k \tag{7.35}$$

where

$$
\begin{aligned}
N_i(f_i, f_{\mathcal{N}_i}) =\ & [\chi_1(f_i), \chi_2(f_i), \cdots, \chi_M(f_i), \\
& \chi(f_i, f_{i_4}) + \chi(f_i, f_{i_8}), \chi(f_i, f_{i_2}) + \chi(f_i, f_{i_6}), \\
& \chi(f_i, f_{i_3}) + \chi(f_i, f_{i_7}), \chi(f_i, f_{i_1}) + \chi(f_i, f_{i_5}), \\
& \chi(f_i, f_{i_2}, f_{i_4}) + \chi(f_i, f_{i_5}, f_{i_6}) + \chi(f_i, f_{i_1}, f_{i_8}), \\
& \chi(f_i, f_{i_2}, f_{i_3}) + \chi(f_i, f_{i_4}, f_{i_6}) + \chi(f_i, f_{i_7}, f_{i_8}), \\
& \chi(f_i, f_{i_2}, f_{i_8}) + \chi(f_i, f_{i_3}, f_{i_4}) + \chi(f_i, f_{i_6}, f_{i_7}), \\
& \chi(f_i, f_{i_1}, f_{i_2}) + \chi(f_i, f_{i_4}, f_{i_5}) + \chi(f_i, f_{i_6}, f_{i_8}), \\
& \chi(f_i, f_{i_1}, f_{i_2}, f_{i_8}) + \chi(f_i, f_{i_2}, f_{i_3}, f_{i_4}) + \\
& \chi(f_i, f_{i_4}, f_{i_5}, f_{i_6}) + \chi(f_i, f_{i_6}, f_{i_7}, f_{i_8})]^T \\
\triangleq\ & [n_1, \ldots, n_K]^T
\end{aligned}
\tag{7.36}
$$

is a vector whose elements count the number of each type of cliques in the 3×3 window. The right-hand side of the first row in the equation above correspond to the potentials weighted by $\alpha_1, \cdots, \alpha_M$, the second row by β_1 and β_2, the third row by β_3 and β_4, the fourth to seventh rows by $\gamma_1, \cdots, \gamma_4$, and the last two rows by ξ_1 (refer to Fig. 2.3).

The conditional likelihood $P(f_i \mid f_{\mathcal{N}_i})$ is related to the potential $U_i(f_i, f_{\mathcal{N}_i}, \theta)$ by

$$P(f_i \mid f_{\mathcal{N}_i}) = \frac{e^{-U_i(f_i, f_{\mathcal{N}_i}, \theta)}}{\sum_{f_i \in \mathcal{L}} e^{-U_i(f_i, f_{\mathcal{N}_i}, \theta)}} \tag{7.37}$$

It can be expressed in terms of the joint probability $P(f_i, f_{\mathcal{N}_i})$ of the labels on the 3×3 window and the joint probability $P(f_{\mathcal{N}_i})$ of the labels at the neighboring sites

$$P(f_i \mid f_{\mathcal{N}_i}) = \frac{P(f_i, f_{\mathcal{N}_i})}{P(f_{\mathcal{N}_i})} \qquad (7.38)$$

The reason for this expression is that $P(f_i, f_{\mathcal{N}_i})$ can be estimated using histogram techniques (see below). Comparing the two equations above yields the equation

$$\frac{e^{-U_i(f_i, f_{\mathcal{N}_i}, \theta)}}{P(f_i, f_{\mathcal{N}_i})} = \frac{\sum_{f_i \in \mathcal{L}} e^{-U_i(f_i, f_{\mathcal{N}_i}, \theta)}}{P(f_{\mathcal{N}_i})} \qquad (7.39)$$

Note that the right-hand side of the above is independent of the value of f_i and therefore, so is the left-hand side.

Consider the left-hand side of (7.39) for any two distinct values, I and I', for f_i. Because the left-hand side is independent of f_i values, therefore we have

$$\frac{e^{-U_i(I, f_{\mathcal{N}_i}, \theta)}}{P(I, f_{\mathcal{N}_i})} = \frac{e^{-U_i(I', f_{\mathcal{N}_i}, \theta)}}{P(I', f_{\mathcal{N}_i})} \qquad (7.40)$$

Rearranging the above and using the relationship (7.35), we obtain

$$\theta^T [N_i(I', f_{\mathcal{N}_i}) - N_i(I, f_{\mathcal{N}_i})] = \ln\left(\frac{P(I, f_{\mathcal{N}_i})}{P(I', f_{\mathcal{N}_i})}\right) \qquad (7.41)$$

For any $I \in \mathcal{L}$ and $f_{\mathcal{N}_i}$, the vector $N_i(I, f_{\mathcal{N}_i})$ can be determined using (7.36) and $P(I, f_{\mathcal{N}_i})$ can be estimated using histogram techniques. Therefore, (7.41) represents a linear equation with unknown θ. One such equation can be obtained for each distinct combination of I, I', and $\mathcal{N}_i$. An overdetermined linear system of equations is obtained from all possible combinations. It can be solved using the standard LS method. This procedure is also used by Nadabar and Jain (1992) for estimating line process parameters in edge detection.

Histogram techniques are proposed to estimate $P(f_i, f_{\mathcal{N}_i})$, for all combinations of $(f_i, f_{\mathcal{N}_i})$, from one or several realizations (images). Assume that there are a total of N distinct 3×3 blocks in the image lattice and that a particular 3×3 configuration $(f_i, f_{\mathcal{N}_i})$ occurs $H(f_i, f_{\mathcal{N}_i})$ times. Then $P(f_i, f_{\mathcal{N}_i}) = \frac{H(f_i, f_{\mathcal{N}_i})}{N}$. The relationship between the histogramming and the ML estimation was established in (Gurelli and Onural 1994), and also used in (Gimel'farb 1996).

A few remarks follow. Any $(f_i, f_{\mathcal{N}_i})$ combination for which $H(f_i, f_{\mathcal{N}_i}) = 0$ cannot be used to obtain a linear equation because of the logarithm in (7.41). Derin and Elliott propose to use only those combinations $(f_i, f_{\mathcal{N}_i})$ for which $H(f_i, f_{\mathcal{N}_i})$ are largest. This selection strategy is quite robust (Derin and Elliott 1987). They also suggest discardinf the three and four cliques (set their potentials to zero), according to some prior or empirical knowledge, to reduce the length of θ and $N_i(f_i, f_{\mathcal{N}_i})$ to simplify the procedure. The normalized form of GRF (Griffeath 1976) described in Section 2.1.5 provides a foundation for the reduction. This will be discussed in Section 7.4.

Figure 7.4 shows estimation results obtained by using the LS fitting method. In this example, the only nonzero parameters are $\beta_1, \ldots, \beta_4$. In the upper row are the texture images generated with specified parameter values. These values are estimated from the images. The estimated values are then used to generate textures in the lower row. Table 7.1 compares the specified values and the estimated values. Note that the estimate for the image in column 4 is numerically not so good; the explanation of Derin and Elliott (1987) for this is that the parameter representation for a realization is not unique.

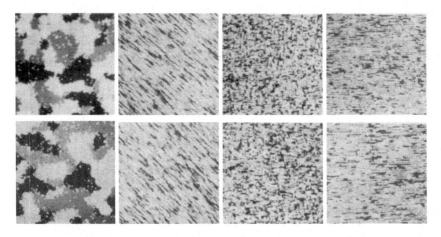

Figure 7.4: Realizations from a Gibbs distribution with specified (upper) and estimated (lower) parameters. See Table 6.1 for parameter values. From (Derin and Elliott 1987) with permission; ©1987 IEEE.

The LS fitting procedure has the following advantages over the coding method and MPL estimation. It does not require numerical maximization, avoiding the need to solve nonlinear equations. Nor is the final solution dependent on the initial estimate. Further, it is more efficient and faster. A disadvantage is that the number of equations (7.41) in the system increases exponentially as the sizes of $\mathcal{S}$ and $\mathcal{L}$ increase. Zero histogram counts also cause problems with the right-hand side of (7.41).

7.1.6 Markov Chain Monte Carlo Methods

Often in the computation or estimation of random models, one needs to draw random samples from a multivariate distribution $P(f) = P(f_1, \ldots, f_m)$ or calculate an expectation of a function of f w.r.t. a distribution $P(f)$. This cannot be done easily by using standard analytical or numerical methods

Table 7.1: Specified and estimated parameters. From (Derin and Elliot 1987) with permission; ©1987 IEEE.

Row-Column	Description	Size	β_1	β_2	β_3	β_4
1-1	specified	128×128	0.3	0.3	0.3	0.3
2-1	estimated	128×128	0.326	0.308	0.265	0.307
1-2	specified	256×256	1.0	1.0	-1.0	1.0
2-2	estimated	256×256	0.752	0.805	-0.762	0.909
1-3	specified	256×256	2.0	2.0	-1.0	-1.0
2-3	estimated	256×256	1.728	1.785	-0.856	-0.851
1-4	specified	128×128	1.0	-1.0	-1.0	-1.0
2-4	estimated	128×128	0.861	0.055	-1.156	-0.254

when $P(f)$ is complex enough. In this case, Markov chain Monte Carlo (MCMC) methods (Smith and Robert 1993; Besag and Green 1993; Gilks et al. 1993; Neal 1993; Besag et al. 1995) provide an effective, tractable means for the sampling and the calculation.

In MCMC, a time homogeneous (ergodic) sequence of states $\{f^{(0)}, f^{(1)}, \ldots\}$, or *Markov chain*, is generated, where a state in a state space $\mathbb{F}$ may correspond to a labeling configuration or parameter vector. At each time $t \geq 0$, the next random state $f^{(t+1)}$ is generated according to a Markov transition probability $P(f^{(t+1)} \mid f^{(t)})$. The transition probability should be constructed in such a way that the chain converges to the target distribution $P(f)$; thus, after a (usually long) "burn-in" period, the samples along the path of the chain can be used as samples from $P(f)$, which enables *Monte Carlo* methods. Sampling according to $P(f^{(t+1)} \mid f^{(t)})$ is much easier than direct sampling from the target $P(f)$, which is the reason for using MCMC techniques.

To make such a Markov chain to converge to the target distribution $P(f)$, it is sufficient to choose the transition probability so that the following time reversibility condition, or detailed balance, is satisfied

$$P(f^{(t)})P(f^{(t+1)} \mid f^{(t)}) = P(f^{(t+1)})P(f^{(t)} \mid f^{(t+1)}) \qquad \forall f^{(t)}, f^{(t+1)} \in \mathbb{F} \tag{7.42}$$

There can be more than one choice of $P(f^{(t+1)} \mid f^{(t)})$ for a given $P(f)$. Different choices lead to different MCMC algorithms, such as Metropolis-Hastings algorithms and the Gibbs sampler, which will be described shortly.

Figure 7.5 shows an MCMC procedure for sampling from $P(f)$. There, $f^{(t)}$ (for $t \leq 0$) is assumed to be from the distribution $P(f)$ already. This is generally not true initially with MCMC starting with an arbitrary state $f^{(0)}$. A solution is to start with an arbitrary state $f^{(0)}$, run the MCMC procedure

---

set $t = 0$; generate $f^{(0)}$ at random;
repeat
      generate next $f^{(t+1)}$ according to $P(f^{(t+1)} \mid f^{(t)})$;
      $t = t + 1$;
until (maximum t reached);
return $\{f^{(t)} \mid t > t_0\}$;

Figure 7.5: MCMC sampler of $P(f)$, where $P(f^{(t+1)} \mid f^{(t)})$ satisfies (7.42).

for an initial "burn-in" period (for $0 \leq t \leq t_0$), and use $\{f^{(t)}\}$ (for $t > t_0$) as samples from $P(f)$ (Besag et al. 1995).

Metropolis-Hastings Algorithm

In the Metropolis algorithm, a new configuration f' is accepted with the acceptance ratio

$$P = \alpha(f' \mid f) = \min\left\{1, \frac{P(f')}{P(f)}\right\} \tag{7.43}$$

Figure 7.6 illustrates the Metropolis algorithm for sampling a Gibbs distribution $P(f) = Z^{-1}e^{-U(f)/T}$ given the T parameter. In each iteration, the next configuration f' is randomly chosen from $\mathcal{N}(f)$ (the vicinity of f) by applying a perturbation on f, for instance, by changing one of the f_i's into a new label f'_i. Random[0,1) is a random number from a uniform distribution in $[0, 1)$. A new configuration f' for which $\Delta U \leq 0$ is accepted for sure, whereas one for which $\Delta U > 0$ is accepted with probability P. The convergence condition "equilibrium is reached" may be judged by "a prescribed number of iterations have been performed".

The Metropolis-Hastings algorithm (Hastings 1970) is a generalization of the Metropolis algorithm in that the acceptance is calculated as

$$\alpha(f' \mid f) = \min\left\{1, \frac{P(f')q(f \mid f')}{P(f)q(f' \mid f)}\right\} \tag{7.44}$$

where $q(f' \mid f)$ is a (arbitrary) density for proposing f' given f. If $q(f \mid f') = q(f' \mid f)$, the Metropolis-Hastings algorithm reduces to the Metropolis algorithm.

The transition probability corresponding to the acceptance probability $\alpha(f' \mid f)$ of (7.44) is

$$P(f' \mid f) = \begin{cases} q(f' \mid f)\alpha(f' \mid f) & \text{if } f' \neq f \\ 1 - \sum_{f''} q(f'' \mid f)\alpha(f'' \mid f) & \text{if } f' = f \end{cases} \tag{7.45}$$

```
initialize f;
repeat
    generate f' ∈ N(f);
    ΔU ← U(f') − U(f);
    P = α(f' | f) = min{1, e^{−ΔU/T}};
    if random[0,1) < P then f ← f';
until (equilibrium is reached);
return f;
```

Figure 7.6: Metropolis sampling of Gibbs distribution.

where $f' \neq f$ means that the proposed f' is accepted and $f' = f$ means f' is rejected. It can be checked that the detailed balance of (7.42) is satisfied by such a transition probability.

Gibbs Sampler

The Gibbs sampler (Geman and Geman 1984) generates the next configuration based on a conditional probability. A candidate f'_i for the next f_i is randomly drawn from the conditional distribution $P(f'_i \mid f_{\mathcal{S}-\{i\}})$. Given $f = \{f_1, \ldots, f_m\}$, the transition to $f' = \{f'_1, \ldots, f'_m\}$ is performed by successively drawing samples from the conditional probabilities as follows:

$$
\begin{array}{lll}
f'_1 & \text{from} & P(f_1 \mid f_2, \ldots, f_m); \\
f'_2 & \text{from} & P(f_2 \mid f'_1, f_3 \ldots, f_m); \\
f'_3 & \text{from} & P(f_3 \mid f'_1, f'_2, f_4, \ldots, f_m); \\
& \vdots & \\
f'_m & \text{from} & P(f_m \mid f'_1, \ldots, f'_{m-1}, f_m);
\end{array}
$$

The transition probability from f to f' is given by

$$P(f' \mid f) = \prod_{i=1}^{m} P(f'_i \mid f'_j, f_k, \ j < l < k) \tag{7.46}$$

The scenario for the Gibbs sampler procedure is similar to that for the Metropolis algorithm illustrated in Fig. 7.6; the difference is in how the next configuration f' is generated.

When f is an MRF, $P(f'_i \mid f_{\mathcal{S}-\{i\}}) = P(f'_i \mid f_{\mathcal{N}_i})$, which is given by

$$P(f_i \mid f_{\mathcal{N}_i}) = \frac{e^{-\left[V_1(f_i)+\sum_{i' \in \mathcal{N}_i} V_2(f_i,f_{i'})\right]/T}}{\sum_{f_i \in \mathcal{L}} e^{-\left[V_1(f_i)+\sum_{i' \in \mathcal{N}_i} V_2(f_i,f_{i'})\right]/T}} \tag{7.47}$$

when up to pairwise cliques are involved. Unlike (2.18), this formula contains a temperature parameter.

Like the Metropolis algorithms, the Gibbs sampler also produces a Markov chain $\{f^{(0)}, f^{(1)}, \ldots\}$ with the target distribution $P(f)$ as the equilibrium (Geman and Geman 1984). Comparatively speaking, the Metropolis sampler is easier to program and runs more efficiently than the Gibbs sampler. However, other considerations also affect the choice between samplers (Besag et al. 1995). In image and vision, Cross and Jain (1983) were the first to use the Metropolis algorithm for sampling MRF's for texture modeling, and Geman and Geman (1984) were the first to use the Gibbs sampler and annealing procedures to find the global estimates of MRF's.

Calculating Expectation Using MCMC

Many statistics, such as the mean and variance of $P(f)$ as well as the normalizing constant Z in a Gibbs distribution, can be expressed as the expectation of a function $\phi(f)$ w.r.t. $P(f)$

$$\overline{a}_{P(f)} = \int a(f)P(f)\mathrm{d}f \tag{7.48}$$

A direct evaluation would be infeasible. MCMC enables one to approximate the value using the following method. Choose the transition probability $P(f' \mid f)$ so that the chain converges to the target distribution $P(f)$. Generate a Markov chain according to $P(f' \mid f)$. Get $f^{(0)}$ after a long enough burn-in period. Collect samples $f^{(0)}, f^{(1)}, \ldots, f^{(N-1)}$. Then use their arithmetic average

$$\overline{a} = \frac{1}{N} \sum_{n=0}^{N-1} a(f^{(n)}) \tag{7.49}$$

to approximate the expectation. The average value converges to the expectation $\overline{a}_{P(f)}$ as $N \to \infty$.

7.1.7 Learning in the FRAME Model

As mentioned in Section 2.6, there are two things to learn in the FRAME model: (1) the potential functions $\theta_I^{(k)}$ and (2) the types of filters $G^{(k)}$.

Given $\overline{H_{samp,\ell}^{(k)}}$ computed from example images f_{samp}, the problem is to estimate the potential functions $\theta_\ell^{(k)}$ of the FRAME model from which f_{samp} is sampled (generated). The problem is posed as one of maximizing (w.r.t. θ) the likelihood $p(f_{samp} \mid \theta)$ of (2.64) or the log-likelihood

$$\ln p(f_{samp} \mid \theta) = -\log Z(\theta) - <\theta, \overline{H_{samp}}>$$ (7.50)

where $Z(\theta) = \sum_f e^{-<\theta, \overline{H_{samp}}>}$. Setting $\frac{\partial \ln p(f_{samp} \mid \theta)}{\partial \theta} = 0$, the minimum θ^* is obtained by solving the equation

$$\sum_f p(f \mid \theta^*) H(f) = \overline{H_{samp}}$$ (7.51)

The left-hand side is the expectation $\overline{H_{p(f|\theta^*)}}$. Because the log-likelihood is strictly convex w.r.t. θ, the minimum solution can be found by applying gradient descent

$$\frac{d\theta_\ell^{(k)}}{dt} = \overline{H_{p(f|\theta),\ell}^{(k)}} - \overline{H_{samp,\ell}^{(k)}} \qquad \forall k, \forall \ell$$ (7.52)

This provides an algorithm for learning the parameters θ of a specific FRAME model from the sample data f_{samp}.

While $\overline{H_{samp,\ell}^{(k)}}$ are calculated from the sample data, $\overline{H_{p(f|\theta),\ell}^{(k)}}$ have to be estimated w.r.t. $p(f \mid \theta)$. Given θ fixed, the estimation may be done by using a Markov chain Monte Carlo (MCMC) procedure (see Section 7.1.6) as follows: (1) Draw N samples (i.e., synthesize N images $f_{syn(1)}, \ldots, f_{syn(N)}$) according to the distribution $p(f \mid \theta)$, using, say, a Gibbs sampler (Geman and Geman 1984). (2) Filter the synthesized images to obtain $H_{syn(1)}^{(k)}, \ldots, H_{syn(N)}^{(k)}$. (3) Assuming that N is large enough, we approximate the expectation using the sample mean, $\overline{H_{p(f|\theta)}^{(k)}} = \frac{1}{N} \sum_{n=1}^{N} H_{syn(n)}^{(k)}$. Refer to Section 7.1.6 for an MCMC approximation of the probabilistic expectation.

Given a fixed model complexity K, the selection of filters $G^{(k)}$ may be done in a stepwise way using the greedy strategy (Zhu et al. 1998). Filters are added to the filter bank one at a time until the number K is reached. Given the current selection of k filters, the next filter $k + 1$ is selected such that the entropy of $p(f)$ is *minimized*. Note, however, that this is not necessarily a globally optimal selection.

The above can be summarized as a three part algorithm: (1) the FRAME algorithm, (2) the Gibbs sampler and (3) the filter selection. More details of the three parts can be found in (Zhu et al. 1998).

7.2 Unsupervised Estimation with Unlabeled Data

The methods described in the previous section are applicable to situations where the data contain either a single MRF or several *segmented* MRF's so that the estimation can be done for each MRF separately. In practice, it is normal for the data to be an observation, such as an image, of the underlying,

unobservable MRF's. There can be several complications compared with the
clean, single-MRF case. First, the data are a transformed version of the true
MRF, for example, due to the additive noise observation model $d_i = f_i + e_i$.
Second, the noise parameters can be unknown and hence also a part of the
parameters to be estimated. In this case, the set of unknown parameters is
composed of two parts, those for the MRF and those due to the noise, collectively denoted as $\theta = \{\theta_f, \theta_d\}$. Third, the data d can contain realizations
of more than one MRF, and the origin of each data point is unknown. Since
the parameters for an MRF should be estimated by using only realizations
of that MRF, the data have to be segmented before being used for the estimation. However, to perform good segmentation, the parameter values of the
MRF's have to be available. This is a dilemma between segmentation and
estimation. The problem of restoration-estimation is similar: The estimation
should be performed using smoothed or restored data; however, to obtain
good restoration, the parameter values of the MRF have to be known. We
consider this chicken-and-egg problem.

One strategy is to perform parameter estimation after labeling (segmentation or restoration); for example, to perform segmentation using clustering
techniques (Silverman and Cooper 1988; Hu and Fahmy 1987). This partitions the input data into blocks which are then merged into clusters according
to some optimization criterion, yielding a segmentation. Each cluster contains
only blocks due to a single MRF (or due to "boundary blocks"). Thereafter,
parameters for each MRF can be estimated separately. However, such a strategy may not yield the best result because optimal labeling relies on correct
parameter values.

In a more sophisticated paradigm, iterative labeling-estimation (Besag
1986; Kelly et al. 1988; Lakshmanan and Derin 1989; Qian and Titterington
1992; Pappas 1992; Won and Derin 1992; Zhang 1992; Bouman and Shapiro
1994) is adopted, where labeling and estimation are performed alternately.
The initial labeling is chosen by using some scheme, possibly with initial
parameters; a parameter estimate is computed based on this labeling, and
is then used to get a hopefully better labeling, a hopefully better estimate
is computed after that, and so on. The expectation-maximization algorithm
(Dempster et al. 1977) provides a justifiable formalism for such labeling-
estimation schemes.

7.2.1   Simultaneous Restoration and Estimation

Let us consider a heuristic procedure for simultaneous image restoration and
parameter estimation (Besag 1986). Our interest is in restoring the underlying
MRF realization, f, from the observed data, d. During this process, two sets
of parameters are to be estimated: a set of the MRF parameters θ_f and a set
of the observation parameters θ_d. The procedure is summarized below.

1. Start with an initial estimate $\hat{f}$ of the true MRF and guess θ_f and θ_d.

2. Estimate θ_d by $\hat{\theta}_d = \arg\max_{\theta_d} P(d \mid f, \theta_d)$.

3. Estimate θ_f by $\hat{\theta}_f = \arg\max_{\theta_f} P(f \mid \theta_f)$.

4. Update $\hat{f}$ by $\hat{f} = \arg\max_f P(f \mid d, \theta_f, \theta_d)$ based on the current $\hat{f}$, θ_f, and θ_d.

5. Go to step 2 for a number of iterations or until $\hat{f}$ approximately converges.

This is illustrated below.

Assume that the data $d = \{d_1, \ldots, d_m\}$ are determined by the additive noise model

$$d_i = f_i + e_i \tag{7.53}$$

where $e_i \sim N(0, \sigma^2)$ is the independent zero-mean Gaussian noise corrupting the true label f_i. Then, for step 2, we can maximize

$$\ln p(d \mid f, \theta_d) = -n \ln(2\pi) - n \ln \sigma - \sum_i \frac{(f_i - d_i)^2}{2\sigma^2} \tag{7.54}$$

where $\theta_d = \sigma^2$. Step 2, which is not subject to the Markovianity of f, amounts to finding the ML estimate of $\theta_d = \sigma^2$, which is, according to (7.5),

$$(\sigma^2)^* = \frac{1}{m} \sum_{i=1}^{m} (f_i - d_i)^2 \tag{7.55}$$

Steps 3 and 4 have to deal with the Markovianity. Consider again the homogeneous and isotropic auto-logistic MRF, characterized by (7.7) and (7.8). For step 3, we can maximize

$$\ln P(f \mid \theta_f) = -\ln Z(\alpha, \beta) + \sum_i \alpha f_i + \sum_i \sum_{j \in \mathcal{N}_i} \beta f_i f_j \tag{7.56}$$

where $\theta_f = \{\alpha, \beta\}$. Heuristics may be used to alleviate difficulties therein. For example, the likelihood function $P(f \mid \theta_d)$ in step 3 may be replaced by the pseudo-likelihood (7.14) or the mean field approximation. A method for computing the maximum pseudo-likelihood estimate of $\theta = \{\alpha, \beta\}$ has been given in Section 7.1.2. The maximization in step 4 may be performed by using a local method such as ICM to be introduced in Section 9.3.1.

Figure 7.7 shows an example. The true MRF, f shown in the upper left, is generated by using a multilevel logistic model, a generalization to the bi-level auto-logistic MRF, with $M = 6$ (given) level labels in $\mathcal{L}$ and the parameters being $\alpha = 0$ (known throughout the estimation) and $\beta = 1.5$. The MRF is corrupted by the addition of Gaussian noise with variance $\sigma^2 = 3.6$, giving the observed data, d in the the upper right. With the correct parameter values $\sigma^2 = 3.6$ and $\beta = 1.5$ known, no parameters are to be estimated and

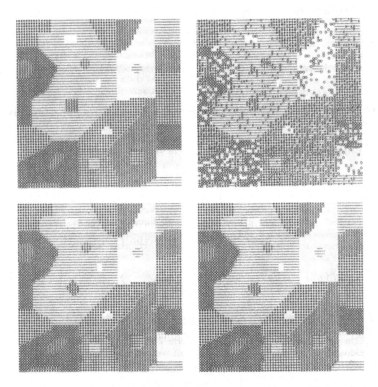

Figure 7.7: Image restoration with unknown parameters. Upper left: True MRF. Upper right: Noisy observation. Lower left: Restoration with the exact parameters. Lower right: Restoration with β estimated. From (Besag 1986) with permission; ©1986 Blackwell Publishing.

only step 3 needs to be done (e.g., using ICM). ICM, with $\beta = 1.5$ fixed, converges after six iterations and the misclassification rate of the restored image is 2.1%. The rate reduces to 1.2% (lower left) after eight iterations when β is gradually increased to 1.5 over the first six cycles. With σ^2 known and β estimated, the procedure consists of all the steps but step 2. The error rate is 1.1% after eight iterations (lower right), with $\beta^* = 1.80$. When σ^2 is also to be estimated, the error rate is 1.2%, with $\beta^* = 1.83$ and $\sigma^2 = 0.366$.

7.2.2 Simultaneous Segmentation and Estimation

In the previous simultaneous restoration and estimation, it was assumed that only a single MRF was present in the data. Segmentation has to be performed when the data contain realizations of more than one MRF. For the reasons noted, there is a need for simultaneous segmentation and estimation. The optimal parameters are estimated during the computation of the optimal segmentation.

Let $\mathcal{L} \in \{1, \dots, M\}$ denote the possible texture types and f represent a labeling or segmentation with $f_i \in \mathcal{L}$ indicating the texture type for pixel i. The segmentation f partitions $\mathcal{S}$ into M types of disjoint regions[3]

$$\mathcal{S}^{(I)}(f) = \{i \in \mathcal{S} \mid I = f_i\}, \quad I \in \mathcal{L} \tag{7.57}$$

(see Equations (3.51)–(3.53). The problem may be formulated in terms of the MAP principle as

$$(f^*, \theta_f^*, \theta_d^*) = \arg\max_{f, \theta_f, \theta_d} P(f, \theta_f, \theta_d \mid d) \tag{7.58}$$

Assume that θ_f and θ_d are independent of each other and that they are uniformly distributed when no prior knowledge about their distributions is available. Then the joint maximization above is equivalent to

$$\begin{aligned} (f^*, \theta_f^*, \theta_d^*) &= \arg\max_{f, \theta_f, \theta_d} P(d \mid f, \theta_d) P(f \mid \theta_f) P(\theta_f) P(\theta_d) \\ &= \arg\max_{f, \theta_f, \theta_d} P(d \mid f, \theta_d) P(f \mid \theta_f) \end{aligned} \tag{7.59}$$

In general, this maximization problem is computationally intractable. However, it is solvable when θ_d can be expressed in closed form as $\theta_d(f, d)$, that is, a function of f and d (Won and Derin 1992).

Assume that the regions are piecewise constant-valued and are governed by an MRF model, and that the observation model $d_i = \varphi(f_i) + e_i$, where e_i is additive identical, independent, zero-mean Gaussian noise and $\varphi(f_i)$ is the gray level for type $I = f_i$ regions. Then, given f and d, the ML estimate of the noise variance for type I MRF regions can be expressed explicitly as a function of f and d

$$\theta_d^*(f, d) = (\sigma_I^2)^* = \frac{1}{\#\mathcal{S}^{(I)}} \sum_{i \in \mathcal{S}^{(I)}} (d_i - \varphi(f_i))^2 \tag{7.60}$$

where $\#\mathcal{S}^{(I)}$ is the number of pixels in type I regions. When θ_d is given as $\theta_d^*(f, d)$, (7.59) is reduced to

$$(f^*, \theta_f^*) = \arg\max_{f, \theta_f} P(d \mid f, \theta_d^*(f, d)) P(f \mid \theta_f) \tag{7.61}$$

However, the minimization over the f-θ_f space is still a difficult one. Therefore, it is divided into the two subproblems

$$f^* = \arg\max_f P(d \mid f, \theta_d^*(f, d)) P(f \mid \theta_f^*) \tag{7.62}$$

$$\theta_f^* = \arg\max_\theta P(d \mid f, \theta_d^*(f^*, d)) P(f^* \mid d, \theta_f) \tag{7.63}$$

---

[3]In completely unsupervised procedures, the number of texture types, M, is also an unknown parameter to be estimated; see section 7.3.

Thus, the procedure iteratively alternates between the two equations. The estimate (f^*, θ_f^*) thus found is suboptimal w.r.t. (7.61).

As in the case of simultaneous restoration and estimation, there are methods for solving (7.62) and (7.63). For example, one may choose a simulated annealing procedure if he can afford to find a good f^* for (7.62) or use heuristic ICM to find a local solution. Pseudo-likelihood, the coding method, or the mean field method may be used to approximate $P(f^* \mid d, \theta_f)$ in (7.63).

Now we turn to the case of textured images. Suppose that the hierarchical Gibbs model of (Derin and Cole 1986; Derin and Elliott 1987) is used to represent textured images. The higher-level MRF corresponds to a region process with distribution $P(f \mid \theta_f)$. When modeled as a homogeneous and isotropic MLL model, $\theta_f = \{\alpha_1, \ldots, \alpha_M, \beta\}$, in which α_I has the effect of controlling the relative percentage of type I pixels and β controls the interaction between neighboring regions. At the lower-level, there are $M = \#\mathcal{L}$ types of MRF's for the filling-in textures with p.d.f. $p(d \mid f, \theta_d)$ with $\theta_d = \{\theta_d^{(I)} \mid I \in \mathcal{L}\}$. When a filling-in is modeled as an auto-normal MRF (Besag 1974), its parameters consist of the mean and the interaction matrix $\theta_d^{(I)} = \{\mu^{(I)}, B^{(I)}\}$. After all, note that the region process is a much "slower varying" MRF than the texture MRF's.

Because it is difficult to compute the true p.d.f. $p(d \mid f, \theta_d)$, we use the pseudo-likelihood, $pl(d \mid f, \theta_d)$, to approximate it.[4] The conditional p.d.f. for d_i takes the form (2.34). When the MRF is also homogeneous, the conditional p.d.f. can be written as

$$p(d_i \mid d_{\mathcal{N}_i}, f, \theta_d) = \frac{1}{\sqrt{2\pi\sigma^2}} e^{-\frac{1}{2\sigma^2}[d_i - \mu - \sum_{i' \in \mathcal{N}_i} \beta_{i,i'}(d_{i'} - \mu)]^2} \quad (7.64)$$

where $I = f_i$. The above formula is a meaningful only when all the sites in the block $\{i, \mathcal{N}_i\}$ have the same texture label. To approximate, we may pretend that the all sites in $\mathcal{N}_i$ have the same label as i even when this is not true; in this case, the formula is only an imprecise one.

The conditional p.d.f. $p(d_i \mid d_{\mathcal{N}_i}, f, \theta_d)$ has a very limited power for texture description because it is about the datum on a single-site i given the neighborhood. Therefore, a scheme based on a block of data may be used to enhance the capability (Won and Derin 1992). Define a block centered at i

$$\mathcal{B}_i = \{i\} \bigcup \mathcal{N}_i \quad (7.65)$$

for each $i \in \mathcal{S} - \partial\mathcal{S}$. Regard the data in $\mathcal{B}_i$ as a vector of $n = \#\mathcal{B}_i$ dependent random variables, denoted $d_{\mathcal{B}_i}$. Assume that the all the sites in $\mathcal{N}_i$ have the same label, $I = f_i$, as i. Then the joint p.d.f. of $d_{\mathcal{B}_i}$ given f and $\theta_d^{(I)}$ is multivariate normal

----

[4]Note that the lowercase letters "pl" are used to denote the PL for the continuous random variables d.

$$p(d_{\mathcal{B}_i} \mid f, \theta_d^{(I)}) = \frac{1}{\sqrt{(2\pi)^n \det(\Sigma)}} e^{-\frac{1}{2}[(d_{\mathcal{B}_i} - \mu^{(I)})^{\mathrm{T}}[\Sigma^{(I)}]^{-1}(d_{\mathcal{B}_i} - \mu^{(I)})]} \quad (7.66)$$

where $\mu^{(I)}$ and $\Sigma^{(I)}$ are the mean vector and covariance matrix of $d_{\mathcal{B}_i}$, respectively.

The parameters $\mu^{(I)}$ and $\Sigma^{(I)}$ represent texture features for type I texture. They can be estimated using the sample mean and covariance. For only a single block centered at i, they are estimated as $\mu_i^{(I)} = d_{\mathcal{B}_i}$ and $\Sigma^{(I)} = (d_{\mathcal{B}_i} - \mu_i^{(I)})^{\mathrm{T}}(d_{\mathcal{B}_i} - \mu_i^{(I)})$. For all blocks in $\mathcal{S}^{(I)}$, they are taken as the averages

$$\hat{\mu}^{(I)} = \frac{1}{\#\mathcal{S}^{(I)}} \sum_{i \in \mathcal{S}^{(I)}} d_{\mathcal{B}_i} \quad (7.67)$$

and

$$\hat{\Sigma}^{(I)} = \frac{1}{\#\mathcal{S}^{(I)}} \sum_{i \in \mathcal{S}^{(I)}} [d_{\mathcal{B}_i} - \mu^{(I)}]^{\mathrm{T}}[d_{\mathcal{B}_i} - \mu^{(I)}] \quad (7.68)$$

Obviously, the estimates have larger errors at or near boundaries, and obviously, $\Sigma^{(I)}$ is symmetric. When $\#\mathcal{S}^{(I)} > n$, which is generally true, $\Sigma^{(I)}$ is positive definite with probability one. Therefore, the validity of $p(d_{\mathcal{B}_i} \mid f, \theta_d^{(I)})$ is generally guaranteed. It is not difficult to establish relationships between the elements in the inverse of the covariance matrix $[\Sigma^{(I)}]^{-1}$ and those in the interaction matrix $B^{(I)}$ (Won and Derin 1992).

Based on the conditional p.d.f.s for the blocks, the pseudo-likelihood for type I texture can be defined as

$$pl^{(I)}(d \mid f, \theta_d^{(I)}) = \prod_{i \in \mathcal{S}^{(I)}} [p(d_{\mathcal{B}_i} \mid f, \theta_d^{(I)})]^{1/n} \quad (7.69)$$

where $\mathcal{S}^{(I)}$ is defined in (3.51), and $1/n = 1/\#\mathcal{S}^{(I)}$ compensates for the fact that $p(d_{\mathcal{B}_i} \mid \cdot)$ is the joint p.d.f. of n elements in $d_{\mathcal{B}_i}$. The overall pseudo-likelihood is then

$$pl(d \mid f, \theta_d) = \prod_{I \in \mathcal{L}} pl^{(I)}(d \mid f, \theta_d^{(I)}) \quad (7.70)$$

Given f, $\theta_d^{(I)}$ can be explicitly expressed as (7.67) and (7.68), and hence $pl(d \mid f, \theta_d)$ is θ_d-free. Then, the problem reduces to finding the optimal f and θ_f.

Given initial values for the parameters θ_f and the segmentation, such a procedure alternately solves the two subproblems

$$\theta_f^{(t+1)} = \arg\max_{\theta_f} \ln P(f^{(t)} \mid \theta_f) \quad (7.71)$$

$$f^{(t+1)} = \arg\max_f \ln P(f \mid \theta_f^{(t+1)}) + \ln pl(d \mid f, \hat{\theta}_d) \quad (7.72)$$

until convergence. In the above, $P(f^{(t)} \mid \theta_f)$ may be replaced by the corresponding pseudo-likelihood. Figure 7.8 shows a segmentation result.

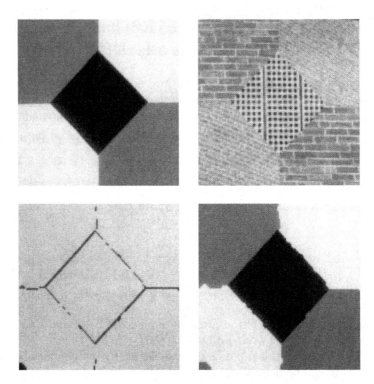

Figure 7.8: Texture segmentation with unknown parameters. Upper left: True regions. Upper right: Observed noisy textured regions. Lower left: Segmentation error. Lower right: Segmentation result. From (Won and Derin 1992) with permission; ©1992 Elsevier/Academic Press.

7.2.3 Expectation-Maximization

The expectation-maximization (EM) algorithm (Dempster et al. 1977) is a general technique for finding maximum likelihood (ML) estimates with *incomplete data*. In the classic ML, the estimate is obtained from the complete data as

$$\theta^* = \arg\max_\theta \ln P(d_{com} \mid \theta) \qquad (7.73)$$

In EM, the complete data are considered to consist of the two parts $d_{com} = \{d_{obs}, d_{mis}\}$, of which only d_{obs} is observed while d_{mis} is missing (or unobservable, and hidden). With only the incomplete data d_{obs}, an EM procedure attempts to solve the following ML estimation problem

$$\theta^* = \arg\max_\theta \ln P(d_{obs} \mid \theta) \qquad (7.74)$$

which is more general than the classic ML.

Intuitively, we may envisage a maximum likelihood procedure that after some initialization for d_{mis} and θ iterates between the following two steps until convergence: (1) Estimate the missing part as $\hat{d}_{mis}$ given the current θ estimate and then use it to augment the observed data $d = d_{obs}$ to form the *complete data set* $d_{com} = \{d_{obs}, \hat{d}_{mis}\}$; and (2) Estimate θ, with d_{com}, by maximizing the complete-data log-likelihood $\ln P(\hat{d}_{mis}, d_{obs} \mid \theta)$. The simultaneous labeling-estimation algorithm described in the previous two subsections is based on such an idea.

However, we cannot work with this log-likelihood directly because it is a random function of the missing variables f (the reason why the procedure above is ad hoc). The idea of the EM algorithm is to use the expectation of the complete-data log-likelihood, $E\left[\ln P(\hat{d}_{mis}, d_{obs} \mid \theta)\right]$, which formalizes the intuitive procedure above.

Related to parameter estimation in MRF models, the missing part corresponds to the unobservable labeling f, $f = d_{mis}$, and the observed part is the given data, $d = d_{obs}$. The complete-data log-likelihood is then denoted by $\ln P(f, d \mid \theta)$. The EM algorithm consists of the following two steps for each iteration:

(1) *The E-step.* Compute the following conditional expectation of the log-likelihood

$$Q(\theta \mid \theta^{(t)}) = E\left[\ln P(f, d \mid \theta) \mid d, \theta^{(t)}\right] \qquad (7.75)$$

(2) *The M-step.* Maximize $Q(\theta \mid \theta(t))$ to obtain the next estimate

$$\theta^{(t+1)} = \arg\max_{\theta} Q(\theta \mid \theta^{(t)}) \qquad (7.76)$$

If in the M-step the next estimate $\theta^{(t+1)}$ is chosen only to ensure

$$Q(\theta^{(t+1)} \mid \theta^{(t)}) > Q(\theta^{(t)} \mid \theta^{(t)}) \qquad (7.77)$$

(not necessarily to maximize Q), then the algorithm is called a generalized EM (GEM).

The E-step computes the conditional expectation of the unobservable labels f given the observed data d and the current estimate $\theta^{(t)}$ and then substitutes the expectations for the labels. The M-step performs maximum likelihood estimation as if there were no missing data (i.e., as if they had been filled in by the expectations).

This describes just one instance of the EM procedure. More generally, the probability $P(f, d \mid \theta)$ can be replaced by any complete-data sufficient statistics. When the prior distribution $p(\theta)$ is known, the M-step can be modified to maximize the expected posterior $E\left[\ln p(\theta, f \mid d) \mid d, \theta^{(t)}\right]$ or $E\left[\ln P(f, d \mid \theta) \mid d, \theta^{(t)}\right] + \ln p(\theta)$, where $p(\theta, f \mid d) = \frac{p(f,d \mid \theta)\, p(\theta)}{P(d)}$, instead of the likelihood.

The EM algorithm is shown to converge to the ML estimates at least locally under some moderate conditions (Wu 1983). Global convergence is also proven for special cases, for example, in (Redner and Walker 1984; Lange 1990). However, the convergence of the EM algorithm can be painfully slow.

Let us be more specific about the expectation $E\left[\ln P(f, d \mid \theta) \mid d, \theta^{(t)}\right]$. For f taking discrete values, which is the case for discrete restoration and for segmentation, the expectation is

$$E\left[\ln P(f, d \mid \theta) \mid d, \theta^{(t)}\right] = \sum_{f \in \mathbb{F}} [\ln P(f, d \mid \theta)] P(f \mid d, \theta^{(t)}) \qquad (7.78)$$

For continuous f, which is for continuous restoration, it is calculated from the p.d.f.s

$$E\left[\ln p(f, d \mid \theta) \mid d, \theta^{(t)}\right] = \int_{\mathbb{F}} [\ln p(f, d \mid \theta)] p(f \mid d, \theta^{(t)}) df \qquad (7.79)$$

Consider the homogeneous and isotropic auto-logistic MRF model described by (7.7) and (7.8). We have

$$\ln P(f \mid \theta) = -\ln Z(\alpha, \beta) + \sum_i \alpha f_i + \sum_i \sum_{j \in \mathcal{N}_i} \beta f_i f_j \qquad (7.80)$$

Assume that the data are obtained according to $d_i = f_i + e_i$ with independent Gaussian noise $e_i \sim N(0, \sigma^2)$. Then

$$\ln p(d \mid f, \theta) = -n \ln(2\pi) - n \ln \sigma - \sum_i \frac{(f_i - d_i)^2}{2\sigma^2} \qquad (7.81)$$

where $\theta = \{\theta_f, \theta_d\} = \{\alpha, \beta, \sigma^2\}$. The expectation is

$$\begin{aligned} Q(\theta \mid \theta^{(t)}) &= E\left[\ln p(f, d \mid \theta) \mid d, \theta^{(t)}\right] \qquad (7.82) \\ &= \sum_{f \in \mathbb{F}} [\ln p(d \mid f, \theta) + \ln p(f \mid \theta)] P(f \mid d, \theta^{(t)}) \end{aligned}$$

A main difficulty in applying EM to MRF parameter estimation is again the complexity of $P(f \mid d, \theta^{(t)})$ in the computation of $Q(\theta \mid \theta^{(t)})$ due to the need to evaluate the partition function for $P(f \mid \theta)$. This difficulty may be overcome by using the pseudo-likelihood method (Chalmond 1989) or the mean field approximation (Zhang 1992; Zhang 1993).

7.2.4 Cross Validation

Lastly, in this section, we describe an analytical technique called "cross validation" for estimating the smoothing parameter λ in the regularization model. Consider the energy of the form

$$E(f) = \sum_{i \in \mathcal{A}} [f(x_i) - d_i]^2 + \lambda \int_a^b [f''(x)]^2 dx \qquad (7.83)$$

(see (1.29)) under additive white noise

$$d_i = f(x_i) + e_i \qquad (7.84)$$

where $e_i \sim N(0, \sigma^2)$ with unknown variance σ^2. It requires no a priori knowledge and is entirely data-driven. The method has been studied in the context of regression and smoothing splines (see Chapter 4 of Wahba (1990)). See also (Geiger and Poggio 1987; Shahraray and Anderson 1989; Thompson et al. 1991; Galatsanos and Katsaggelos 1992; Reeves 1992) for its applications in image and vision processing.

The idea of cross validation for parameter estimation is the following. Divide the data into an estimation subset and a validation subset. The former subset is used to obtain a parameter estimate, and the latter is used to validate the performance under the estimate. However, cross validation does not use one subset exclusively for one purpose (estimation or validation); it allows all the data to be used for both purposes. For instance, we can divide the data into m subsets, compute an estimate from all the subsets but one, and validate the estimate of the left-out subset. Then, we perform the estimation-validation with a different subset left-out. We repeat this m times.

Ordinary cross validation (OCV) uses the "leaving-out-one" strategy. Each point is left out in turn and its value is predicted from the rest of the data using the fitted covariance function. The prediction errors are summed. The best λ minimizes the summed error. This is expressed as follows.

Let $f_\lambda^{[k]}$ be the minimizer of the energy

$$E^{[k]}(f) = \sum_{i \neq k} \frac{1}{m} [f(x_i) - d_i]^2 + \lambda \int_a^b [f''(x)]^2 dx \qquad (7.85)$$

in which d_k is left out. The OCV function is defined as

$$V_o(\lambda) = \frac{1}{m} \sum_{k=1}^m [f_\lambda^{[k]}(x_k) - d_k]^2 \qquad (7.86)$$

In the above, $[f_\lambda^{[k]}(x_k) - d_k]^2$ measures the error incurred by using $f_\lambda^{[k]}(x_k)$ to predicate d_k. Therefore, $V_o(\lambda)$ is the average prediction error. The OCV estimate of λ is the minimizer of $V_o(\lambda)$.

Let f_λ^* be the minimizer of the complete energy (7.83) under the parameter value λ. The relationship between f_λ^* and $f_\lambda^{[k]}$ can be derived. Define the $m \times m$ influence matrix $A(\lambda)$ as satisfying

$$f_\lambda^* = A(\lambda)d \qquad (7.87)$$

where f_λ^* and d are $m \times 1$ vectors. The OCV function can be expressed in terms of $A(\lambda)$ as (Craven and Wahba 1979)

$$V_o(\lambda) = \frac{1}{m} \sum_{k=1}^{m} \frac{[f_\lambda^*(x_k) - d_k]^2}{(1 - a_{kk})^2} \tag{7.88}$$

where $f_\lambda^*(x_k) = \sum_{j=1}^{m} a_{kj} d_j$. By comparing (7.86) with (7.88), one obtains

$$[f_\lambda^{[k]}(x_k) - d_k]^2 = \frac{[f_\lambda^*(x_k) - d_k]^2}{1 - a_{kk}(\lambda)} \tag{7.89}$$

OCV is modified to generalized cross validation (GCV) to achieve certain desirable invariance properties that do not generally hold for OCV. Let Γ be any $m \times m$ orthogonal matrix, and consider a new vector of data $\tilde{d} = \Gamma d$. In general, the OCV estimate can give a different value of λ; in contrast, the GCV estimate is invariant under this (rotation) transformation (Golub et al. 1979).

The GCV function is a weighted version of $V_o(\lambda)$

$$V(\lambda) = \frac{1}{m} \sum_{k=1}^{m} [f_\lambda^{[k]}(x_k) - d_k]^2 w_k(\lambda) \tag{7.90}$$

where the weights

$$w_k(\lambda) = \left[\frac{1 - a_{kk}(\lambda)}{1 - \frac{1}{m} \sum_{j=1}^{m} a_{jj}} \right]^2 \tag{7.91}$$

give the relative effect of leaving out d_k. It is obtained by replacing a_{kk} in (7.88) by $\frac{1}{m} \sum_{j=1}^{m} a_{jj}$. The GCV estimate λ_{GCV} is the minimizer of $V(\lambda)$.

7.3 Estimating the Number of MRF's

In the previous discussions on parameter estimation of multiple MRF's, the number of distinct MRF models (regions, textures, etc.), here denoted by K, is assumed to be known. In completely unsupervised estimation and clustering schemes, this number is an unknown parameter and also has to be estimated along with other parameters. The selection of the model number is generally a difficult issue in clustering analysis (Jain and Dubes 1988). Various criteria have been proposed for the estimation of K. They can be classified as two broad categories: heuristic approaches and statistical approaches. An example of the former category is the bootstrap technique (Jain and Moreau 1987); there, the value of K that provides the most stable partitions (in terms of a heuristic criterion) is the estimate of the number of clusters in the data set. In the following, a particular class of statistically based criteria are discussed.

7.3.1 Akaike Information Criterion (AIC)

Let K be the number to be chosen from the set $\mathbf{K} = \{1, \ldots, K_{\max}\}$, $\hat{\theta}^{[K]}$ the ML estimate of the model parameter vector of the K models, $L(K) = p_K(d \mid \hat{\theta}^{[K]})$ the maximized likelihood, $n(K)$ the number of independent adjustable parameters in the K models, and N the sample size. Various maximization criteria take the form (Sclove 1987)

$$G(K) = \ln[L(K)] - a(N)\, n(K) - b(K, N) \qquad (7.92)$$

Basically, the value of $G(K)$ increases with the maximized likelihood $L(K)$ but, due to other terms, decreases with the value M. Criteria $G(K)$ differ in the choices of $a(N)$ and $b(K, N)$.

For $a(N) = b(K, N) = 0$, the K that maximizes $G(K)$ is reduced to the maximum likelihood estimate. Such an estimate has been shown to be biased (Sakamoto et al. 1987). The Akaike information criterion (AIC) (Akaike 1974) is obtained as an expected entropy, giving $a(N) = 1$ and $b(K, N) = 0$. The AIC has been shown to be asymptotically unbiased (Sakamoto et al. 1987). From the Bayes viewpoint (Schwartz 1987; Kashyap 1988), $G(K)$ can be obtained by expanding the posterior probability $P(K \mid data)$. In (Schwartz 1987), $a(N) = \frac{1}{2}\ln(N)$ and $b(K, N) = 0$. Kashyap has the same $a(N)$ as Schwarts but with $b(K, N) = \frac{1}{2}\ln[\det B(K, N)]$ where $B(K, N)$ is the Hessian of $L(K)$ evaluated at the maximum likelihood estimates of K. Rissanen (1978) formulates $G(K)$ in terms of minimum description length and gives $a(N) = \frac{1}{2}\ln(\frac{N+1}{24})$ and $b(K, N) = \ln(K + 1)$. Some results comparing the performance of the criteria are presented in (Sclove 1987), showing that Schwartz's method gives better results than the AIC.

Sclove (1983) used the AIC to verify the number of region classes in image segmentation. Zhang and Modestino (1990) present a cluster validation scheme based on AIC (Akaike 1974) and argue that the AIC approach is advantageous over other heuristic schemes for cluster validation. Some modified criteria based on (7.92) have also been proposed. Bouman and Liu (1991) adopt a modified criterion for texture segmentation where a texture is modeled as a Gaussian AR random field (Jain 1981) with a number of H nonzero prediction coefficients. They replace the likelihood function with the joint function $p_K(d, f \mid \theta^{[K]})$ where f is the labeling that determines the partition of $\mathcal{S}$ into regions, $G(\theta^{[K]}, f, K) = \ln[p_K(d, f \mid \theta^{[K]})] - K(H + 3) + 1$. The criterion is optimized w.r.t. the parameters θ, the partition f and the number of textures K. Won and Derin (1992) propose $G(K) = \ln[L(K)] - N^c n(K)$, where c is a prespecified constant, for determining the number of texture regions.

7.3.2 Reversible Jump MCMC

Reversible jump MCMC, previously studied as the jump-diffusion sampler (Grenander and Miller 1994) and then elaborated by Green (1995), is an

MCMC method adapted for general state spaces of variable dimensionality. Assume that there are a number of $n(K)$ parameters to estimate when the model number is K, where K may be constrained to take a value in $\mathbb{K} = \{1, \ldots, K_{\max}\}$. Assuming that every parameter is a real number, then the parameter subspace for $\omega = (K, \theta^{[K]})$ is $\Omega_K = \{K\} \times \mathbb{R}^{n(K)}$. With variable K, the entire parameter space is the collection of all subspaces, $\Omega = \cup_{K \in \mathbb{K}} \Omega_K$. In order to implement sampling over Ω, a reversible Markov chain is constructed so that the sampling is allowed to jump between parameter subspaces of variable dimensionality.

A reversible jump MCMC procedure here is aimed at sampling a target joint distribution, which could be a likelihood $p(d \mid K, \theta^{[K]})$ or joint posterior $p(K, \theta^{[K]} \mid d)$ over Ω. A Markov chain is generated according to a transition probability $p(\omega' \mid \omega)$ designed in such a way that the detailed balance equation of (7.42) is satisfied and the chain has the target distribution as its equilibrium.

If a move from ω to ω' does not jump so that the dimensionality of $\theta^{[K]}$ remains unchanged, the reversible jump MCMC reduces to the conventional MCMC (to be specific, the Motroplis-Hastings algorithm). For dimension-changing moves, something extra should be done for switching between subspaces.

Consider moves between two subspaces $\Omega_1 = \{K_1\} \times \mathbb{R}^{n_1}$ and $\Omega_2 = \{K_2\} \times \mathbb{R}^{n_2}$. A move from Ω_1 to Ω_2 could be implemented by drawing a vector of random variables $u^{[1]}$ of length m_1 from some distribution $p_1(u^{[1]})$, independently of $\theta^{[1]} \in \mathbb{R}^{n_1}$, and then setting $\theta^{[2]} \in \mathbb{R}^{n_2}$ to some deterministic function of $\theta^{[1]}$ and $u^{[1]}$. Similarly, to move back, $u^{[2]}$ of length m_2 will be generated from $p_2(u^{[2]})$, and $\theta^{[1]}$ is set to some function of $\theta^{[2]}$ and $u^{[2]}$. The dimensions must match (i.e. $n_1 + m_1 = n_2 + m_2$) and there must be a bijection between $(\theta^{[1]}, u^{[1]})$ and $(\theta^{[2]}, u^{[2]})$ – that is, $(\theta^{[2]}, u^{[2]}) = \phi_2(\theta^{[1]}, u^{[1]})$ and $(\theta^{[1]}, u^{[1]}) = \phi_2^{-1}(\theta^{[2]}, u^{[2]}) = \phi_1(\theta^{[2]}, u^{[2]})$.

The acceptance probability for a move from $f = (K_1, \theta^{[1]})$ to $f' = (K_2, \theta^{[2]})$ can be derived as (Green 1995)

$$\alpha(f' \mid f) = \min \left\{ 1, \frac{P(K_2, \theta^{[2]} \mid d) \, p_2(u^{[2]}) \, q(K_2, \theta^{[2]})}{P(K_1, \theta^{[1]} \mid d) \, p_1(u^{[1]}) \, q(K_1, \theta^{[1]})} \left| \frac{\partial \phi_2(\theta^{[1]}, u^{[1]})}{\partial(\theta^{[1]}, u^{[1]})} \right| \right\} \tag{7.93}$$

where $P(K, \theta \mid d)$ is the target distribution of states $(K, \theta) \in \Omega$ given data d, $q(K, \theta)$ is the probability of choosing the move to state $f = (K, \theta)$, and $\left| \frac{\partial \phi_2(\theta^{[2]}, u^{[2]})}{\partial(\theta^{[1]}, u^{[1]})} \right|$ is the Jacobian of the mapping ϕ_2. The acceptance probability for the reverse move can be formulated similarly.

In (Richardson and Green 1997) and (Barker and Rayner 1997), the reversible jump MCMC is incorporated into simultaneous segmentation and parameter estimation procedures. A joint posterior distributions $p(K, \theta^{[K]} \mid d)$ is formulated, where K represents the number of distinct regions and $\theta^{[K]}$ is the collection of parameters of the MRF models for the regions and observation noise. Given K, $p(K, \theta^{[K]} \mid d)$ may be formulated (Richardson and

Green 1997; Barker and Rayner 1997), and the formulation is in principle similar to what is detailed in Section 7.2.2.

7.4 Reduction of Nonzero Parameters

This is an issue in MRF representation that affects parameter estimation. If an MRF can be represented using fewer parameters, then algorithms for it can be more efficient. It is known that Gibbs clique potential parameter representations for an MRF are not unique (Griffeath 1976; Kindermann and Snell 1980). For the sake of economy, it is preferred to use a representation that requires the smallest possible number of nonzero clique parameters. The normalized clique potential parameterization described in Section 2.1.5 can be used for this purpose.

Given a countable set $\mathcal{L}$ of labels and a set $\mathcal{C}$ of all cliques, one may choose an arbitrary label $\ell^0 \in \mathcal{L}$ and let $V_c(f) = 0$ whenever $f_i = \ell^0$ for some $i \in c$. Then one can specify nonzero clique parameters for the other configurations to define an MRF.

Nadabar and Jain (1991) use a normalized clique potential representation for MRF-based edge detection. Assuming the 4-neighborhood system on the image lattice, there can be eight different types of edge cliques, whose shapes are shown in the top part of Fig. 7.9. The label at each edge site takes one of the two values, nonedge or edge, in the label set $\mathcal{L}=\{\text{NE,E}\}$. This gives 35 {edge-label, clique-type} combinations. In the normalized representation, clique potentials $V_c(f)$ are set to zero whenever f_i takes a particular value, say f_i=NE for some $i \in c$. In this way, only 8 clique potentials are nonzero; thus the number of nonzero parameters is reduced from 35 to 8.

In (Zhang 1995), parameter reduction is performed for the compound Gauss-Markov model of Jeng and Wood (1990) and Jeng and Wood (1991). The model contains line process variables on which clique potential parameters are dependent. By imposing reasonable symmetry constraints, the 80 interdependent parameters of the model are reduced to 7 independent ones. This not only reduces the parameter number but also guarantees the consistency of the model parameters.

After the normalized representation is chosen, the parameter estimation can be formulated using any method, such as one of those described in the previous sections. In (Jain and Nadabar 1990), for example, the Derin-Elliott LS method is adopted to estimate line process parameters for edge detection. The estimated normalized clique potentials are shown in Fig. 7.9 in comparison with an ad hoc choice made in their previous work (Jain and Nadabar 1990). The corresponding results of edges detected using various schemes are shown in Fig. 7.10.

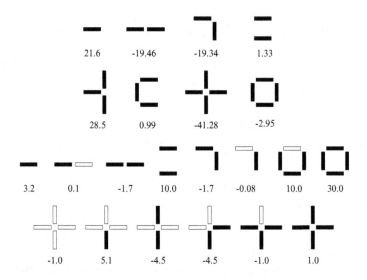

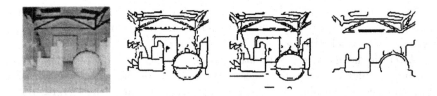

Figure 7.9: Normalized nonzero clique potentials estimated by using the least squares method (top) and nonzero clique potentials chosen ad hoc (bottom). Filled bars are edges and nonfilled bars nonedges.

Figure 7.10: Step edge detection using various schemes. From left to right: the input image; edges detected by using estimated canonical clique potentials; and edges obtained using the ad hoc clique potentials; edges detected using the Canny edge detector. From (Nadabar and Jain 1992) with permission; ©1992 IEEE.

Chapter 8

Parameter Estimation in Optimal Object Recognition

Object recognition systems involve parameters such as thresholds, bounds, and weights. These parameters have to be tuned before the system can perform successfully. A common practice is to choose such parameters manually on an ad hoc basis, which is a disadvantage. This chapter[1] presents a theory of parameter estimation for optimization-based object recognition where the optimal solution is defined as the global minimum of an energy function. The theory is based on supervised learning from training examples. *Correctness* and *instability* are established as criteria for evaluating the estimated parameters. A correct estimate enables the labeling implied in each example configuration to be encoded in a unique global energy minimum. The instability is the ease with which the minimum is replaced by a nonexample configuration after a perturbation. The optimal estimate minimizes the instability. Algorithms are presented for computing correct and minimal-instability estimates. The theory is applied to the parameter estimation for MRF-based recognition, and promising results are obtained.

8.1 Motivation

Object recognition systems almost inevitably involve parameters such as thresholds, bounds, and weights (Grimson 1990). In optimization-based object recognition, where the optimal recognition solution is explicitly defined as the global extreme of an objective function, these parameters can be part of the definition of the objective function by which the global cost (or gain)

---

[1]This chapter is based on Li (1997b).

of the solution is measured. The selection of the parameters is crucial for a system to perform successfully.

Among all the admissible parameter estimates, only a subset of them lead to the desirable or correct solutions to the recognition. Among all the correct estimates, a smaller number of them are better in the sense that they lead to correct solutions for a larger variety of data sets. One of them may be optimal in the sense that it makes the vision procedure the most stable to uncertainties and the least prone to local optima in the search for the global optimum.

The manual method performs parameter estimation in an ad hoc way by trial and error: A combination of parameters is selected to optimize the objective function, and then the optimum is compared with the desirable result in the designer's perception and the selection is adjusted. This process is repeated until a satisfactory choice, that makes the optimum consistent with the desirable result, is found. This is a process of *supervised learning from examples*. When the objective function takes a right functional form, a correct manual selection may be made for a small number of data sets. However, there is no reason to believe that the manual selection is an optimal or even a good one. Such empirical methods have been criticized for their ad hoc nature.

This chapter aims to develop an automated optimal approach for parameter estimation[2] applied to optimization-based object recognition schemes. A theory of parameter estimation based on supervised learning is presented. The learning is "supervised" because a training set of examples is given. Each example represents a desirable recognition result where a recognition result is a labeling of the scene in terms of the model objects. *Correctness* and *optimality* are proposed as the two-level criteria for evaluating parameters estimates.

A correct selection of parameters enables the configuration given by each example to be embedded as a unique global energy minimum. In other words, if the selection is incorrect, the example configuration will not correspond to a global minimum. While a correct estimate can be learned from the examples, it is generally not the only correct solution. *Instability* is defined as the measure of the ease with which the global minimum is replaced by a nonexample labeling after a perturbation to the input. The optimality minimizes the *instability* so as to maximize the ability to generalize the estimated parameters to other situations not directly represented by the examples.

Combining the two criteria gives a constrained minimization problem: minimize the instability subject to the correctness. A nonparametric algorithm is presented for learning an estimate which is optimal as well as correct. It does not make any assumption about the distributions and is useful for cases where the size of the training example set is small and where the

---

[2]In this chapter, parameter "selection", "estimation" and "learning" are used interchangeably.

underlying parametric models are not accurate. The estimate thus obtained is optimal w.r.t. the training data.

The theory is applied to a specific model of MRF recognition proposed in (Li 1994a). The objective function in this model is the posterior energy of an MRF. The form of the energy function has been derived, but it involves parameters that have to be estimated. The optimal recognition solution is the maximum a posteriori (MAP) configuration of an MRF. Experiments conducted show very promising results in which the optimal estimate serves well for recognizing other scenes and objects.

A parametric method based on *maximum likelihood* is also described for computing the optimal parameter estimate under the Gaussian-MRF assumption. It takes advantage of the assumption and may be useful when the size of the training data is sufficiently large. The parameter estimate thus computed is optimal w.r.t. the assumption.

Although automated and optimal parameter selection for object recognition in high-level vision is an important and interesting problem that has existed for a long time, reports on this topic are rare. Works have been done in related areas. In (Poggio and Edelman 1990), to recognize 3D objects from different viewpoints, a function mapping any viewpoint to a standard view is learned from a set of perspective views. In (Weng et al. 1993), a network structure is introduced for automated learning to recognize 3D objects. In (Pope and Lowe 1993), a numerical graph representation for an object model is learned from features computed from training images. In (Pelillo and Refice 1994) a procedure is proposed for learning compatibility coefficients for relaxation labeling by minimizing a quadratic error function. Automated and optimal parameter estimation for low-Level problems has achieved significant progress. MRF parameter selection has been dealt with in statistics (Besag 1974; Besag 1975) and in applications such as image restoration, reconstruction, and texture analysis (Cross and Jain 1983; Cohen and Cooper 1987; Derin and Elliott 1987; Qian and Titterington 1989; Zhang 1988; Nadabar and Jain 1992). The problem is also addressed from the regularization viewpoint (Wahba 1980; Geiger and Poggio 1987; Shahraray and Anderson 1989; Thompson et al. 1991).

The chapter is organized as follows. Section 8.2 presents the theory. Section 8.3 applies the theory to an MRF recognition model. Section 8.4 presents the experimental results. Finally, conclusions are made in Section 8.5.

8.2 Theory of Parameter Estimation for Recognition

In this section, correctness, instability, and optimality are proposed for evaluating parameter estimates. Their relationships to nonparametric pattern recognition are discussed, and nonparametric methods for computing correct and optimal estimates are presented. Before preceeding, necessary notations for optimization-based object recognition are introduced.

8.2.1   Optimization-Based Object Recognition

In optimization-based recognition, the optimal solution is explicitly defined as the extreme of an objective function. Let f be a configuration representing a recognition solution. The cost of f is measured by a global objective function $E(f \mid \theta)$, also called the energy. The definition of E is dependent on f and a number of $K + 1$ parameters $\theta = [\theta_0, \theta_1, \ldots, \theta_K]^T$. As the optimality criterion for model-based recognition, it also relates to other factors such as the observation, denoted $\mathcal{G}$, and model references, denoted $\mathcal{G}'$. Given $\mathcal{G}$, $\mathcal{G}'$, and θ, the energy maps a solution f to a real number by which the cost of the solution is evaluated. The optimal solution corresponds to the global energy minimum, expressed as

$$f^* = \arg \min_f E(f \mid \theta) \qquad (8.1)$$

In this regard, it is important to formulate the energy function so that the "correct solution" is embedded as the global minimum. The energy may also serve as a guide to the search for a minimal solution. In this respect, it is desirable that the energy should differentiate the global minimum from other configurations as much as possible.

The energy function may be derived using one of the following probabilistic approaches: fully parametric, partially parametric, and nonparametric. In the fully parametric approach, the energy function is derived from probability distributions in which all the parameters involved are known. Parameter estimation is a problem only in the partially parametric and nonparametric cases.

In the partially parametric case, the forms of distributions are given but some parameters involved are unknown. One example is the Gaussian distribution with an unknown variance, and another is the Gibbs distribution with unknown clique potential parameters. In this case, the problem of estimating parameters in the objective function is related to estimating parameters in the related probability distributions.

In the nonparametric approach, no assumptions are made about distributions and the form of the objective function is obtained based on experiences or prespecified "basis functions" (Poggio and Edelman 1990). This also applies to situations where the data set is too small to have statistical significance.

An important form for E in object recognition is the weighted sum of various terms, expressed as

$$E(f \mid \theta) = \theta^T U(f) = \sum_{k=0}^{K} \theta_k U_k(f) \qquad (8.2)$$

where $U(f) = [U_0(f), U_1(f), \ldots, U_K(f)]^T$ is a vector of *potential functions*. A potential function is dependent on f, $\mathcal{G}$, and $\mathcal{G}'$, where the dependence can be nonlinear in f, and often measures the violation of a certain constraint incurred by the solution f. This linear combination of (nonlinear) potential functions is not an unusual form. It has been used in many matching and recognition works; see (Duda and Hart 1973; Fischler and Elschlager 1973; Davis 1979; Ghahraman et al. 1980; Jacobus et al. 1980; Shapiro and Haralick 1981; Oshima and Shirai 1983; Bhanu and Faugeras 1984; Wong and You 1985; Fan et al. 1989; Nasrabadi et al. 1990; Wells 1991; Weng et al. 1992; Li 1994a). Note that when $E(f \mid \theta)$ takes the linear form, multiplying θ by a positive factor $\kappa > 0$ does not change the minimal configuration

$$\arg\min_f E(f \mid \theta) = \arg\min_f E(f \mid \kappa\theta) \qquad (8.3)$$

Because of this equivalent, an additional constraint should be imposed on θ for the uniqueness. In this work, θ is confined to having a unit Euclidean length

$$\|\theta\| = \sqrt{\sum_{k=0}^{K} \theta_k^2} = 1 \qquad (8.4)$$

Given an observation $\mathcal{G}$, a model reference $\mathcal{G}'$, and the form of $E(f \mid \theta)$, it is the θ value that completely specifies the energy function $E(f \mid \theta)$ and thereby defines the minimal solution f^*. It is desirable to learn the parameters from examples so that the minimization-based recognition is performed correctly. The criteria for this purpose are established in the next subsection.

8.2.2 Criteria for Parameter Estimation

An example is specified by a triple $(\bar{f}, \mathcal{G}, \mathcal{G}')$, where $\bar{f}$ is the example configuration (recognition solution) telling how the scene ($\mathcal{G}$) should be labeled or interpreted in terms of the model reference ($\mathcal{G}'$). The configuration $\bar{f}$ may be a structural mapping from $\mathcal{G}$ to $\mathcal{G}'$. Assume that there are L model objects; then at least L examples have to be used for learning to recognize the L object. Let the instances be given as

$$\{(\bar{f}^\ell, \mathcal{G}^\ell, \mathcal{G}'^\ell) \mid \ell = 1, \ldots, L\} \qquad (8.5)$$

We propose two-level criteria for learning θ from examples:

1. *Correctness*. This defines a parameter estimate that encodes constraints into the energy function in a correct way. A correct estimate, denoted $\theta_{correct}$, should embed each $\bar{f}^\ell$ into the minimum of the corresponding energy $E(f^\ell \mid \theta_{correct})$, that is

$$\bar{f}^\ell = \arg\min_f E^\ell(f \mid \theta_{correct}) \qquad \forall \ell \qquad (8.6)$$

where the definition of $E^\ell(f \mid \theta)$ is dependent on the given scene $\mathcal{G}^\ell$ and the model $\mathcal{G}'^\ell$ of a particular example $(\bar{f}^\ell, \mathcal{G}^\ell, \mathcal{G}'^\ell)$ as well as θ. Briefly, a correct θ is one that makes the minimal configuration f^* defined in (8.1) coincide with the example configuration $\bar{f}^\ell$.

2. *Optimality.* This is aimed at maximizing the generalizability of the parameter estimate to other situations. A measure of *instability* is defined for the optimality and is to be minimized.

The correctness criterion is necessary for a vision system to perform correctly and is of fundamental importance. Only when this is met does the MRF model make correct use of the constraints. The optimality criterion is not necessary in this regard but makes the estimate most generalizable.

Correctness

In (8.6), a correct estimate $\theta_{correct}$ enables the $\bar{f}$ to be encoded as the global energy minimum for this particular pair $(\mathcal{G}, \mathcal{G}')$. Therefore, it makes any energy change due to a configuration change from $\bar{f}$ to $f \neq \bar{f}$ positive; that is,

$$
\begin{aligned}
\Delta E^\ell(f \mid \theta) &= E^\ell(f \mid \theta) - E^\ell(\bar{f}^\ell \mid \theta) &\qquad (8.7)\\
&= \sum_k \theta_k [U_k(f) - U_k(\bar{f}^\ell)] > 0 &\qquad \forall f \in \mathcal{F}^\ell
\end{aligned}
$$

where

$$
\mathcal{F}^\ell = \{f \mid f \neq \bar{f}^\ell\} \qquad (8.8)
$$

is the set of all non-$\bar{f}^\ell$ configurations. Let

$$
\Theta_{correct} = \{\theta \mid \Delta E^\ell(f \mid \theta) > 0, \forall f \in \mathcal{F}^\ell, \ell = 1, \ldots, L\} \qquad (8.9)
$$

be the set of all correct estimates. The set, if non-empty, usually comprises not just a single point but a region in the allowable parameter space. Some of the points in the region are better in the sense of stability, to be defined below.

Instability

The value of the energy change $\Delta E^\ell(f \mid \theta) > 0$ can be used to measure the (local) stability of $\theta \in \Theta_{correct}$ w.r.t. a certain configuration $f \in \mathcal{F}^\ell$. Ideally, we want $E^\ell(\bar{f}^\ell \mid \theta)$ to be very low and $E^\ell(f \mid \theta)$ to be very high, such that $\Delta E^\ell(f \mid \theta)$ is very large, for all $f \neq \bar{f}^\ell$. In such a situation, $\bar{f}^\ell$ is expected to be a stable minimum where the stability is w.r.t. perturbations in the observation and w.r.t. the local minimum problem with the minimization algorithm.

The smaller $\Delta E^\ell(f \mid \theta)$ is, the larger is the chance with which a perturbation to the observation will cause $\Delta E^\ell(f \mid \theta_{correct})$ to become negative to violate the correctness. When $\Delta E^{\ell}(f \mid \theta_{correct}) < 0$, $\bar{f}^{\ell}$ no longer corresponds to the global minimum. Moreover, we assume that configurations f whose energies are slightly higher than $E^{\ell}(\bar{f}^{\ell} \mid \theta)$ are possibly *local energy minima* at which an energy minimization algorithm is most likely to get stuck.

Therefore, the energy difference (i.e., the local stabilities) should be enlarged. One may define the global stability as the sum of all $\Delta E^{\ell}(f \mid \theta)$. For reasons to be explained later, *instability*, instead of stability, is used for evaluating θ.

The local instability for a correct estimate $\theta \in \Theta_{correct}$ is defined as

$$c^{\ell}(\theta, f) = \frac{1}{\Delta E^{\ell}(f \mid \theta)} \tag{8.10}$$

where $f \in \mathcal{F}^{\ell}$. It is "local" because it considers only one $f \in \mathcal{F}^{\ell}$. It is desirable to choose θ such that the value of $c^{\ell}(\theta, f)$ is small *for all* $f \in \mathcal{F}^{\ell}$. Therefore, we defined the global *p-instability* of θ

$$C_p^{\ell}(\theta) = \left\{ \sum_{f \in \mathcal{F}^{\ell}} [c^{\ell}(\theta, f)]^p \right\}^{1/p} \tag{8.11}$$

where $p \geq 1$. The total global p-instability of θ is

$$C_p(\theta) = \sum_{\ell=1}^{L} C_p^{\ell}(\theta) \tag{8.12}$$

In the limit as $p \to \infty$, we have that[3]

$$C_{\infty}^{\ell}(\theta) = \max_{f \in \mathcal{F}^{\ell}} c^{\ell}(\theta, f) = \frac{1}{\min_{f \in \mathcal{F}^{\ell}} \Delta E^{\ell}(f \mid \theta)} \tag{8.13}$$

is due solely to f having the smallest $c^{\ell}(\theta, f)$ or largest $\Delta E^{\ell}(f \mid \theta)$ value.

Unlike the global stability definition, the global instability treats each item in the following manner: Those f having smaller $\Delta E^{\ell}(f \mid \theta)$ (larger $c^{\ell}(\theta, f)$) values affect $C_p^{\ell}(\theta)$ in a more significant way. For $p = 2$, for example, the partial derivative is

$$\frac{\partial C_2^{\ell}(\theta)}{\partial \theta_k} = \frac{\partial C_2^{\ell}(\theta)}{\partial \Delta E^{\ell}(f \mid \theta)} \frac{\partial \Delta E^{\ell}(f \mid \theta)}{\partial \theta_k} \tag{8.14}$$

$$= \frac{1}{[\Delta E^{\ell}(f \mid \theta)]^3} [U_k(f) - U_k(\bar{f}^{\ell})] \tag{8.15}$$

where $E^{\ell}(f \mid \theta)$ takes the linear form (8.2). The smaller the $\Delta E^{\ell}(f \mid \theta)$ is, the more it affects θ. This is desirable because such f are more likely than the others to violate the correctness, because their $\Delta E^{\ell}(f \mid \theta)$ values are small, and should be more influential in determining θ.

---

[3]This is because, for the p-norm defined by $\|y\|_p = (|y_1|^p + |y_2|^p + \cdots + |y_n|^p)^{1/p}$, we have $\lim_{p \to \infty} \|y\|_p = \max_j |y_j|$.

Optimality

The optimal estimate is defined as the one in $\Theta_{correct}$ that minimizes the instability

$$\bar{\theta} = \arg \min_{\theta \in \Theta_{correct}} C_p(\theta) \qquad (8.16)$$

Obviously, $C_p(\theta)$ is positive for all $\theta \in \Theta_{correct}$, and hence the minimal solution always exists. The minimal solution tends to increase $\Delta E(f \mid \theta)$ values in the global sense and thus maximizes the extent to which an example configuration $\bar{f}$ remains to be the global energy minimum when the observation d is perturbed. It is also expected that with such a $\bar{\theta}$, local minima corresponding to some low-energy-valued f are least likely to occur in minimization.

The correctness in (8.7), instability in (8.11), and optimality in (8.16) are defined without specifying the form of the energy $E(f \mid \theta)$. Therefore, the principle established so far is general for any optimization-based recognition models. Minimizing the instability with the constraint of the correctness is a nonlinear programming problem when the instability is nonlinear in θ.

For conciseness, in the following, the superscript ℓ will be omitted most of the time unless necessary.

8.2.3   Linear Classification Function

In this work, we are interested in cases where the $E(f \mid \theta)$ is linear in θ. Assume that, with a nonparametric or partially parametric modeling method, the energy is derived to take the linear form (8.2). With the linear form, the energy change can be written as

$$\Delta E(f \mid \theta) = \theta^T x(f) \qquad (8.17)$$

where

$$x(f) = [x_0(f), x_1(f), \ldots, x_K(f)]^T = U(f) - U(\bar{f}) \qquad (8.18)$$

is the potential change. Denote the set of all potential changes by

$$\mathcal{X} = \{x(f) \mid f \in \mathcal{F}\} \qquad (8.19)$$

Note that $\mathcal{X}$ *excludes* $x(\bar{f})$, the vector of zeros. The set $\mathcal{X}$ will be used as the *training data set*. When there are $L > 1$ examples,

$$\mathcal{X} = \{x^\ell(f) = U^\ell(f) - U^\ell(\bar{f}^\ell) \mid f \in \mathcal{F}^\ell, \forall \ell\} \qquad (8.20)$$

contains training data from all the instances. $\mathcal{X}$ is the $K+1$-dimensional data space in which x's are points.

The correctness (8.7) can be written as

$$\theta^T x(f) > 0 \qquad \forall x \in \mathcal{X} \qquad (8.21)$$

In pattern recognition, $\theta^T x$ is called a *linear classification function* when considered as a function of x; $\theta^T x(f)$ is called a *generalized linear classification*

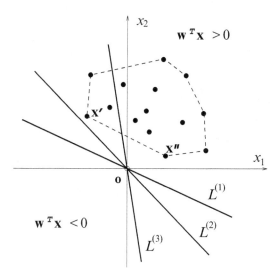

Figure 8.1: Correct and incorrect parameter estimates. A case where the C_∞-optimal hyperplane is parallel to $\overline{x'x''}$. From (Li 1997b) with permission; ©1997 Kluwer.

function when considered as a function of f. There exist useful theories and algorithms for linear pattern classification functions (Duda and Hart 1973).

The equation $\theta^T x = 0$ is a hyperplane in the space $\mathcal{X}$, passing through the origin $x(\bar{f}) = \mathbf{0}$. With a correct θ, the hyperplane divides the space into two parts, with all $x(f)$ ($f \in \mathcal{F}$) on one side, more exactly the "positive" side, of the hyperplane. The Euclidean distance from x to the hyperplane is equal to $\theta^T x / \|\theta\|$, a *signed* quantity. After the normalization (8.4), the point-to-hyperplane distance is just $\theta^T x(f)$.

The correctness can be judged by checking the minimal distance from the point set $\mathcal{X}$ to the hyperplane. We define the "separability" as the smallest distance value

$$S(\theta) = \min_f \theta^T x(f) / \|\theta\| \tag{8.22}$$

and it can also be considered as the stability of the system with a given θ. The correctness is equivalent to the positivity of the separability.

It is helpful to visually illustrate the optimality using C_∞. With C_∞, the minimal-instability solution is the same as the minimax solution

$$\bar{\theta} = \arg \min_{\theta \in \Theta_{correct}} C_\infty(\theta) = \arg \min_{\theta \in \Theta_{correct}} \left[\max_{f \in \mathcal{F}} \Delta E(f \mid \theta) \right] \tag{8.23}$$

In this case, $S(\theta) = 1/C_\infty(\theta)$ and the above is maximal separability.

Figure 8.1 qualitatively illustrates correct/incorrect and maximal separability parameters. It is an example in two-dimensional space where an energy

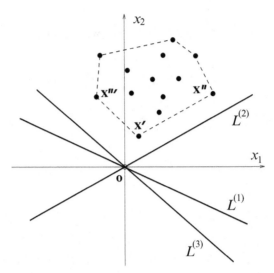

Figure 8.2: A case where the optimal hyperplane is perpendicular to $\overline{ox'}$. From (Li 1997b) with permission; ©1997 Kluwer.

change takes the form of $\Delta E(f \mid \theta) = \theta^T x(f) = \theta_1 x_1 + \theta_2 x_2$. The point $x(\bar{f}) = \mathbf{0}$ coincides with the origin of the x_1–x_2 space. Data $x(f)$ $(f \in \mathcal{F})$ are shown as filled dots. The three lines $L^{(1)}$, $L^{(2)}$, and $L^{(3)}$ represent three hyperplanes, corresponding to three different estimates of parameters θ. Parameter estimates for $L^{(1)}$ and $L^{(2)}$ are correct ones because they make all the data points on the positive side of the hyperplane and thus satisfy (8.7). However, $L^{(3)}$ is not a correct one. Of the two correct estimates in Fig. 8.1, the one corresponding to $L^{(1)}$ is better than the other in terms of C_∞.

The separability determines the range of disturbances in x within which the example configuration $\bar{f}$ remains minimal. Refer to point $x' = x(f')$ in the figure. Its distance to $L^{(2)}$ is the smallest. The point may easily deviate across $L^{(2)}$ to the negative half space due to some perturbation in the observation d. When this happens, $\Delta E(f \mid \theta) < 0$, causing $E(f' \mid \theta)$ to be lower than $E(\bar{f} \mid \theta)$. This means that $\bar{f}$ is no longer the energy minimum. If the parameters are chosen as those corresponding to $L^{(1)}$, the separability is larger and violation of the correctness is less likely to happen.

The dashed polygon in the figure forms the convex hull (polytope) of the data set. Only those data points that form the hull affect the minimax solution whereas those inside the hull are ineffective. The ineffective data points inside the hull can be removed and only the effective points, the number of which may be small compared with the whole data set, need be considered in the solution-finding process. This increases the efficiency.

There are two possibilities for the orientation of the maximal-separability hyperplane w.r.t. the polytope. The first is that the optimal hyperplane is parallel to one of the sides (faces in cases of 3D or higher-dimensional parameter space) of the polytope, which is the case in Fig. 8.1. The other is that the optimal hyperplane is perpendicular to the line linking the origin ∅ and the nearest data point, which is the case in Fig. 8.2. This is a property we may use to find the minimax solution. When the points constituting the polytope are identified, all the possible solutions can be enumerated; when there are only a small number of them, we can find the minimax solution by an exhaustive comparison. In Fig. 8.2, $L^{(2)}$ is parallel to $\overline{x'x''}$ and $L^{(3)}$ is parallel to $\overline{x'x'''}$, but $L^{(1)}$ is perpendicular to $\overline{0x'}$. Let θ_1, θ_2 and θ_3 be the sets of parameters corresponding to $L^{(1)}$, $L^{(2)}$ and $L^{(3)}$. Suppose that in the figure the following relations hold: $S(\theta_1) > S(\theta_2)$ and $S(\theta_1) > S(\theta_3)$. Then θ_1 is the best estimate because it maximizes the separability.

The minimax solution (8.23) with $p = \infty$ was used in the above only for illustration. With $p = \infty$, a continuous change in x may lead to a discontinuous change in the C_∞-optimal solution. This is due to the decision-making nature of the definition which may cause discontinuous changes in the minimum. We use other p-instability definitions, with $p = 1$ or 2 for example, because they give more stable estimates.

8.2.4   A Nonparametric Learning Algorithm

Consider the case where $p = 2$. Because $\theta^T x \geq 0$, minimizing $C_2(\theta)$ is equivalent to minimizing its square,

$$[C_2(\theta)]^2 = \sum_{x \in \mathcal{X}} \frac{1}{(\theta^T x)^2} \qquad (8.24)$$

The problem is given formally as

$$\begin{array}{ll} \text{minimize} & \sum_{x \in \mathcal{X}} \frac{1}{(\theta^T x)^2} \\ \text{subject to} & \|\theta\| = 1 \\ & \theta^T x > 0 \quad \forall x \in \mathcal{X} \end{array} \qquad (8.25)$$

This is a nonlinear programming problem and can be solved using the standard techniques. In this work, a gradient-based, nonparametric algorithm is used to obtain a numerical solution.

The perceptron algorithm (Rosenblatt 1962) has already provided a solution for learning a correct parameter estimate. It iterates on θ to increase the objective of the form $\theta^T x$ based on the gradient information, where the gradient is simply x. In one cycle, all $x \in \mathcal{X}$ are tried in turn to adjust θ. If $\theta^T x > 0$, meaning x is correctly classified under θ, no action is taken. Otherwise, if $\theta^T x \leq 0$, θ is adjusted according to the gradient-ascent rule

$$\theta \longleftarrow \theta + \mu x \qquad (8.26)$$

where μ is a small constant. The update is followed by a normalization operation

---

Algorithm learning(θ, $\mathcal{X}$)
/* *Learning correct and optimal θ from the training data $\mathcal{X}$* /
Begin Algorithm
      initialize(θ);
      do {
            $\theta_{last} \leftarrow \theta$;
            if ($\theta^T x > 0 \;\; \exists x \in \mathcal{X}$) {
                correct(θ);
                $\theta \leftarrow \theta/\|\theta\|$;
            }
            $\theta \longleftarrow \theta + \mu \sum_{x \in \mathcal{X}} \frac{x}{(\theta^T x)^3}$;
            $\theta \leftarrow \theta/\|\theta\|$;
      } until ($\|\theta_{last}^T - \theta\| < \epsilon$);
      return($\bar{\theta} = \theta$);
End Algorithm

Figure 8.3: Algorithm for finding the optimal combination of parameters. From (Li 1997b) with permission; ©1997 Kluwer.

$$\theta \leftarrow \theta/\|\theta\| \tag{8.27}$$

when $\|\theta\| = 1$ is required. The cycle is repeated until a correct classification is made for all the data and thus a correct θ is learned. With the assumption that the solution exists, also meaning that the data set $\mathcal{X}$ is linearly separable from the origin of the data space, the algorithm is guaranteed to converge after a finite number of iterations (Duda and Hart 1973).

The objective function $[C_2(\theta)]^2 = \sum_{x \in \mathcal{X}} \frac{1}{(\theta^T x)^2}$ can be minimized in a similar way. The gradient is

$$\nabla [C_2(\theta)]^2 = -2 \sum_{x \in \mathcal{X}} \frac{x}{(\theta^T x)^3} \tag{8.28}$$

An update goes as

$$\theta \longleftarrow \theta + \mu \sum_{x \in \mathcal{X}} \frac{x}{(\theta^T x)^3} \tag{8.29}$$

where μ is a small constant. If θ were unconstrained, the above might diverge when $(\theta^T x)^3$ becomes too small. However, the update is again followed by the normalization $\|\theta\| = 1$. This process repeats until θ converges to $\bar{\theta}$.

In our implementation, the two stages are combined into one procedure as shown in Fig. 8.3. In the procedure, initialize(θ) sets θ at random, correct(θ)

learns a correct θ from $\mathcal{X}$, and ($\|\theta_{last}^T - \theta\| < \epsilon$), where $\epsilon > 0$ is a small number, verifies the convergence. The algorithm is very stable.

The amount of change in each minimization iteration is $\mu \sum_{x \in \mathcal{X}} \frac{x}{(\theta^T x)^3}$. The influence from x is weighted by $1/(\theta^T x)^3$. This means that those x with smaller $\theta^T x$ values (closer to the hyperplane) have bigger force in pushing the hyperplane away from themselves, whereas those with big $\theta^T x$ values (far away from the hyperplane) have small influence. This effectively stabilizes the learning process.

When $p = \infty$, we are facing the minimax problem (8.23). An algorithm for solving this is the "generalized portrait technique" (Vapnik 1982), which is designed for constructing hyperplanes with maximum separability. It is extended by Boser, Guyon, and Vapnik (1992) to train classifiers of the form $\theta^T \varphi(x)$, where $\varphi(x)$ is a vector of functions of x. The key idea is to transform the problem into the dual space by means of the Lagrangian. This gives a quadratic optimization with constraints. The optimal parameter estimate is expressed as a linear combination of supporting patterns, where the supporting patterns correspond to the data points nearest to the hyperplane. Two benefits are gained from this method: There are no local minima in the quadratic optimization and the maximum separability obtained is insensitive to small changes of the learned parameters.

The θ computed using the nonparametric procedure is optimal w.r.t. the training data $\mathcal{X}$. It is the best result that can be obtained from $\mathcal{X}$ for generalization to other data. Better results may be obtained, provided that more knowledge about the training data is available.

8.2.5 Reducing Search Space

The data set $\mathcal{X}$ in (8.20) may be very large because there are a combinatorial number of possible configurations in $\mathcal{F} = \{f \neq \bar{f}\}$. In principle, all $f \neq \bar{f}$ should be considered. However, we assume that the configurations thay are neighboring $\bar{f}$ have the largest influence on the selection of θ. Define the neighborhood of $\bar{f}$ as

$$\mathcal{N}_{\bar{f}} = \{f = (f_1, \ldots, f_m) \mid f_i \neq \bar{f}_i, f_i \in \mathcal{L}, \exists^1 i \in \mathcal{S}\} \tag{8.30}$$

where $\exists^1$ reads "one and only one exists" and $\mathcal{L}$ is the set of admissible labels for every f_i. The $\mathcal{N}_{\bar{f}}$ consists of all $f \in \mathcal{F}$ that differ from $\bar{f}$ by one and only one component. This confinement reduces the search space to an enormous extent. After the configuration space is confined to $\mathcal{N}_{\bar{f}}$, the set of training data is computed as

$$\mathcal{X} = \{x = U(f) - U(\bar{f}) \mid f \in \mathcal{N}_{\bar{f}}\} \tag{8.31}$$

which is much smaller than the $\mathcal{X}$ in (8.20).

8.3 Application in MRF Object Recognition

The theory is applied to the parameter estimation for MRF object recognition where the form of the energy is derived based on MRF's. MRF modeling provides one approach to optimization-based object recognition (Modestino and Zhang 1989; Cooper 1990; Baddeley and van Lieshout 1992; Kim and Yang 1992; Li 1994a). The MAP solution is usually sought. The posterior distribution of configurations f is of Gibbs type

$$P(f \mid d) = Z^{-1}e^{-E(f \mid \theta)} \qquad (8.32)$$

where Z is a normalizing constant called the partition function and $E(f \mid \theta)$ is the posterior energy function measuring the global cost of f. In the following, $E(f \mid \theta)$ is defined and converted to the linear form $\theta^T U(f)$.

8.3.1 Posterior Energy

An object or a scene is represented by a set of features where the features are attributed by their properties and constrained to one another by contextual relations. Let a set of m features (sites) in the scene be indexed by $\mathcal{S} = \{1, \ldots, m\}$, a set of M features (labels) in the considered model object by $\mathcal{L} = \{1, \ldots, M\}$, and everything in the scene not modeled by labels in $\mathcal{L}$ by $\{0\}$ which is a virtual NULL label. The set union $\mathcal{L}^+ = \mathcal{L} \cup \{0\}$ is the augmented label set. The structure of the scene is denoted by $\mathcal{G} = (\mathcal{S}, d)$ and that of the model object by $\mathcal{G}' = (\mathcal{L}, D)$, where d denotes the visual constraints on features in $\mathcal{S}$ and D describes the visual constraints on features in $\mathcal{L}$, where the constraints can be, for example, properties and relations between features.

Let object recognition be posed as assigning a label from $\mathcal{L}^+$ to each of the sites in $\mathcal{S}$ so as to satisfy the constraints. The labeling (configuration) of the sites is defined by $f = \{f_1, \ldots, f_m\}$, in which $f_i \in \mathcal{L}^+$ is the label assigned to i. A pair $(i \in \mathcal{S}, f_i \in \mathcal{L}^+)$ is a match or correspondence. Under contextual constraints, a configuration f can be interpreted as a mapping from the structure of the scene $\mathcal{G} = (\mathcal{S}, d)$ to the structure of the model object $\mathcal{G}' = (\mathcal{L}, D)$. Therefore, such a mapping is denoted as a triple $(f, \mathcal{G}, \mathcal{G}')$.

The observation $d = (d_1, d_2)$, which is the feature extracted from the image, consists of two sources of constraints, unary properties d_1 for single-site features, such as color and size, and binary relations d_2 for pair-site features, such as angle and distance. More specifically, each site $i \in \mathcal{S}$ is associated with a set of K_1 properties $\{d_1^{(k)}(i) \mid k = 1, \ldots, K_1, i \in \mathcal{S}\}$ and each pair of sites with a set of K_2 relations $\{d_2^{(k)}(i, i') \mid k = 1, \ldots, K_2; i, i' \in \mathcal{S}\}$. In the model object library, we have model features $\{D_1^{(k)}(I) \mid k = 1, \ldots, K_1, I \in \mathcal{L}\}$ and $\{D_2^{(k)}(I, I') \mid k = 1, \ldots, K_2; I, I' \in \mathcal{L}\}$ (note that $\mathcal{L}$ excludes the NULL label). According to (4.14), under the labeling f, the observation d is a noise-contaminated version of the corresponding model features D

$$d_1(i) = D_1(f_i) + e(i), \qquad d_1(i, i') = D_2(f_i, f_{i'}) + e(i, i') \qquad (8.33)$$

where $f_i, f_{i'} \neq 0$ are nonNULL matches and e is a white Gaussian noise; that is, $d_1(i)$ and $d_2(i, i')$ are white Gaussian distributions with conditional means $D_1(f_i)$ and $D_2(f_i, f_{i'})$, respectively.

The posterior energy $E(f) = U(f \mid d)$ takes the form shown in (4.18), rewritten as

$$
\begin{aligned}
E(f) = \quad & \sum_{i \in \mathcal{S}} V_1(f_i) + \\
& \sum_{i \in \mathcal{S}} \sum_{i' \in \mathcal{N}_i} V_2(f_i, f_{i'}) + \\
& \sum_{i \in \mathcal{S}: f_i \neq 0} V_1(d_1(i) \mid f_i) + \\
& \sum_{i \in \mathcal{S}: f_i \neq 0} \sum_{i' \in \mathcal{S} - \{i\}: f_{i'} \neq 0} V_2(d_2(i, i') \mid f_i, f_{i'})
\end{aligned}
\qquad (8.34)
$$

The first and second summations are due to the joint prior probability of the MRF labels f; the third and fourth are due to the conditional p.d.f. of d or the likelihood of f, respectively. Refer to (4.11), (4.12), (4.16), and (4.17).

8.3.2 Energy in Linear Form

The parameters involved in $E(f)$ are the noise variances $[\sigma_n^{(k)}]^2$ and the prior penalties v_{n0} ($n = 1, 2$). Let the parameters be denoted uniformly by $\theta = \{\theta_n^{(k)} \mid k = 0, \ldots, K_n, n = 1, 2\}$. For $k = 0$,

$$\theta_n^{(0)} = v_{n0} \qquad (8.35)$$

and for $k \geq 1$,

$$\theta_n^{(k)} = (2[\sigma_n^{(k)}]^2)^{-1} \qquad (8.36)$$

Note all $\theta_n^{(k)} \geq 0$. Let the different energy components be uniformly denoted by $U = \{U_n^{(k)} \mid k = 0, \ldots, K_n, n = 1, 2\}$. For $k = 0$,

$$U_1^{(0)}(f) = N_1 = \#\{f_i = 0 \mid i \in \mathcal{S}\} \qquad (8.37)$$

is the number of NULL labels in f and

$$U_2^{(0)}(f) = N_2 = \#\{f_i = 0 \text{ or } f_{i'} = 0 \mid i \in \mathcal{S}, i' \in \mathcal{N}_i\} \qquad (8.38)$$

is the number of label pairs, at least one of which is NULL . For $k \geq 1$, $U_n^{(k)}(f)$ relates to the likelihood energy components; They measure how much the observations $d_n^{(k)}$ deviate from the values $D_n^{(k)}$ that should-be true under f:

$$U_1^{(k)}(f) \triangleq U_1^{(k)}(d \mid f) = \sum_{i \in \mathcal{S}, f_i \neq 0} [d_1^{(k)}(i) - D_1^{(k)}(f_i)]^2 / \{2[\sigma_1^{(k)}]^2\} \qquad (8.39)$$

and

$$U_2^{(k)}(f) \stackrel{\triangle}{=} U_2^{(k)}(d \mid f) \tag{8.40}$$

$$= \sum_{i \in \mathcal{S}, f_i \neq 0} \sum_{i' \in \mathcal{S}, i' \neq i, f_{i'} \neq 0} [d_2^{(k)}(i,i') - D_2^{(k)}(f_i, f_{i'})]^2 / \{2[\sigma_2^{(k)}]^2\}$$

After some manipulation, the energy can be written as

$$E(f \mid \theta) = \sum_{n=1}^{2} \sum_{k=0}^{K_n} \theta_n^{(k)} U_n^{(k)}(f) = \theta^T U(f) \tag{8.41}$$

where θ and $U(f)$ are column vectors of $K_1 + K_2 + 2$ components. Given an instance $(\bar{f}, \mathcal{G}, \mathcal{G}')$, the $U(\bar{f})$ is a *known* vector of real numbers. The θ is the vector of *unknown* weights to be determined. The stability follows immediately as

$$\Delta E(f \mid \theta) = \theta^T x(f) \tag{8.42}$$

where $x(f) = U(f) - U(\bar{f})$.

Some remarks on θ, $U(f)$, and E are in order. Obviously, all $\theta_n^{(k)}$ and $U_n^{(k)}$ are nonnegative. In the ideal case of exact (possibly partial) matching, all $U_n^{(k)}(f)$ $(k \geq 1)$ are zeros because $d_1^{(k)}(i)$ and $D_1^{(k)}(f_i)$ are exactly the same and so are $d_2^{(k)}(i,i')$ and $D_2^{(k)}(f_i, f_{i'})$. In the general case of inexact matching, the sum of the $U_n^{(k)}$ should be as small as possible for the minimal solution. The following are some properties of E:

- Given f, $E(f \mid \theta)$ is linear in θ. Given θ, it is linear in $U(f)$.

- For $\kappa > 0$, θ and $\kappa\theta$ are equivalent, as has been discussed.

- The values of $\theta_n^{(0)}$ relative to those of $\theta_n^{(k)}$ $(k \geq 1)$ affect the rate of NULL labels. The higher the penalties $\theta_n^{(0)}$ are, the more sites in $\mathcal{S}$ will be assigned nonNULL labels and vice versa.

The first property enables us to use the results we established for linear classifiers in learning the correct and optimal θ. According to the second property, a larger $\theta_n^{(k)}$ relative to the rest makes the constraints $d_n^{(k)}$ and $D_n^{(k)}$ play a more important role. Useless and misleading constraints $d_n^{(k)}$ and $D_n^{(k)}$ should be weighted by 0. Using the third property, one can decrease $\theta_n^{(0)}$ values to increase the number of the NULL labels. This is because for the minimum energy matching, a lower cost for NULL labels makes more sites labeled NULL , which is equivalent to discarding more not so reliable nonNULL labels into the NULL bin.

8.3.3 How the Minimal Configuration Changes

The following analysis examines how the minimum $f^* = \arg \min_f E(f \mid \theta)$ changes as the observation changes from $d0$ to $d = d0 + \delta d$, where δd is a perturbation. In the beginning, when $\|\delta d\|$ is close to 0, f^* should remain as the minimum for a range of such small δd. This is simply because $E(f \mid \theta)$ is continuous w.r.t. d. When the perturbation becomes larger and larger, the minimum has to give way to another configuration.

When should a change happen? To see the effect more clearly, assume the perturbation is in observation components related to only a particular i so that the only changes are $d0_1^{(k)}(i) \rightarrow d_1^{(k)}(i)$ and $d0_2^{(k)}(i, i') \rightarrow d_2^{(k)}(i, i')$, $\forall i' \in \mathcal{N}_i$. First, assume that f_i^* is a nonNULL label ($f_i^* \neq 0$) and consider such a perturbation δd that incurs a *larger* likelihood potential. Obviously, as the likelihood potential (conditioned on $\{f_{i'}^* \neq 0 \mid i' \in \mathcal{N}_i\}$)

$$V(d \mid f_i^*) = \sum_{k=1}^{K_1} \theta_1^{(k)} (d_1^{(k)}(i) - D_1^{(k)}(f_i^*))^2 + \tag{8.43}$$

$$\sum_{i' \in \mathcal{S}, i' \neq i, f_{i'}^* \neq 0} \sum_{k=1}^{K_2} \theta_2^{(k)} (d_2^{(k)}(i, i') - D_2^{(k)}(f_i^*, f_{i'}^*))^2$$

increases, it will eventually become cheaper for $f_i^* \neq 0$ to change to $f_i = 0$. More accurately, this should happen when

$$V(d \mid f_i^*) > \theta_1^{(0)} + \sum_{i' \in \mathcal{N}_i, f_{i'}^* \neq 0} \theta_2^{(0)} = \theta_1^{(0)} + N_2^i \theta_2^{(0)} \tag{8.44}$$

where

$$N_2^i = \#\{i' \in \mathcal{N}_i \mid f_{i'}^* \neq 0\} \tag{8.45}$$

is the number of nonNULL labeled sites in $\mathcal{N}_i$ under f^*.

Next, assume $f_i^* = 0$ and consider such a perturbation that incurs a *smaller* likelihood potential. The perturbation has to be such that d and D more closely resemble each other. As the conditional likelihood potential

$$V(d \mid f_i) = \sum_{k=1}^{K_1} \theta_1^{(k)} (d_1^{(k)}(i) - D_1^{(k)}(f_i))^2 + \\ \sum_{i' \in \mathcal{S}, i' \neq i, f_{i'}^* \neq 0} \sum_{k=1}^{K_2} \theta_2^{(k)} (d_2^{(k)}(i, i') - D_2^{(k)}(f_i, f_{i'}^*))^2 \tag{8.46}$$

decreases, it will eventually become cheaper for $f_i^* = 0$ to change to one of the nonNULL labels, $f_i \neq 0$. More accurately, this should happen when

$$V(d \mid f_i) < \theta_1^{(0)} + \sum_{i' \in \mathcal{N}_i, f_{i'}^* \neq 0} \theta_2^{(0)} = \theta_1^{(0)} + N_2^i \theta_2^{(0)} \tag{8.47}$$

The analysis above shows how the minimal configuration f^* adjusts as d changes when θ is fixed. On the other hand, the f^* can be maintained unchanged by adjusting θ; this means a different encoding of constraints into E.

8.3.4   Parametric Estimation under Gaussian Noise

Assuming the functional form of the noise distribution is known, then we can take advantage of (partial) parametric modeling for the estimation. When the noise is additive white Gaussian with unknown $[\sigma_n^{(k)}]^2$, the estimate can be obtained in closed form. The closed form estimation is performed in two steps. (1) Estimate the noise variances $[\bar{\sigma}_n^{(k)}]^2$ ($k \geq 1$) and then compute the weights $\bar{\theta}_n^{(k)}$ ($k \geq 1$) using the relationship in (8.36). (2) Then compute the allowable $\bar{\theta}_n^{(0)}$ relative to $\bar{\theta}_n^{(k)}$ ($k \geq 1$) to satisfy the correctness in (8.7). Optimization like (8.16) derived using a nonparametric principle may not be applicable in this case.

Given d, D, and $\bar{f}$, the Gaussian noise variances can be estimated by maximizing the joint likelihood function $p(d \mid \bar{f})$ (ML estimation). The ML estimates are simply

$$[\bar{\sigma}_1^{(k)}]^2 = \frac{1}{N_1'} \sum_{i \in \mathcal{S}, \bar{f}_i \neq 0} [d_1^{(k)}(i) - D_1^{(k)}(\bar{f}_i)]^2 \tag{8.48}$$

and

$$[\bar{\sigma}_2^{(k)}]^2 = \frac{1}{N_2'} \sum_{i \in \mathcal{S}, \bar{f}_i \neq 0} \sum_{i' \in \mathcal{S}, i' \neq i, \bar{f}_{i'} \neq 0} [d_2^{(k)}(i, i') - D_2^{(k)}(\bar{f}_i, \bar{f}_i)]^2 \tag{8.49}$$

where

$$N_1' = \#\{i \in \mathcal{S} \mid f_i \neq 0\} \tag{8.50}$$

is the number of nonNULL labels in f and

$$N_2' = \#\{(i, i') \in \mathcal{S}^2 \mid i' \neq i, \ f_i \neq 0, \ f_{i'} \neq 0\} \tag{8.51}$$

is the number of label pairs of which neither is the NULL . The optimal weights for $k \geq 1$ can be obtained immediately by

$$\bar{\theta}_n^{(k)} = 1/2[\bar{\sigma}_n^{(k)}]^2 \tag{8.52}$$

So far, only the example configurations $\bar{f}$, not others, are used in computing the $\bar{\theta}_n^{(k)}$.

Now the remaining problem is to determine $\bar{\theta}_n^{(0)}$ to meet the correctness (8.7). Because $\theta_n^{(0)} = v_{n0}$, this is done to estimate the MRF parameters v_{n0} in the prior distributions implied in the given examples. There may be a *range* of $\bar{\theta}_n^{(0)}$ under which each $\bar{f}$ is correctly encoded. The range is determined by the lower and upper bounds.

In doing so, only those configurations in $\mathcal{N}_{\bar{f}}$ that reflect transitions from a nonNULL to the NULL label and the other way around are needed; the other configurations, which reflect transitions from one nonNULL label to another, are not. This subset is obtained by changing each of the nonNULL labels in $\bar{f}$ to the NULL label or changing each of the NULL labels to a nonNULL label.

First, consider label changes from a nonNULL label to the NULL label. Assume a configuration change from $\bar{f}$ to f is due to the change from $\bar{f}_i \neq 0$ to $f_i = 0$ for just one $i \in \mathcal{S}$. The corresponding energy change is given by

$$\tfrac{1}{2}\Delta E(f \mid \theta) = \theta_1^{(0)} - \sum_{k=1}^{K_1} \bar{\theta}_1^{(k)}[d_1^{(k)}(i) - D_1^{(k)}(\bar{f}_i)]^2 + \\ \sum_{i' \in \mathcal{N}_i, \bar{f}_{i'} \neq 0} \theta_2^{(0)} - \\ \sum_{i' \in \mathcal{S}, i' \neq i, \bar{f}_{i'} \neq 0} \sum_{k=1}^{K_2} \bar{\theta}_2^{(k)}[d_2^{(k)}(i, i') - D_2^{(k)}(\bar{f}_i, \bar{f}_{i'})]^2$$

(8.53)

The change above must be positive, $\Delta E(f \mid \theta) > 0$. Suppose there are N nonNULL labeled sites under $\bar{f}$ and therefore $\bar{f}$ has N such neighboring configurations. Then N such inequalities of $\Delta E(f \mid \theta) > 0$ can be obtained. The two unknowns, $\theta_1^{(0)}$ and $\theta_2^{(0)}$, can be solved for and used as the lower bounds $(\theta_1^{(0)})_{min}$ and $(\theta_2^{(0)})_{min}$.

Similarly, the upper bounds can be computed by considering label changes from the NULL label to a nonNULL label. The corresponding energy change due to a change from $\bar{f}_i = 0$ to $f_i \neq 0$ is given by

$$\tfrac{1}{2}\Delta E(f \mid \theta) = \sum_{k=1}^{K_1} \bar{\theta}_1^{(k)}[d_1^{(k)}(i) - D_1^{(k)}(f_i)]^2 - \theta_1^{(0)} \\ \sum_{i' \in \mathcal{S}, \bar{f}_{i'} \neq 0} \sum_{k=1}^{K_2} \bar{\theta}_2^{(k)}[d_2^{(k)}(i, i') - D_2^{(k)}(f_i, f_{i'})]^2 - \\ \sum_{i' \in \mathcal{N}_i, \bar{f}_{i'} \neq 0} \theta_2^{(0)}$$

(8.54)

The change above must also be positive, $\Delta E(f \mid \theta) > 0$. Suppose there are N NULL labeled sites under $\bar{f}$ and recall that there are M possible nonNULL labels in $\mathcal{L}$. Then $N \times M$ inequalities can be obtained. The two unknowns, $\theta_1^{(0)}$ and $\theta_2^{(0)}$, can be solved and used as the upper bounds $(\theta_1^{(0)})_{max}$ and $(\theta_2^{(0)})_{max}$. If the example configurations $\bar{f}$ are minimal for the corresponding $\mathcal{G}$ and $\mathcal{G}'$, then the solution must be consistent; that is, $(\theta_n^{(0)})_{min} < (\theta_n^{(0)})_{max}$, for each instance.

Now, the space of all correct parameters is given by

$$\Theta_{correct} = \{\theta_n^{(k)} \mid \theta_n^{(0)} \in [(\theta_n^{(0)})_{min}, (\theta_n^{(0)})_{max}]; \ \theta_n^{(k)} = \bar{\theta}_n^{(k)}, k \geq 1\} \quad (8.55)$$

A correct θ makes $\theta^T x > 0$ for all $x \in \mathcal{X}$. The hyperplane $\theta^T x = 0$ partitions $\mathcal{X}$ into two parts, with all $x \in \mathcal{X}$ on the positive side of it. The value for $\bar{\theta}_n^{(0)}$ may simply be set to the average $[(\theta_n^{(0)})_{min} + (\theta_n^{(0)})_{min}]/2$.

When there are $L > 1$ instances, $\theta_n^{(k)}$ ($k \geq 1$) are obtained from the data set computed from all the instances. Given the common $\theta_n^{(k)}$, L correct ranges can be computed. The correct range for the L instances as a whole is the intersection of the L ranges. As a result, the overall $(\theta_n^{(0)})_{min}$ is the maximum of all the lower bounds and $(\theta_n^{(0)})_{max}$ the minimum of all the upper bounds. Although each range can often be consistent (i.e., $(\theta_n^{(0)})_{min} < (\theta_n^{(0)})_{max}$ for each n), there is less of a chance to guarantee that they, as a whole, are consistent for all $\ell = 1, \ldots, L$: The intersection may be empty when $L > 1$.

This inconsistency means a correct estimate does not exist for all the instances as a whole. There are several reasons for this. First of all, the assumptions, such as the model being Gaussian, are not verified by the data set, especially when the data set is small. In this case, the noise in different instances has different variances; when the ranges are computed under the assumption that the ML estimate is common to all instances, they may not be consistent with each other. This is the most direct reason for the inconsistency. Second, $\bar{f}$ in some examples cannot be embedded as the minimal energy configuration to satisfy the given constraints. Such instances are misleading and also cause inconsistency.

8.4 Experiments

The following experiments demonstrate: (i) the computation (learning) of the optimal parameter $\bar{\theta}$ from the examples given in the form of a triplet $(\bar{f}, \mathcal{G}, \mathcal{G}')$, and (ii) the use of the learned estimate $\bar{\theta}$ to recognize other scenes and models. The nonparametric learning algorithm is used because the data size is too small to assume a significant distribution. The convergence of the learning algorithm is demonstrated.

8.4.1 Recognition of Line Patterns

This experiment performs the recognition of simulated objects of line patterns under 2D rotation and translation. There are six possible model objects shown in Fig. 8.4. Figure 8.5 gives an example used for parameter estimation. The scene is given in the dotted and dashed lines, which are generated as follows. (1) Take a subset of lines from each of the three objects in Fig. 8.4(a)–(c); (2) rotate and translate each of the subsets; (3) mix the transformed subsets; (4) randomly deviate the positions of the endpoints of the lines, which results in the dotted lines; and (5) add spurious lines, shown as the dashed lines. The scene generated consists of several subsets of model patterns plus spurious lines, as shown in Fig. 8.5(b). The example configuration $\bar{f}$ is shown in Fig. 8.5(a). It maps the scene to one of the models given in Fig. 8.4. The alignment between the dotted lines of the scene and the solid lines of the model gives the nonNULL labels of $\bar{f}$, whereas the unaligned lines of the scene are labeled as NULL .

The following four types of bilateral relations are used with $n = 2$ and $K_2 = 4$):

(1) $d_2^{(1)}(i, i')$: the angle between lines i and i';

(2) $d_2^{(2)}(i, i')$: the distance between the mid-points of the lines;

(3) $d_2^{(3)}(i, i')$: the minimum distance between the endpoints of the lines; and

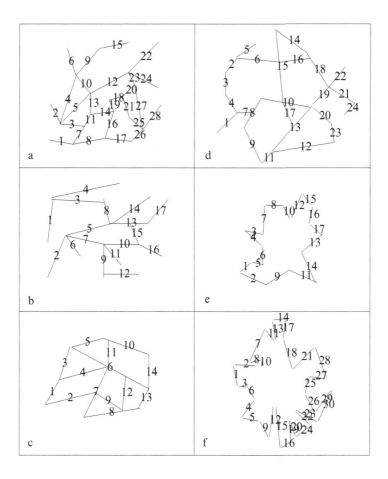

Figure 8.4: The six objects of line patterns in the model base. From (Li 1997b) with permission; ©1997 Kluwer.

(4) $d_2^{(4)}(i, i')$: the maximum distance between the endpoints of the lines.

Similarly, there are four model relations $D_2^{(k)}(I, I')$ ($k = 1, \ldots, 4$) of the same type. No unary properties are used ($K_1 = 0$). The $\mathcal{G}$ and $\mathcal{G}'$ are composed of these four relational measurements. Therefore, there are five components ($k = 0, 1, \ldots, 4$) in x and θ.

The C_2-optimal parameters are computed as $\bar{\theta} = \{\bar{\theta}_2^{(0)}, \bar{\theta}_2^{(1)}, \ldots, \bar{\theta}_2^{(4)}\} = \{0.58692, 0.30538, 0.17532, 0.37189, 0.62708\}$, which satisfies $\|\theta\| = 1$. The computation takes a few seconds on an HP series 9000/755 workstation. To be used for recognition, $\bar{\theta}$ is multiplied by a factor of $0.7/\bar{\theta}_2^{(0)}$, yielding the final

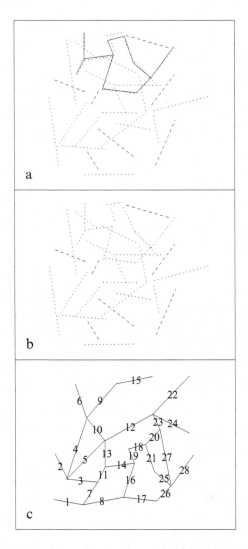.

236   8. *Parameter Estimation in Optimal Object Recognition*

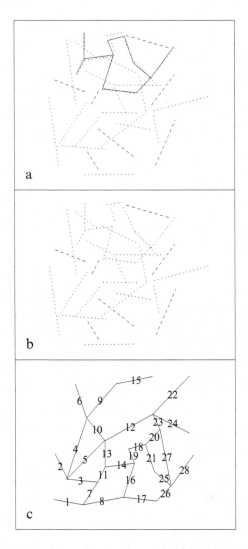

Figure 8.5: An exemplary instance consisting of (a) exemplary configuration $\bar{f}$, (b) scene $\mathcal{G}$, and (c) model $\mathcal{G}'$. From (Li 1997b) with permission; ©1997 Kluwer.

weights $\theta^* = \{0.70000,\ 0.36422,\ 0.20910,\ 0.44354,\ 0.74789\}$ (our recognition system requires $(\theta_2^{(0)})^* = 0.7$).

The θ^* is used to define the energy for recognizing other objects and scenes. The recognition results are shown in Fig. 8.6. There are two scenes, one in the upper row and the other in the lower row, composed of the dotted

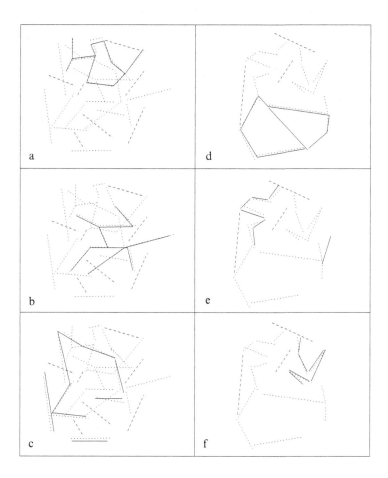

Figure 8.6: The optimal parameter estimate learned from the example is used to recognize other scenes and models (see the text). From (Li 1997b) with permission; ©1997 Kluwer.

and dashed lines. The upper one was used in the example, whereas the lower scene contains subparts of the three model objects in Fig. 8.4(d)–(f). Each scene is matched against the six model objects. The optimally matched object lines are shown as solid lines aligned with the scenes. The objects in the scenes are correctly matched to the model objects.

8.4.2   Recognition of Curved Objects

This experiment deals with jigsaw objects under 2D rotation, translation and uniform scaling. There are eight model jigsaw objects shown in Fig. 4.10. In this case of curved objects, the features of an object correspond to the corner points of its boundary. For both the scene and the models, the boundaries are extracted from the images using the Canny detector followed by hysteresis and edge linking. Corners are detected after that. No unary relations are used ($K_1 = 0$). Denoting the corners by $p_1, \ldots, p_m$, the following five types of bilateral relations are used ($n = 2; K_2 = 5$) based on a similarity-invariant curve representation of curves (Li 1993):

(1) $d_2^{(1)}(i, i')$: ratio of curve arc length $\widehat{p_i p_{i'}}$ and chord length $\overline{p_i p_{i'}}$;

(2) $d_2^{(2)}(i, i')$: ratio of curvature at p_i and $p_{i'}$;

(3) $d_2^{(3)}(i, i')$: invariant coordinate vector;

(4) $d_2^{(4)}(i, i')$: invariant radius vector; and

(5) $d_2^{(5)}(i, i')$: invariant angle vector.

They are computed using information about both the boundaries and the corners. Similarly, there are five model relations $D_2^{(k)}(I, I')$ ($k = 1, \ldots, 5$) of the same types. Therefore, there are six components ($k = 1, \ldots, 5$) in each x and θ, one for the NULL and five for the relational quantities above.

Figure 8.7 gives the example used for parameter estimation. The scene in Fig. 8.7(b) contains rotated, translated and scaled parts of one of the model jigsaw objects. Some objects in the scene are considerably occluded. The alignment between the model jigsaw object (the highlighted curve in (a)) and the scene gives nonNULL labels of $\bar{f}$, whereas the unaligned boundary corners of the scene are labeled as NULL .

The C_2-optimal parameters are computed as $\bar{\theta} = \{\bar{\theta}_2^{(0)}, \bar{\theta}_2^{(1)}, \ldots, \bar{\theta}_2^{(6)}\} = \{0.95540, 0.00034, 0.00000, 0.06045, 0.03057, 0.28743\}$, which satisfies $\|\theta\| = 1$. It takes a few seconds on the HP workstation. Note that the weight $\bar{\theta}_2^{(2)}$ for $d_2^{(2)}(i, i')$ and $D_2^{(2)}(I, I')$ (ratio of curvature) are zero. This means that this type of feature is not reliable enough to be used. Because our recognition system has a fixed value of $\theta_2^{(0)} = 0.7$, $\bar{\theta}$ is multiplied by a factor of $0.7/\bar{\theta}_2^{(0)}$, yielding the final weights $\theta^* = \{0.70000, 0.00025, 0.00000, 0.04429, 0.02240, 0.21060\}$.

The θ^* is used to define the energy function for recognizing other objects and scenes. The recognition results are shown in Fig. 8.8. There are two scenes, one in the left column and the other in the right column. The scene on the left was the one used in the example and the one on the right is a new scene. The optimally matched model objects are shown in the highlighted curves aligned with the scenes. The same results can also be obtained using the C_∞-optimal estimate, which is $\{0.7, 0.00366, 0.00000, 0.09466, 0.00251\}$.

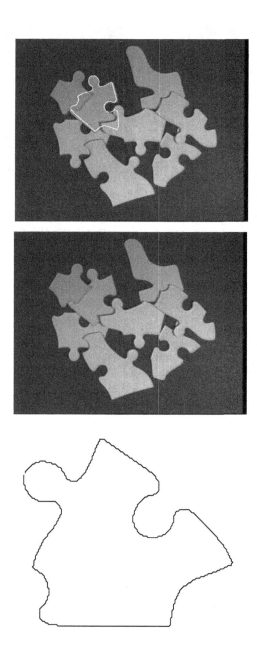

Figure 8.7: An example consisting of (a) example configuration $\bar{f}$, (b) scene $\mathcal{G}$ and (c) model $\mathcal{G}'$. From (Li 1997b) with permission; ©1997 Kluwer.

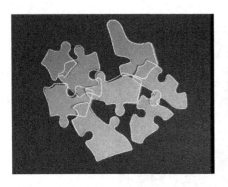

Figure 8.8: The learned estimate is used to recognize other scenes and models. The matched model jigsaw objects are aligned with the scene. From (Li 1997b) with permission; ©1997 Kluwer.

8.4.3   Convergence

The parameter estimation algorithm is very stable and has a nice convergence property. Figure 8.9 shows how the global instability measure C_2 and one of the learned parameters $\bar{\theta}_2^{(3)}$ evolve given different starting points. The values stabilize after hundreds of iterations, and different starting points converge to the same point.

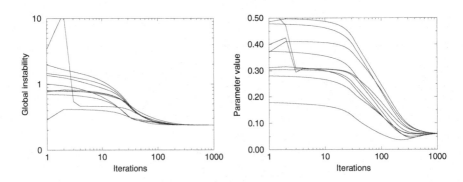

Figure 8.9: Convergence of the algorithm from different starting points. Left: Trajectories of C_2. Right: Trajectories of $\bar{\theta}_2^{(3)}$. From (Li 1997b) with permission; ©1997 Kluwer.

8.5 Conclusion

While manual selection is a common practice in object recognition systems, this chapter has presented a novel theory for automated optimal parameter estimation in optimization-based object recognition. The theory is based on learning from examples. Mathematical principles of correctness and instability are established and defined for the evaluation of parameter estimates. A learning algorithm is presented for computing the optimal (i.e., minimal-instability) estimate. An application to MRF-based recognition is given. Experiments conducted show very promising results. Optimal estimates automatically learned from examples can be well generalized for recognizing other scenes and objects.

The training examples are given to reflect the designer's judgment of desirable solutions. However, a recognizer with a given functional form cannot be trained by arbitrary examples. The example should be selected properly to reflect the correct semantics, in other words, they should be consistent with the constraints with which the functional form is derived. Assuming the form of the objective function is correct and the training set contains useful information, then the more examples are used for training, the more generalizable the learned parameter estimate will be.

The learning procedure also provides a means for checking the validity of the energy function derived from mathematical models. An improper mathematical model leads to an improper functional form. If no correct parameter estimates can be learned, it is a diagnostic symptom that the assumptions used in the model are not suitable for modeling the reality of the scene. The procedure also provides useful information for feature selection. Components of the optimal parameter estimate will be zero or near zero for unstable features.

Chapter 9

Minimization – Local Methods

After the energy function $E(f)$, including both the functional form and the parameters involved, is given and thus the optimal solution $f^* = \min_f E(f)$ is entirely defined, the remaining problem is to find the solution. It is most desirable to express the solution in closed form, but this is generally very difficult in vision problems due to the complexity caused by interactions between labels. Therefore, optimal solutions are usually computed by using some iterative search techniques. This chapter describes techniques for finding local minima and discusses related issues.

9.1 Problem Categorization

A minimization problem can be categorized as continuous or combinatorial according to whether the label set $\mathcal{L}$ is continuous or discrete; it can be further categorized as constrained or unconstrained according to whether f is constrained within a subspace of the entire search space. A vision problem may also be modeled as a combination of several minimization processes; for example, continuous restoration with a line process is a combination of continuous and combinatorial minimization processes (See Chapters 3 and 5). As far as computational algorithms are concerned, a view of local versus global methods is important when there are multiple local minima in the solution space.

Because classical unconstrained, continuous minimization methods are most mature, it is sometimes desirable to convert combinatorial and constrained problems into forms suitable for such classical methods. A combinatorial problem may be converted into a continuous one, for example, using the notion of relaxation labeling (Rosenfeld et al. 1976) or mean field approximation (Peterson and Soderberg 1989). A constrained minimization may be

S.Z. Li, *Markov Random Field Modeling in Image Analysis*,
Advances in Pattern Recognition, DOI: 10.1007/978-1-84800-279-1_9,
© Springer-Verlag London Limited 2009

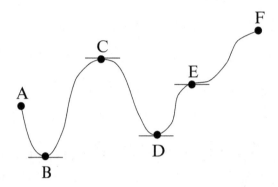

Figure 9.1: Local and global extrema.

converted into an unconstrained one, for example, using classical methods such as the penalty method or the Lagrange multiplier method (Fletcher 1987). This is not only for mathematical convenience but also brings about possibilities for analog implementations.

Figure 9.1 illustrates extremal points of a continuous function in $\mathbb{R}$. A, C, and F are local maxima, but only F is the global maximum. Points B and D are local minima of which B is the global one. B, C, D, and E are stationary points where the gradient vanishes, but E is an inflection point (or saddle point in higher-dimensional cases) rather than an extremum. A gradient-descent search algorithm will stop at B or D. Theoretically, it can also stop at a saddle point or even a local maximum because the gradient there is exactly zero. But such a chance is small in numerical computation because of the quantization, and the algorithm is likely to pass by these points.

The localness and globalness of a minimum f are given w.r.t. a neighborhood system $\mathcal{N} = \{\mathcal{N}(f) \mid f \in \mathbb{F}\}$ where $\mathcal{N}(f)$ is the set of neighboring configurations of f in the solution space. For $\mathbb{F} = \mathbb{R}^{\mathcal{S}}$, the neighborhood of f can be suitably defined as

$$\mathcal{N}_\epsilon(f) = \{x \mid x \in \mathbb{F}, 0 < \|x - f\| \leq \epsilon\} \tag{9.1}$$

where $\|\cdot\|$ is a norm and $\epsilon > 0$ is a real number. For a discrete problem where $\mathcal{L}_d$ is a discrete label set and $\mathbb{F}_d = \mathcal{L}_d^{\mathcal{S}}$, we may define the "$k$-neighborhood" of $f \in \mathbb{F}_d$ as

$$\mathcal{N}^k(f^*) = \{x \in \mathbb{F}_d \mid x \text{ differs from } f^* \text{ by at most } k \text{ labels}\} \tag{9.2}$$

For example, assuming that $\mathcal{L} = \{a, b, c\}$ is the set of allowable labels for every f_i ($i = 1, \ldots, 5$), then $\{a, b, a, a, a\}$ is a 1-neighbor of $\{a, a, a, a, a\}$ and

$\{b, a, a, a, c\}$ is a 2-neighbor. Note that $\{a, b, a, a, a\}$ is also a 2-neighbor of $\{a, a, a, a, a\}$ (due to the phrase "at most" in the definition) but not vice versa.

A point f^* is a *local minimum* of E w.r.t. $\mathcal{N}$ if

$$E(f^*) \leq E(f) \quad \forall f \in \mathcal{N}(f^*) \tag{9.3}$$

A local minimum is also a *global minimum* if the neighborhood is defined to include all the other configurations,

$$\mathcal{N}(f^*) = \{f \mid \forall f \in \mathbb{F}, f \neq f^*\} = \mathbb{F} - \{f^*\} \tag{9.4}$$

When $E(f^*)$ is strictly the lowest ($E(f^*) < E(f)$ for all $f \neq f^*$) f^* is the unique global minimum. Multiple global minima occur if there exist more than one point $\xi \in \mathbb{F}$ for which $E(\xi)$ take the globally minimum value $E(f^*)$.

Local search is the basis for many minimization algorithms. Figure 9.2 describes the idea of a deterministic local search. The idea is very simple and natural: At a point $f \in \mathbb{F}$, we look for where to go within the neighborhood $\mathcal{N}(f)$. If f' in the neighborhood leads to an improvement, $E(f') < E(f)$, we replace f by f'. This process continues until no further improvement can be made. When $E(f)$ is lower bounded, the algorithm converges to a local minimum.

Set $f = f(0)$;
Repeat
      Generate f' from $\mathcal{N}(f)$ such that $E(f') < E(f)$;
      $f \leftarrow f'$;
Until $E(f') \geq E(f)$ for all $f' \in \mathcal{N}(f)$;

Figure 9.2: Deterministic local search algorithm.

There are different ways of generating (or searching for) a candidate f'. In the continuous case, f' may be chosen based on the gradient: $f' = f - \mu \nabla E(f)$, where $\mu > 0$ is a step size. This leads to a zero gradient point, which is necessary for the minimization. In the discrete case, some enumerative method may be used to generate $f' \in \mathcal{N}(f)$ for which $E(f') < E(f)$.

The local search discussed so far is deterministic. In a stochastic local search, f' is generated at random. It is not necessary that $E(f')$ be lower than $E(f)$. Whether to accept f' for which $E(f')$ is higher than $E(f)$ is decided according to some probabilistic rules.

9.2 Classical Minimization with Continuous Labels

Let $\mathcal{L} = [X_l, X_h]$ be a continuous label set. Consider the prototypical problem of (piecewise) continuous restoration in which the energy takes the form

$$E(f) = \sum_{i \in \mathcal{S}} \chi_i (f_i - d_i)^2 + \lambda \sum_{i \in \mathcal{S}} \sum_{i' \in \mathcal{N}_i} g(f_i - f_{i'}) \qquad (9.5)$$

where $\chi_i \in \{1, 0\}$ indicates the presence or absence of the data d_i and $g(\eta)$ is the potential function. Because a minimum is necessarily a stationary point, the gradient $\nabla E(f)$ must be a zero vector at the point. Therefore, the following system of equations must be satisfied by f^*:

$$\frac{1}{2} \frac{\partial E}{\partial f_i} = \chi_i (f_i - d_i) + \lambda \sum_{i' \in \mathcal{N}_i} (f_i - f_{i'}) h(f_i - f_{i'}) = 0, \quad \forall i \in \mathcal{S} \qquad (9.6)$$

where $g'(\eta) = 2\eta h(\eta)$ is assumed (see (5.24)). The system is nonlinear if $h(\eta)$ is a nonlinear function. Higher-order derivatives have to be examined in order to determine whether such a stationary point is a local minimum, maximum, or a saddle point. We assume that it is a local minimum.

Solving the system of equations (9.6) is a classical problem (Dahlquist and Bjorck 1974). When h is a linear function, the system is linear and f can be solved for by using matrix operations. For a nonlinear system, a numerical iterative method is generally utilized. In a simple gradient-based algorithm, for example, a stationary point satisfying (9.6) is computed by iteratively updating f using the rule

$$\begin{aligned} f_i^{(t+1)} \leftarrow f_i^{(t)} \quad &- \quad 2\mu \{ \, \chi_i[f_i^{(t)} - d_i] \\ &+ \quad \lambda \sum_{i' \in \mathcal{N}_i} (f_i^{(t)} - f_{i'}^{(t)}) \, h(f_i^{(t)} - f_{i'}^{(t)}) \, \} \end{aligned} \qquad (9.7)$$

where μ is a small constant.

The fixed-point iteration method provides another method for solving the problem. Rearrange (9.6) into the form

$$f = \varphi(f) \qquad (9.8)$$

which is composed of m fixed-point equations $f_i = \varphi_i(f)$ $(i = 1, \ldots, m)$. The following is called fixed-point iteration:

$$f^{(t+1)} \leftarrow \varphi(f^{(t)}) \qquad (9.9)$$

Let $J_\varphi(f)$ be the Jacobian matrix of φ and $\mathbb{D}$ be a region in $\mathbb{R}^{\mathcal{S}}$. Suppose (1) φ is a continuous mapping, $\varphi : \mathbb{D} \to \mathbb{D}$, and (2) φ is a contraction mapping (that is, $\|J_\varphi(f)\| \le L < 1$ for all $f \in \mathbb{D}$, where $\|\cdot\|$ is a norm and

L is a constant). Then the fixed-point iteration (9.8) converges to a unique fixed-point $f^* = \lim_{t \to \infty} f^{(t)}$ for every $f^{(0)} \in \mathbb{D}$. The way for constructing fixed-point function φ is not unique; some choices lead to fast convergence whereas others may diverge. The rate of convergence depends linearly on L.

To solve (9.6) using fixed-point iteration, we may construct a mapping like

$$\varphi_i(f) = \frac{1}{2\lambda C + \chi_i}\{\chi_i d_i + \lambda C(f_{i-1} + f_{i+1}) + \\ \lambda[g'(f_{i-1} - f_i) + g'(f_{i+1} - f_i) - C(f_{i-1} - 2f_i + f_{i+1})]\} \tag{9.10}$$

where C is a constant. It is not difficult to verify that $f_i = \varphi_i(f)$ is equivalent to (9.6). It can be shown that (9.10) is a contraction mapping when $C = \max_\eta g''(\eta)$ and $\chi_i = 1$ ($\forall i$). In this case, the fixed-point iteration (9.9) converges to a unique fixed point for which $f^* = \varphi(f^*)$, where $\varphi = (\varphi_1, \ldots, \varphi_m)$.

9.3 Minimization with Discrete Labels

When the label set is discrete, the minimization of the posterior energy is combinatorial. In this section, we consider methods for minimizing $E(f)$ when each f_i takes a value in $\mathcal{L} = \{1, \ldots, M\}$.

9.3.1 Iterated Conditional Modes

Since it is difficult to maximize the joint probability of an MRF, Besag (1986) proposed a deterministic algorithm called iterated conditional modes (ICM) that maximizes local conditional probabilities sequentially. The ICM algorithm uses the "greedy" strategy in the iterative local maximization. Given the data d and the other labels $f_{\mathcal{S}-\{i\}}^{(k)}$, the algorithm sequentially updates each $f_i^{(k)}$ into $f_i^{(k+1)}$ by maximizing $P(f_i \mid d, f_{\mathcal{S}-\{i\}})$, the conditional (posterior) probability, w.r.t. f_i.

Two assumptions are made in calculating $P(f_i \mid d, f_{\mathcal{S}-\{i\}})$. First, the observation components $d_1, \ldots, d_m$ are conditionally independent given f, and each d_i has the same known conditional density function $p(d_i \mid f_i)$ dependent only on f_i. Thus

$$p(d \mid f) = \prod_i p(d_i \mid f_i) \tag{9.11}$$

The second assumption is that f depends on the labels in the local neighborhood, which is the Markovianity. From the two assumptions and the Bayes theorem, it follows that

$$P(f_i \mid d, f_{\mathcal{S}-\{i\}}) \propto p(d_i \mid f_i) \, P(f_i \mid f_{\mathcal{N}_i}) \tag{9.12}$$

Obviously, $P(f_i \mid d_i, f_{\mathcal{N}_i}^{(k)})$ is much easier to maximize than $P(f \mid d)$, which is the point of ICM.

Maximizing (9.12) is equivalent to minimizing the corresponding posterior potential using the rule

$$f_i^{(k+1)} \leftarrow \arg\min_{f_i} V(f_i \mid d_i, f_{\mathcal{N}_i}^{(k)}) \qquad (9.13)$$

where

$$V(f_i \mid d_i, f_{\mathcal{N}_i}^{(k)}) = \sum_{i' \in \mathcal{N}_i} V(f_i \mid f_{i'}^{(k)}) + V(d_i \mid f_i) \qquad (9.14)$$

For example, for the discrete restoration formulated in Section 3.2.2, the posterior potential for (3.19) is

$$V(f_i \mid d_i, f_{\mathcal{N}_i}) = (f_i - d_i)^2/\sigma + v_{20} \sum_{i' \in \mathcal{N}_i} [1 - \delta(f_i - f_{i'})] \qquad (9.15)$$

where $\sum_{i' \in \mathcal{N}_i} [1 - \delta(f_i - f_{i'})] = \#\{f_{i'} \neq f_i \mid i' \in \mathcal{N}_i\}$ is the number of neighboring sites whose labels $f_{i'}$ differ from f_i.

For discrete $\mathcal{L}$, $V(f_i \mid d_i, f_{\mathcal{N}_i})$ is evaluated with each $f_i \in \mathcal{L}$ and the label causing the lowest $V(f_i \mid d_i, f_{\mathcal{N}_i})$ value is chosen as the value for $f_i^{(k+1)}$. When applied to each i in turn, the above defines an updating cycle of ICM. The iteration continues until convergence. The convergence is guaranteed for the serial updating and is rapid (Besag 1986).

The result obtained by ICM depends very much on the initial estimator $f^{(0)}$, as is widely reported. Currently, it is not known how to set the initialization properly to obtain a good solution. A natural choice for $f^{(0)}$ is the maximum likelihood estimate

$$f^{(0)} = \arg\max_f p(d \mid f) \qquad (9.16)$$

when the noise is identically, independently distributed Gaussian.

ICM can also be applied to problems where f_i takes a continuous value. In minimizing (3.22) for continuous restoration, for example, one needs to maximize

$$V(f_i \mid d_i, f_{\mathcal{N}_i}) = (f_i - d_i)^2 + \lambda \sum_{i' \in \mathcal{N}_i} g(f_i - f_{i'}) \qquad (9.17)$$

for each $i \in \mathcal{S}$. To do this, it is necessary to solve $\frac{dV(f_i \mid d_i, f_{\mathcal{N}_i})}{df_i} = 0$.

In a genuine MRF algorithm, no two neighboring sites should be updated simultaneously. The "coding method" (Besag 1974) may be incorporated into ICM to parallelize the iteration. Using codings, $\mathcal{S}$ are partitioned into several sets such that no two sites in one set are neighbors (see Section 7.1.3). Therefore, all f_i on a single coding can be updated in parallel.

9.3.2   Relaxation Labeling

In this section, the *relaxation labeling* (RL) method is described as an approach for minimizing an energy $E(f)$ over a discrete space, where $E(f)$ may be, for example, the posterior energy $U(f \mid d)$ for the MAP-MRF restoration of a multilevel image. RL was originally proposed as a class of parallel iterative numerical procedures which uses contextual constraints to reduce ambiguities in image analysis (Rosenfeld et al. 1976). It has been widely used in the image and vision community. It is known that RL has polynomial complexity yet empirically produces good results.

Formally, Faugeras and Berthod (1981) define a class of global criteria in terms of transition probabilities. Peleg (1980) interprets RL in terms of Bayes analysis. Haralick (1983) illustrates RL as minimizing the expected loss from a Bayes viewpoint. Hummel and Zucker (1983) formulate the probabilistic relaxation approach as a global optimization problem in which a so-called average local consistency is defined and maximized. They show that, computationally, finding consistent labeling is equivalent to solving a variational inequality. Using the notion of continuous RL, the combinatorial minimization is converted into a real minimization subject to linear constraints, and the combinatorial minimization is performed by constrained real minimization.

A development in relaxation labeling is the use of contextual constraints about the observed data. Li formulates an objective function for relaxation labeling in which contextual constraints from both the prior knowledge and the observation are considered (Li 1992c; Li 1992a; Li 1992b). Kittler, Christmas, and Petrou (1993) later justified the formulation from a probabilistic viewpoint.

Representation of Continuous RL

In continuous RL, the labeling state for each site is represented as a $\#\mathcal{L}$ position vector; for i, it is

$$p_i = [p_i(I) \mid I \in \mathcal{L}] \tag{9.18}$$

subject to the *feasibility* constraints

$$\begin{cases} \sum_{I \in \mathcal{L}} p_i(I) = 1 & \forall i \\ p_i(I) \geq 0 & \forall i, I \end{cases} \tag{9.19}$$

This was interpreted as a fuzzy assignment model (Rosenfeld et al. 1976). The real value $p_i(I) \in [0, 1]$ reflects the strength with which i is assigned label I. Each p_i lies in a hyperplane in the nonnegative quadrant portion (simplex) of the multidimensional real space $\mathbb{R}^{\mathcal{L}}$, as shown in the shaded area of Fig. 9.3. The feasible labeling assignment space is

$$\mathbb{P}_i = \left\{ p_i(I) > 0 \mid I \in \mathcal{L}, \sum_{I \in \mathcal{L}} p_i(I) = 1 \right\} \tag{9.20}$$

The set $p = \{p_i \mid i \in \mathcal{S}\}$ is called a *labeling assignment*. The feasible space for labeling assignments is the product

$$\mathbb{P} = \mathbb{P}_1 \times \mathbb{P}_2 \cdots \times \mathbb{P}_m \qquad (9.21)$$

The final solution p^* must be *unambiguous*

$$p_i^*(I) = \{0, 1\} \qquad \forall i, I \qquad (9.22)$$

with $p_i^*(I) = 1$ meaning that i is unambiguously labeled I. This gives the unambiguous spaces for p_i as

$$\mathbb{P}_i^* = \left\{ p_i(I) \in \{0, 1\} \mid I \in \mathcal{L}, \sum_{I \in \mathcal{L}} p_i(I) = 1 \right\} \qquad (9.23)$$

So every p_i in $\mathbb{P}_i^*$ is one of the vectors $(1, 0, \ldots, 0)$, $(0, 1, \ldots, 0)$, and $(0, 0, \ldots, 1)$ corresponding to the "corners" of the simplex $\mathbb{P}_i$ (see Fig. 9.3). The unambiguous labeling assignment space is

$$\mathbb{P}^* = \mathbb{P}_1^* \times \mathbb{P}_2^* \cdots \times \mathbb{P}_m^* \qquad (9.24)$$

which consists of the corners of the space $\mathbb{P}$. Noncorner points in the continuous space $\mathbb{P}$ provide routes to the corners of $\mathbb{P}$ (i.e., the points in the discrete space $\mathbb{P}^*$). An unambiguous labeling assignment is related to the corresponding discrete labeling by

$$f_i = I \qquad \text{if} \quad p_i^*(I) = 1 \qquad (9.25)$$

The unambiguity may be satisfied automatically upon convergence of some iterative algorithms; if not, it has to be enforced, for example, using a maximum selection (winner-take-all) operation. However, the forced unambiguous solution is not necessarily a local minimum. It has been largely ignored in the literature how to ensure the result of a continuous RL algorithm to be an unambiguous solution.

Maximization Formulation

Although RL can be used for both minimization and maximization, it is sometimes algorithmically more suitable for maximization. Therefore, in practice, we convert the minimization of the MRF energy into the maximization of a corresponding gain function. The gain is the sum of compatibility functions.

The RL compatibility functions can be defined based on the Gibbs clique potential functions as follows. The unary compatibility function is defined by

$$r_i(I) = Const_1 - V_1(I \mid d) \qquad (9.26)$$

where $V_1(I \mid d)$ is the function of potentials incurred by single-site cliques. The binary compatibility function is defined by

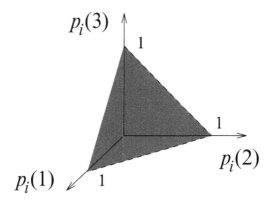

Figure 9.3: The labeling space for p_i with three labels. The feasible space is shown as the shaded area (simplex), and the unambiguous space consists of the three corner points.

$$r_{i,i'}(I, I') = Const_2 - V_2(I, I' \mid d) \qquad (9.27)$$

where $V_2(I, I' \mid d)$ is the function of potentials incurred by pair-site cliques. The constants $Const_1$ and $Const_2$ are chosen so that all the compatibility functions are nonnegative.

The gain with the labeling assignment representation can now be written as

$$G(p) = \sum_{i \in \mathcal{S}} \sum_{I \in \mathcal{L}} r_i(I) \, p_i(I) + \qquad (9.28)$$

$$\sum_{i \in \mathcal{S}} \sum_{I \in \mathcal{L}} \sum_{i' \in \mathcal{S}, i' \neq i} \sum_{I' \in \mathcal{L}} r_{i,i'}(I, I') \, p_i(I) \, p_{i'}(I')$$

which is to be *maximized*. The above is the standard form for objective functions in RL formulations. Obviously, there is one–one correspondence between the maxima of $G(f)$ and the minima of $E(f)$ because of the relationship $G(f) \propto Const - E(f)$ where $Const$ is some constant.

Iterative Updating Equations

In maximizing $G(f)$ under the feasibility and unambiguity constraints, an RL algorithm iteratively updates p based on the gradient $q = -\nabla E(f) = \{q_i(I) \mid i \in \mathcal{S}, I \in \mathcal{L}\}$. The gradient components are given by

$$q_i(I) = \frac{\partial G}{\partial p_i(I)} = r_i(I) + 2 \sum_{i' \in \mathcal{S}} \sum_{I' \in \mathcal{L}} r_{i,i'}(I, I') \, p_{i'}(I') \qquad (9.29)$$

for symmetric $r_{i,i'}(I, I')$. Let $p = p^{(t)}$ represent the labeling state at time t. The initial assignment is set to $p^{(0)} \in \mathbb{P}$. A current $p^{(t)}$ is updated into a new state $p^{(t+1)}$ based on the information in $p^{(t)}$ and $q^{(t)}$

$$p^{(t+1)} \leftarrow \Phi(p^{(t)}, q^{(t)}) \qquad (9.30)$$

where Φ is an updating operator. RL algorithms differ from one another in the choice of Φ. Various update equations are described in the following.

Rosenfeld, Hummel, and Zucker (1976) proposed to use the fixed-point iteration

$$p_i^{(t+1)}(I) \leftarrow \frac{p_i^{(t)}(I)[1 + q_i^{(t)}(I)]}{\sum_J p_i^{(t)}(J)[1 + q_i^{(t)}(J)]} \qquad (9.31)$$

Obviously, the feasibility constraints (9.19) are satisfied by the normalization. The iteration equation above is only one of many choices. For example, when the compatibility coefficients are nonnegative, we have $q_i^{(t)}(I) \geq 0$; in this case, the following updating scheme also maintains the feasibility while performing gradient ascent

$$p_i^{(t+1)}(I) \leftarrow \frac{p_i^{(t)}(I)q_i^{(t)}(I)}{\sum_J p_i^{(t)}(J)q_i^{(t)}(J)} \qquad (9.32)$$

In the continuous RL algorithms by Faugeras and Berthod (1981) and Hummel and Zucker (1983), the feasibility is maintained by using a technique called *gradient projection* (GP) (Rosen 1960). GP is an extended steepest descent procedure for solving programming problems under linear constraints. It finds a feasible direction u from the gradient q and the current $p^{(t)}$, for the updating scheme

$$p^{(t+1)} \leftarrow p^{(t)} + \mu u \qquad (9.33)$$

where $\mu > 0$ is a scalar factor. The vector $u = \frac{1}{\mu}[p^{(t+1)} - p^{(t)}]$ must satisfy the condition

$$\sum_{I \in \mathcal{L}} u_i(I) = \frac{1}{\mu} \sum_{I \in \mathcal{L}} [p_i^{(t+1)}(I) - p_i^{(t)}(I)] = \frac{1}{\mu}(1 - 1) = 0 \qquad (9.34)$$

in order to maintain the feasibility constraints. If $p_i^{(t)}(I) = 0$, then there must be $u_i(I) \geq 0$ so that $p_i^{(t+1)}(I)$ would not be negative. For convenience, a normalization of u may be imposed so that $\|u\| = 1$. Then, the feasible space for u is defined by

$$\mathbb{U} = \left\{ u \in \mathbb{R}^{S \times \mathcal{L}} \mid \sum_{I \in \mathcal{L}} u_i(I) = 0, \|u\| = 1, \text{ and } u_i(I) \geq 0 \text{ when } p_i(I) = 0 \right\} \qquad (9.35)$$

The maximum value for μ is chosen so that the updated vector $p^{(t+1)}$ still lies in the space $\mathbb{P}$. A detailed implementation of such a projection operator used in (Hummel and Zucker 1983) is described in (Mohammed et al. 1983). A discussion on a parallel distributed implementation of the Hummel-Zucker algorithm in a multiprocessor architecture environment is given in (McMillin and Ni 1989).

Local versus Global Solutions

Gradient-based optimization methods guarantee only local solutions when the cost function is nonconvex. RL algorithms are local minimizers by nature. Given an RL iteration scheme and the data, two factors affect the solution: the initial assignment of labels and the compatibility function (Rosenfeld et al. 1976). However, it is shown that for certain choices of compatibility function, an RL algorithm can have a *single* convergence point regardless of the initial assignment (Bozma and Duncan 1988; O'Leary and Peleg 1983).

In some earlier RL works (Rosenfeld et al. 1976; Peleg and Rosenfeld 1978), the dependence of solutions on initializations is regarded as a desirable property. There, compatibility functions are defined *without* referring to the observations and therefore should be considered as containing a priori information only. Constraints due to the observation are encoded in the initial assignment. Therefore, traditional RL solutions rely greatly on the initialization; convergence to a solution regardless of the initialization is considered undesirable. However, the posterior MRF energy already contains both sources of information, i.e., on the prior and the observation. Therefore the global energy minimum is the desirable solution (Geman and Geman 1984). An "annealing labeling" algorithm, a GNC-like algorithm for the MAP-MRF matching will be described in Section 10.3.2.

9.3.3 Belief Propagation

Belief propagation (BP) can be used as an efficient way of solving inference problems formulated as maximizing marginal or posterior probabilities defined in terms of up to second-order relationships, such as with pairwise MRF's and Bayesian networks (see also Chapter 2). The BP uses the idea of passing local massages around the nodes through edges (Pearl 1988). The standard BP works for graphs without loops and guarantees the convergence. Several variants of BP algorithms also exist. BPs specific to MRF's and Bayesian networks are equivalent and thus one can can choose one of the algorithms without loosing generality (Yedidia et al. 2003). The BP algorithms described here are in the context of MRF's.

Standard BP

Let $\mathcal{S}$ be the set of sites with the neighborhood system $\mathcal{N}$, $\mathcal{L}$ be the set of discrete labels, and $\mathcal{G} = (\mathcal{S}, \mathcal{N})$ be the associated graph. Let d_i be the observation on the site i. For each i, let $r_i(f_i)$ be some unary compatibility functions describing dependency between f_i and d_i, often call the evidence for f_i. For each pair $(i, i') \in \mathcal{N}$ (an edge), define a compatibility function $r_{i,i'}(f_i, f_{i'})$. Then, the joint probability can be rewritten as

$$P(f) = Z^{-1} \prod_{i \in \mathcal{S}} r_i(f_i) \prod_{i \in \mathcal{S}} \prod_{i' \in \mathcal{N}_i} r_{i,i'}(f_i, f_{i'}) \qquad (9.36)$$

See also (2.19) and (Yedidia et al. 2003). The marginal probability of f_i is defined as

$$P_i(f_i) = \sum_{(f_1, \ldots, f_{i-1}, f_{i+1}, \ldots, f_M)} P(f_1, \ldots, f_M) \qquad (9.37)$$

The belief, denoted $b_i(f_i)$, is an approximation of the marginal probability $P_i(f_i)$ for i to be labeled f_i.

The BP works by iteratively passing local messages. This is similar to RL described previously. Let $m_{i,i'}(f_{i'})$ be the message that sites i sends to its neighbors i' to support for i' to be labeled $f_{i'}$. By the probability rule, the beliefs and the messages are related by the belief equation (Yedidia et al. 2003)

$$\hat{P}_i(f_i) = Z_i^{-1} r_i(f_i) \prod_{i' \in \mathcal{N}_i} m_{i',i}(f_i) \qquad (9.38)$$

where the product term is the collective massage coming into i and Z_i^{-1} is a normalizing constant so that $\sum_{f_i} \hat{P}_i(f_i) = 1$.

However, different than RL where the assignments p is updated, BP algorithms update the messages $m_{i,i'}(f_{i'})$ at each iteration. The standard BP has two versions of the message update rule, sum-product and max-product. In the sum-product BP, the message update rule is

$$m_{i,i'}(f_{i'}) \leftarrow \alpha_{i,i'} \sum_{f_i} r_i(f_i) r_{i,i'}(f_i, f_{i'}) \prod_{i'' \in \mathcal{N}_i - \{i'\}} m_{i'',i}(f_i) \qquad (9.39)$$

where $\mathcal{N}_i - \{i'\}$ is the set difference. The constant α is a normalization constant which does not really influence the final beliefs but affects the numerical stability. The max-product BP uses the message update rule

$$m_{i,i'}(f_{i'}) \leftarrow \alpha_{i,i'} \max_{f_i} r_i(f_i) r_{i,i'}(f_i, f_{i'}) \prod_{i'' \in \mathcal{N}_i - \{i'\}} m_{i'',i}(f_i) \qquad (9.40)$$

A message update rule is applied iteratively in a synchronous or asynchronous way. In the synchronous schedule, all sites send the messages at the same time. In the asynchronous schedule, a site send its messages immediately after $m_{i,i'}(f_{i'})$ is calculated. The latter usually has faster convergence.

Generalizations

The standard BP can be generalized in several ways. The first generalization deals with graphs with loops. Loopy BP (Smyth et al. 1997) applies the standard BP to graphs with loops. Although loopy BP does not guarantee the convergence, it works well for most application problems empirically (Mackay and Neal 1995; Weiss 1997; Frey 1997) but fail for others (Murphy et al. 1999). The reasons are complicated, and little has been found to explain it.

The tree re-weighted (TRW) message passing algorithm (Wainwright 2002; Kolmogorov and Wainwright 2005) is a modified version of the standard BP. It converts a graph with loops into spanning trees and then applies the standard BP on the trees to obtain an optimal or an approximately optimal configuration. When the graph is a tree, these algorithms are equivalent to the standard BP. Theoretical results on the convergence of the TRW are provided in (Kolmogorov and Wainwright 2005).

While energies with single and pairwise cliques can be represented by trees and graphs, representations of higher-order cliques need some generalization. The generalized BP (GBP) has been proposed for this (Yedidia et al. 2000).

9.3.4 Convex Relaxation

The labeling assignment problem of maximizing (9.29) is equivalent to quadratic programming (QP) (Korte and Vygen 2006) with linear constraints (Wierschin and Fuchs 2002) and can be approximately solved by using a convex relaxation algorithm.

Quadratic Programming Reformulation

Let us first represent the constrained QP in matrix form. Consider $p = [p_i(I) \mid i \in \mathcal{S}, I \in \mathcal{L}]$ defined in (9.18). Let $j = I + (i-1)M$ so as to denote $p_i(I)$ by $p(j)$. By this, we express p as a vector $p = [p(j) \mid j = 1, \cdots, m \times M]$. The constraint (9.19) can be rewritten as

$$\begin{cases} \sum_{I \in \mathcal{L}} p(I + (i-1)M) = 1 & \forall i \in \mathcal{S} \\ p(j) \geq 0 & j = 1, \cdots, m \times M \end{cases} \tag{9.41}$$

Similarly, vectorize $r = [r_i(I)]$ in (9.29) into a row vector $[r(1), \ldots, r(m \times M)]$. Then the first term of (9.29) can be rewritten as

$$\sum_{i \in \mathcal{S}} \sum_{I \in \mathcal{L}} r_i(I)\, p_i(I) = rp \tag{9.42}$$

Further, represent $[r_{i,i'}(I, I')]$ in (9.29) as an $mM \times mM$ matrix

$$R = \begin{bmatrix} R(1,1) & \cdots & R(1,m) \\ \vdots & \ddots & \vdots \\ R(m,1) & \cdots & R(m,m) \end{bmatrix} \tag{9.43}$$

where each block $R(i, i')$ is an $M \times M$ matrix

$$R(i, i') = \begin{bmatrix} r_{i,i'}(1,1) & \cdots & r_{i,i'}(1, M) \\ \vdots & \ddots & \vdots \\ r_{i,i'}(M, 1) & \cdots & r_{i,i'}(M, M) \end{bmatrix} \qquad (9.44)$$

Then the second term of (9.29) can be rewritten as

$$\sum_{i \in \mathcal{S}} \sum_{I \in \mathcal{L}} \sum_{i' \in \mathcal{S}, i' \neq i} \sum_{I' \in \mathcal{L}} r_{i,i'}(I, I')\, p_i(I)\, p_{i'}(I') = p^T R p \qquad (9.45)$$

With these notations, the gain function (9.29) can be rewritten as the quadratic form

$$G(p) = rp + p^T R p \qquad (9.46)$$

The labeling assignment problem can be solved by maximizing the quadratic gain subject to the constraints (9.41) such that

$$\begin{aligned} \max \quad & rp + p^T R p \\ \text{s.t.} \quad & \sum_{I \in \mathcal{L}} p(I + (i-1)M) = 1, \quad \forall i \in \mathcal{S} \qquad (9.47) \\ & p(j) \in [0, 1] \end{aligned}$$

The solution of this QP relaxation is equivalent to the original MAP estimation (Ravikumar and Lafferty 2006).

Several convex relaxation algorithms have been proposed to obtain an approximation solution, such as linear programming (LP) relaxation (Schlesinger 1976; Werner 2007), semi-definite programming (SDP) relaxation (Goemans and Williamson 1995; Schellewald and Schnorr 2003), and second-order cone programming (SOCP) relaxation (Muramatsu and Suzuki 2003; Kumar and Hebert 2006).

LP Relaxation

LP relaxation converts the quadratic function (9.46) into a linear function by introducing new variables. Note that the term $p^T R p$ is a scalar, so $p^T R p = \text{trace}(p^T R p)$. Because $\text{trace}(AB) = \text{trace}(BA)$, we have

$$p^T R p = \text{trace}(R p p^T) \qquad (9.48)$$

Letting $Q = p p^T$, the equation above can be written as

$$p^T R p = \text{trace}(RQ) = R \bullet Q \qquad (9.49)$$

where $R \bullet Q$ is the Frobenius dot product of the matrices. Then the optimization problem (9.47) can be reformulated as the constrained linear programming problem

$$
\begin{aligned}
\text{max} \quad & rp + R \bullet Q \\
\text{s.t.} \quad & \sum_{I \in \mathcal{L}} p(I + (i-1)M) = 1, \quad \forall i \in \mathcal{S} \\
& Q = pp^T \\
& p(j) \in [0, 1]
\end{aligned}
\tag{9.50}
$$

While the constraint $Q = pp^T$ is nonconvex, the simplest LP relaxation removes this constraint and relaxes the problem (9.50) to

$$
\begin{aligned}
\text{max} \quad & rp + R \bullet Q \\
\text{s.t.} \quad & \sum_{I \in \mathcal{L}} p(I + (i-1)M) = 1, \quad \forall i \in \mathcal{S} \\
& p(j) \in [0, 1]
\end{aligned}
\tag{9.51}
$$

To get a better solution, the constraint $Q = pp^T$ can be substituted by the following two constraints (Koster et al. 1998; Chekuri et al. 2001):

$$
Q = Q^T \tag{9.52}
$$

$$
\sum_{j'} Q_{j,j'} = mp(j), \quad \forall j \tag{9.53}
$$

The first part is obvious because Q is symmetric. The left hand side of the second part, the sum of the j'th row of Q, can be rewritten as

$$
\sum_{j'} Q_{j,j'} = p(j) \left(\sum_{j'=1}^{m \times M} p(j') \right) \tag{9.54}
$$

Under the feasibility constraint (9.19), we have $\sum_{j'=1}^{m \times M} p(j') = m$ and the second part.

From these, we obtain the following constrained LP:

$$
\begin{aligned}
\text{max} \quad & rp + R \bullet Q \\
\text{s.t.} \quad & \sum_{I \in \mathcal{L}} p(I + (i-1)M) = 1, \quad \forall i \in \mathcal{S} \\
& \sum_{j'} Q_{j,j'} = mp(j), \quad \forall j \\
& Q = Q^T \\
& p(i) \in [0, 1]
\end{aligned}
\tag{9.55}
$$

Various LP algorithms, such as the simplex algorithm, interior-point algorithm and ellipse algorithm, can be employed to solve the problem above.

SDP Relaxation

In LP relaxation, the nonconvex constraint $Q = pp^T$ is substituted by two convex constraints, but they are not exactly equivalent. $Q = pp^T$ is mathematically equivalent to the following two conditions (Kim and Kojima 2001):

$$Y = \begin{bmatrix} 1 & p^T \\ p & Q \end{bmatrix} \geq 0 \tag{9.56}$$

$$\operatorname{rank}(Y) = 1 \tag{9.57}$$

The first is positive semi-definite and hence convex. The second, called the rank 1 constraint, is not a convex constraint. The SDP relaxation simply removes the rank 1 constraint. This method is also-called lift-and-project SDP relaxation because the dimension of Y is one greater than Q.

Under these conditions, the problem (9.50) can be relaxed as the following SDP:

$$\max \quad rp + R \bullet Q$$

$$\text{s.t.} \quad \sum_{I \in \mathcal{L}} p(I + (i-1)M) = 1, \quad \forall i \in \mathcal{S}$$

$$Y = \begin{bmatrix} 1 & p^T \\ p & Q \end{bmatrix} \geq 0 \tag{9.58}$$

$$Q = Q^T$$

$$p(j) \in [0,1]$$

SDP relaxation provides a good approximation theoretically but is computationally prohibited. The second-order cone programming (SOCP) relaxation (Muramatsu and Suzuki 2003; Kumar et al. 2006) provides primal-dual algorithms for finding solutions to the problem.

9.3.5 Highest Confidence First

Highest confidence first (HCF) (Chou and Brown 1990; Chou et al. 1993) is a deterministic algorithm for combinatorial minimization in which the label set is discrete, $\mathcal{L} = \{1, \ldots, M\}$. Its feature is the introduction of a special *uncommitted* label and the strategy for "committing" a site.

Denote the *uncommitted* label by 0. The original label set is augmented by this label into $\mathcal{L}^+ = \{0, 1, \ldots, M\}$. A label f_i is said to be uncommitted if $f_i = 0$ or committed if $f_i \in \mathcal{L}$. Initially, all labels are set uncommitted, $f = \{0, \ldots, 0\}$. A rule is that once a site is committed, its label cannot go back to 0 but can change to another value in $\mathcal{L}$.

The commitment and label changes are based on a stability measure defined in the following. The two assumptions made in ICM are also made in HCF. Therefore we have the conditional posterior potential

$$E_i(f_i) = V(f_i \mid d_i, \hat{f}_{\mathcal{N}_i}) = V(d_i \mid f_i) + \sum_{c:i \in c} V_c(f_i \mid f_{\mathcal{N}_i}) \tag{9.59}$$

where $c : i \in c$ means any clique c containing site i. After augmenting the label set, we define the conditional potential for a *committed* label $f_i \in \mathcal{L}$ as[1]

$$E_i(f_i) = V(f_i \mid d_i, \hat{f}_{\mathcal{N}_i}) = V(d_i \mid f_i) + \sum_{c:i\in c} V'_c(f_i \mid f_{\mathcal{N}_i}) \qquad (9.60)$$

where

$$V'_c = \begin{cases} 0 & \text{if } f_j = 0 \; \exists j \in c \\ V_c & \text{otherwise} \end{cases} \qquad (9.61)$$

Therefore, a site has no effect on its neighbors unless it has committed. When no neighbor is active, the local energy measure reduces to the likelihood of the label. A remark is that the *uncommitted* label is different from the NULL label described in Chapter 4 for object recognition in that any clique involving a site labeled NULL incurs a nonzero penalty.

The stability of i w.r.t. f is defined as

$$S_i(f) = \begin{cases} -\min_{l\in\mathcal{L}, l\neq l_{\min}} [E_i(l) - E_i(l_{\min})] & \text{if } f_i = 0 \\ \min_{l\in\mathcal{L}, l\neq f_i} [E_i(l) - E_i(f_i)] & \text{otherwise} \end{cases} \qquad (9.62)$$

where $l_{\min} = \arg\min_{l\in\mathcal{L}} E_i(l)$. The stability S of an uncommitted site is the negative difference between the lowest and the second lowest local energies (conditional potentials). The stability S of a committed site is the difference between the current local energy $E_i(f_i)$ and the lowest possible energy due to any other label. Note that the range of a stability is $-\infty < S_i < +\infty$. A negative stability means there is room for improving. All uncommitted sites have nonpositive S_i. The magnitude of S_i is equal to the change in energy due to the change in the label f_i. A lower value of $S_i(f)$ indicates a more stable configuration, and a negative value with larger magnitude gives us more confidence to change f to a new configuration.

The following rule is imposed for deciding the order of update. At each step, only the least stable site is allowed to change its label or to make its commitment. Suppose that $k = \arg\max_i S_i(f)$ is the least stable site. Then, if $f_k = 0$, change f_k to

$$f'_k = \arg\min_{l\in\mathcal{L}} E_k(l) \qquad (9.63)$$

Otherwise, change f_k to

$$f'_k = \arg\min_{l\in\mathcal{L}, l\neq f_k} [E_k(l) - E_k(f_k)] \qquad (9.64)$$

Therefore, the first committed label f_i is that which yields the maximum local likelihood $V(d_i \mid f_i)$. The HCF algorithm is described in Fig. 9.4. Create_Heap creates a heap in which the S-values are sorted and the least stable

---

[1]Our definition has a different appearance from Chou's $E_s(l)$ in Equation (14) of Chou and Brown (1990), but is equivalent to it. Note that the definition of $E_s(l)$ given in (Chou et al. 1993) contains a bug.

```
f = {0,...,0};
Create_Heap;
while (S_top(f) < 0) {
      k = top;
      Change_State(k);
      Update_S(k);
      Adjust_Heap(k);
      for each j ∈ N_k {
            Update_S(j);
            Adjust_Heap(j);
      }
}
return(f* = f)
```

Figure 9.4: The highest confidence first algorithm.

site is placed at the *top*. Change_State(k) changes f_k to f'_k according to the rule described earlier. Update_S(k) updates the stability value of k using the current f. Adjust_Heap(k) adjusts the heap according to the current S-values.

A parallel version of HCF, called "local HCF", is described in (Chou et al. 1993). In local HCF, the update on each iteration is performed according to the following rule: For each site in parallel, change the state of the site if its stability is negative and lower than the stabilities of its neighbors.

Some results, comparing HCF with other algorithms including simulated annealing and ICM, are given in (Chou et al. 1993). The results show that HCF is, on the whole, better in terms of the minimized energy $E(f^*)$. The initialization is not a problem in HCF because it is always set to zeros.

9.3.6   Dynamic Programming

Dynamic programming (DP) (Bellman and Dreyfus 1962) is an optimization technique for problems where not all variables (labels in MRF's) are interrelated simultaneously. This is useful for the computation of MRF's because contextual constraints therein, causal or noncausal, are localized. The technique is based on the *principle of optimality*. Bellman states it as follows: "An optimal policy has the property that whatever the initial state and the initial decision are, the remaining decisions must constitute an optimal policy with regard to the state resulting from the first decision."

Suppose that the global energy can be decomposed into the form

$$E(f_1,\ldots,f_m) = E_1(f_1,f_2) + E_2(f_2,f_3) + \cdots + E_{m-1}(f_{m-1},f_m) \quad (9.65)$$

DP generates a sequence of functions of one variable

$$D_1(f_2) = \min_{f_1} E_1(f_1, f_2)$$
$$D_2(f_3) = \min_{f_2}[D_1(f_2) + E_2(f_2, f_3)] \tag{9.66}$$

$$\vdots$$

The first function is obtained as follows: For each f_2 value, choose the minimum of $E_1(f_1, f_2)$ over all f_1 values. The sequence can be written in the recursive form: For $k = 1, \ldots, m - 1$,

$$D_k(f_{k+1}) = \min_{f_k}[D_{k-1}(f_k) + E_k(f_k, f_{k+1})] \tag{9.67}$$

with $D_0(f_1) = 0$. The minimal energy solution is obtained by

$$\min_f E(f_1, \ldots, f_m) = \min_{f_m} D_{m-1}(f_m) \tag{9.68}$$

This is a valid formula if the problem can really be decomposed into the form of (9.65).

With regard to the validity, we have the relationship

$$\min_{f_1, f_2, f_3} [E_1(f_1, f_2) + E_2(f_2, f_3)] \le \min_{f_3} \left\{ \min_{f_2} \left[\min_{f_1} E_1(f_1, f_2) + E_2(f_2, f_3) \right] \right\} \tag{9.69}$$

If the equality holds, it means the DP algorithm gives the true minimum of E. The condition for the equality to hold is that $E_2(f_2, f_3)$ be independent of f_1. If this is not true, then the DP solution is not a global one. This conclusion can be generalized to situations with more variables.

If each f_k takes on M discrete values in $\mathcal{L}$, then to compute $D_{k-1}(f_k)$ for each f_k value, one must evaluate the minimand $[D_{k-2}(f_{k-1})+E_{k-1}(f_{k-1}, f_k)]$ for the M different f_{k-1} values in order to determine the minimal value of the minimand. Therefore, the overall minimization involves $(m - 1)M^2 + M$ evaluations of the minimands. This is an enormous reduction from the exhaustive minimization; for the latter case, $|\mathcal{L}|^{|S|}$ evaluations of $E(f_1, \ldots, f_m)$ have to be made, where $|\cdot|$ denotes the number of elements in the set.

Applications of DP in vision include curve detection (Ballard and Brown 1982), matching (Fischler and Elschlager 1973), and energy minimization for MRF-based image segmentation (Derin and Elliott 1987) and active contours (Amini et al. 1990). The Derin-Elliot algorithm is described below.

The algorithm is designed for the MLL region model with additive white Gaussian noise. The MRF is defined on an image lattice $\mathcal{S} = \{(i, j) \mid 1 \le i, j \le n\}$. Each true label $f_{i,j}$ takes a value in $\mathcal{L} = \{\ell_1, \ldots, \ell_M\}$. The posterior energy is expressed as

$$E(f) = U(f \mid d) = \sum_{c \in \mathcal{C}} V_c(f) + \sum_{I \in \mathcal{L}} \sum_{(i,j) \in \mathcal{S}^{(I)}} \frac{1}{\sigma^2}(d_{i,j} - \ell_I)^2 \tag{9.70}$$

where V_c are the MLL clique potentials on the 8-neighborhood system given in Fig. 2.3, $\mathcal{S}^{(I)} = \{(i,j) \in \mathcal{S} \mid f_{i,j} = \ell_I\}$ is the set of sites whose labels take the value $\ell_I \in \mathcal{L}$, and σ^2 is the noise variance. Similar to (9.65), a decomposition of energy is done as

$$E(f) = \sum_{k=1}^{m-1} E_k(f_k, f_{k+1}) \tag{9.71}$$

where f_k denotes the kth row. Function $E_k(f_k, f_{k+1})$ concerns only the two rows

$$E_k(f_k, f_{k+1}) = \sum_{c \in \mathcal{C}^{k,k+1}} V_c(f) + \sum_{I \in \mathcal{L}} \sum_{\mathcal{S}_I^k} \frac{1}{\sigma^2} (d_{k,j} - \ell_I)^2 \tag{9.72}$$

where

$$
\begin{aligned}
\mathcal{C}^{k,k+1} &= \{c \mid c \text{ is a clique with pixels only in columns } k \text{ and } k+1\} \\
\mathcal{S}_I^k &= \{(k,j) \mid f_{k,j} = \ell_I, 1 \le i \le m\}
\end{aligned}
\tag{9.73}
$$

In this decomposed form, $E(f)$ can be minimized using the DP technique just described. Because (9.72) is a valid formula for the MLL model on the 8-neighborhood system, the DP solution should give an exact MAP estimate.

When the observation model is due to textures, the likelihood function cannot be decomposed into independent terms for the application of DP. Derin and Elliott (1987) make some assumptions to simplify the conditional probabilities involved before applying the DP algorithm. This gives a suboptimal solution.

9.4   Constrained Minimization

In constrained minimization, extra constraints, either equalities or inequalities, are imposed on the solution f. The problem with K equality constraints is stated as

$$
\begin{aligned}
&\min_f &&E(f) \\
&\text{subject to} &&C_k(f) = 0 \quad k = 1, \dots, K
\end{aligned}
\tag{9.74}
$$

where $C_k(f) = 0$ denotes the equality constraints. For an inequality constraint of the form $D(f) \le 0$, we can introduce a nonnegative slack function of one variable (e.g., y^2) and convert it into an equality constraint $D(f) + y^2 = 0$, so only problems with equality constraints are considered here. The constraints define the feasible region $\mathbb{F}$ for f.

The problem is termed linear programming when the energy function and the constraint functions are all linear in f; otherwise it is nonlinear. Methods for linear and nonlinear programming have quite different features. Because our energy functions $E(f)$ can be nonlinear in f, our discussion will be on nonlinear programming problems.

When the constraint functions are linear, the gradient projection method (Rosen 1960) can be used to find feasible directions for the steepest descent. Given that $f \in \mathbb{F}$ is in the feasible region, the gradient $\nabla E(f)$ is computed. Direct gradient descent gives $f - \mu \nabla E(f)$. But this vector may no longer be in $\mathbb{F}$. To keep the feasibility, the new configuration is obtained by projecting $f - \mu \nabla E(f)$ onto $\mathbb{F}$. Using this, the updated configuration still lies in the feasible region. The computation is not difficult when the constraints are linear such that $\mathbb{F}$ corresponds to a hyperplane. This is the GP operator used in RL described earlier.

In classical methods for nonlinear programming, constrained minimization problems are converted into unconstrained ones (e.g., by using penalty functions and Lagrange multipliers). In RL, the feasibility constraints are respected throughout the updating process. However, in the numerical computation in penalty and Lagrange methods, this may not necessarily be so.

9.4.1 Penalty Functions

Using penalty functions, we can force f to stay inside or close to the feasible region determined by $C_k(f) = 0$ while minimizing $E(f)$. A simply penalty function for $C_k(f) = 0$ can be taken as $C_k^2(f)$. Adding them to the original energy yields a new energy

$$E(f, \beta) = E(f) + \sum_k \beta_k C_k^2(f) \qquad (9.75)$$

where $\beta_k > 0$ determines the amount of the penalty. Minimizing the above w.r.t. f for a fixed β is an unconstrained problem. The minimum in the limit $\beta_k \to \infty$ ($\forall k$)

$$f^* = \arg \lim_{\beta_k \to \infty (\forall k)} \min_f E(f, \beta) \qquad (9.76)$$

will be one satisfying the constraints $C_k(f) = 0$. This suggests an algorithm that solves a sequence of unconstrained minimization problems:

1. Choose a fixed sequence $\beta^{(0)}, \beta^{(1)}, \dots$ with $\beta_k^{(t)} < \beta_k^{(t+1)}$ and $\beta_k^{(t)} \to \infty$.

2. For each $\beta^{(t)}$, find a local minimum $f^*(\beta^{(t)}) = \arg \min_f E(f, \beta^{(t)})$.

3. Terminate when $C_k^2(f^*(\beta^{(t)}))$ is small enough.

In practice, step (2) is usually done numerically using an iterative method. Moreover, f^* obtained previously with $\beta^{(t)}$ is used as the initial point for the next minimization with $\beta^{(t+1)}$.

Consider the problem of pose estimation from a set of corresponding points discussed in Section 4.4.1. Assume $\mathbb{F}^0$ is the unconstrained solution space in which a pose f consists of an arbitrary rotation matrix and a translation vector in 3D. The energy $E(f)$ without constraints measures squared errors. If poses are constrained to be Euclidean, as assumed, any rotation matrix must be orthogonal. The penalty function for orthogonality was given in (4.43) as the single-site potential function. A similar penalty function has been used by (Haralick et al. 1989).

In Section 3.3.2 where boundary detection was formulated as constrained minimization, certain edge patterns, such as isolated edges, sharp turns, quadruple junctions, and "small" structures, are forbidden in forming desirable boundaries. These are expressed as $U^E(f^E) = 0$. In the penalty method, the penalty term $U^E(f^E)$ is added to the energy $E(f^P, f^E)$, yielding $E'(f^P, f^E) = E(f^P, f^E) + \lambda_E U^E(f^E)$. Sending λ_E to $+\infty$ in the minimization will rule out occurrences of forbidden edge configurations.

To overcome the problem of local minima, an annealing process can be incorporated with the penalty method. In constrained simulated annealing (Geman et al. 1990), an annealing parameter T is introduced into E, yielding

$$E(f, \beta, T) = \left[E(f) + \sum_k \beta_k C_k^2(f) \right] / T \qquad (9.77)$$

Using a stochastic sampling scheme such as the Metropolis algorithm or Gibbs sampler (see Section 10.1), one can generate a Gibbs distribution f with energy functions $E(f, \beta, T)$. With an annealing scheme in which $\beta^{(t)} \nearrow \infty$ and $T^{(t)} \searrow 0$ at a suitably coupled rate, $f^{(t)}$ will converge to the global minimum of $E(f)$ with probability approaching one. Although this is computationally more demanding than ICM (Besag 1986), it invariably arrives at a better labeling.

The penalty method has a number of advantages: It is easy to use, it allows inexact constraints (constraints are not necessary to fulfill exactly), and it converges to a feasible solution when $\beta_k \to \infty$. However, allowing inexact constraints is disadvantageous for those problems in which the exact fulfillment of constraints is required. Moreover, when β_i are very large, the system becomes *stiff* and the problem becomes ill-conditioned (Fletcher 1987). The Lagrange multiplier method may be better in this regard.

9.4.2   Lagrange Multipliers

The Lagrange multiplier method converts the constrained minimization into an unconstrained minimization of the Lagrange function of $m + K$ variables

$$L(f, \gamma) = E(f) + \sum_k \gamma_k C_k(f) \qquad (9.78)$$

where γ_k are called Lagrange multipliers. For f^* to be a local minimum subject to the constraints, it is necessary that (f^*, γ^*) be a stationary point of the Lagrange function:

$$\begin{aligned} \nabla_f L(f^*, \gamma^*) &= 0 \\ \nabla_\gamma L(f^*, \gamma^*) &= 0 \end{aligned} \qquad (9.79)$$

If (f^*, γ^*) is a saddle point for which

$$L(f^*, \gamma) \leq L(f^*, \gamma^*) \leq L(f, \gamma^*) \qquad (9.80)$$

then f^* is a local minimum of $E(f)$ satisfying $C_k(f^*) = 0$ (Gottfried 1973). The following dynamics can be used to find such a saddle point:

$$\frac{\mathrm{d}f_i}{\mathrm{d}t} = -\frac{\partial L(f, \gamma)}{\partial f_i} = -\frac{\partial E(f)}{\partial f_i} - \sum_k \gamma_k \frac{\partial C_k(f)}{\partial f_i}$$

$$\frac{\mathrm{d}\gamma_k}{\mathrm{d}t} = +\frac{\partial L(f, \gamma)}{\partial \gamma_k} = +C_k(f) \qquad (9.81)$$

It performs energy descent on f but ascent on γ. Convergence results of this system have been obtained by Arrow, Hurwicz, and Uzawa (1958). The dynamics has been used for neural computing (Platt and Barr 1988).

The penalty terms in the penalty function (9.75) can be added to the Lagrange function (9.78) to give an augmented Lagrange function (Powell 1969; Hestenes 1969)

$$L(f, \gamma, \beta) = E(f) + \sum_k \gamma_k C_k(f) + \sum_k \beta_k [C_k(f)]^2 \qquad (9.82)$$

where β_k are *finite* weights. The addition of the penalty terms does not alter the stationary point and sometimes helps damp oscillations and improve convergence. Because β_k are finite, the ill-conditioning of the problem in the penalty method when $\beta_k \to \infty$ is alleviated. After the penalty term is added, the dynamics for f_i is

$$\frac{\mathrm{d}f_i}{\mathrm{d}t} = -\frac{\partial E(f)}{\partial f_i} - \sum_k \gamma_k \frac{\partial C_k(f)}{\partial f_i} - 2 \sum_k \beta_k C_k(f) \frac{\partial C_k(f)}{\partial f_i} \qquad (9.83)$$

9.4.3 Hopfield Method

The graded Hopfield neural network (HNN) (Hopfield 1984) provides another way of applying penalty functions. Let us use the constrained minimization in RL as an example to illustrate this point. There, the feasibility (9.19) must be satisfied by the labeling assignment at any time and the unambiguity (9.22) should be satisfied by the final labeling assignment. The unambiguity constraint can be imposed by the term

$$E_a(f) = \sum_i \sum_I \int_0^{f_i(I)} \psi_T^{-1}(f) df \qquad (9.84)$$

where ψ_T^{-1} is the inverse of a function ψ_T to be illustrated below. A local state $f_i(I) \in [0, 1]$ is related to an *internal variable* $u_i(I) \in (-\infty, +\infty)$ by

$$f_i(I) = \psi_T(u_i(I)) \qquad (9.85)$$

where $\psi_T(u)$ is usually a sigmoid function

$$\psi_T(u) = 1/[1 + e^{-u/T}] \tag{9.86}$$

controlled by the parameter T. In very high gains when $T \to 0^+$, $f_i(I)$ is forced to be 0 or 1 depending on whether $u_i(I)$ is positive or negative (Hopfield 1984). The term $E_a(f)$ reaches the minimum of zero only when all $f_i(I)$'s are either 0 or 1. This means that minimizing this term with $T \to 0^+$ in effect leads to unambiguous labeling.

On the other hand, the feasibility can be imposed by

$$E_b(f) = \sum_i \left[\sum_I f_i(I) - 1 \right]^2 = 0 \tag{9.87}$$

This term has its minimum value of zero when the feasibility is satisfied. Now the constrained optimization minimizes the functional

$$E'(f \mid T) = E(f) + aE_a(f \mid T) + bE_b(f) \tag{9.88}$$

where a and b are weights. A minimal solution which is feasible and unambiguous is given by

$$f^* = \arg \min_f \lim_{T \to 0^+} E'(f \mid T) \tag{9.89}$$

To derive a dynamic system for the minimization, introduce a time variable into f such that $f = f^{(t)}$. The energy change due to the state change $\frac{\mathrm{d}f_i(I)}{\mathrm{d}t}$ is

$$\frac{\mathrm{d}E'}{\mathrm{d}t} = -\sum_{i,I} \frac{\mathrm{d}f_i(I)}{\mathrm{d}t} \left\{ q_i(I) - a\, u_i(I) - 2b \left[\sum_I f_i(I) - 1 \right] \right\} \tag{9.90}$$

where

$$q_i(I) = \frac{\mathrm{d}E}{\mathrm{d}f_i(I)} = V_1(I \mid d_1(i)) + 2 \sum_{i',I'} V_2(I',I \mid d_2(i,i'))\, f_{i'}(I') \tag{9.91}$$

is a gradient component. The following dynamics minimizes the three-term energy E':

$$C\frac{\mathrm{d}u_i(I)}{\mathrm{d}t} = q_i(I) - a\, u_i(I) - 2b \left[\sum_I f_i(I) - 1 \right] \tag{9.92}$$

where the capacitance $C > 0$ controls the convergence of the system. With the dynamics above, the energy change

$$\frac{\mathrm{d}E'}{\mathrm{d}t} = -C \sum_{i,I} \frac{\mathrm{d}f_i(I)}{\mathrm{d}t} \frac{\mathrm{d}u_i(I)}{\mathrm{d}t} - C \sum_{i,I} \frac{\mathrm{d}\psi_T^{-1}(f_i(I))}{\mathrm{d}f_i(I)} \left[\frac{\mathrm{d}f_i(I)}{\mathrm{d}t} \right]^2 \tag{9.93}$$

is nonpositive since ψ_T^{-1} is a monotonically increasing function and C is positive. This update rule will lead to a minimal solution that is both feasible and unambiguous when $T \to 0$, $a \to \infty$ and $b \to \infty$.

To solve combinatorial optimization, Wacholder, Han, and Mann (1989) developed an algorithm that combines the HNN of Hopfield (1984) and the Lagrange multiplier dynamics of Platt and Barr (1988). Good convergence to feasible solutions was reported. In Section 9.5, we will describe a method that combines the HNN and the augmented Lagrangian function to perform relaxation labeling.

The HNN approach has been used to convert binary-valued line process variables to real ones for low-Level processes (Koch et al. 1986; Yuille 1987) (see Section 5.1.1). It has also been used as an alternative to relaxation labeling for matching; this is seen as symbolic image labeling (Jamison and Schalkoff 1988), subgraph isomorphism for object recognition (Nasrabadi et al. 1990), and matching to multiple view models (Lin et al. 1991).

Comparing RL and HNN for combinatorial minimizations such as object matching, our experience is that the former performs better. Due to the influence of the extra penalty function added in the HNN approach, the modified energy may no longer solve the original matching problem (Horn 1988). Moreover, it easily leads to unfavorable local solutions. The RL algorithm is much more careful in this regard. It uses gradient projection to choose the best direction and magnitude for the state to evolve. In other words, the problem of local optima is more significant in the HNN approach. Computationally, the RL needs a smaller number of iterations before convergence.

9.5 Augmented Lagrange-Hopfield Method

A combinatorial optimization can often be converted into a constrained real optimization with equality and inequality constraints. The penalty and the Lagrange multiplier methods can be used for coping with equality constraints and the barrier method can be used for coping with inequality constraints. However, the penalty method suffers from ill-conditioning, and the Lagrange method suffers from the zigzagging problem (Fletcher 1987). The augmented Lagrange (AL) method (Powell 1969) combines both the Lagrange and the penalty methods and effectively overcomes the associated problems. In AL, the relative weight for the penalty terms need not be infinitely large. This not only overcomes the ill-conditioning problem but is also beneficial for obtaining better-quality solutions because the relative importance of the original objective function is emphasized more; at the same time, its use of quadratic penalty terms "convexifies" and hence stabilizes the system, overcoming the zigzagging problem (Fletcher 1987).

Mean field annealing (MFA) (Peterson and Soderberg 1989) provides still another continuous method. Assuming that the minima of the original energy and the corresponding mean field effective energy coincide, the MFA aims to approximate the global minimum of the original energy by tracking that of the effective energy with decreasing temperature. An analysis shows that the effective energy of MFA is identical to a combination of the original energy, a particular barrier term and a standard Lagrange term (Yuille and Kosowsky 1994).

Here, another deterministic method, called the augmented Lagrange-Hopfield (ALH) method, is presented for the combinatorial optimization in MAP-MRF image restoration and segmentation. In solving the converted constrained real optimization, the ALH method uses the augmented Lagrangian multiplier method (Powell 1969) to satisfy the equality constraints and the Hopfield network encoding (Hopfield 1984) to impose the inequality constraints. The use of AL effectively overcomes instabilities inherent in the penalty method and the Lagrange multiplier method. The resulting algorithm solves a system of differential equations. Experimental results in both image restoration and segmentation have compared the ALH method with the ICM, HCF, and SA. The results show that the ALH outperforms ICM and HCF and is comparable to SA in terms of the solution quality; it quickly yields a good solution after a dozen of iterations, a number similar to that required by ICM and HCF but much smaller than for SA. A discussion on MFA results is also provided.

9.5.1   MAP-MRF Estimation as Constrained Optimization

The underlying image signal is denoted $f = \{f_i \mid i \in \mathcal{S}\}$, where $\mathcal{S} = \{1, \ldots, m\}$ indexes the set of sites corresponding to image pixels. Each pixel takes on a discrete value f_i in the label set $\mathcal{L} = \{1, \ldots, M\}$; f is an underlying image to be restored or a segmented map to be computed. The spatial relationship of the sites, each of which is indexed by a single number in $\mathcal{S}$, is determined by a neighborhood system $\mathcal{N} = \{\mathcal{N}_i \mid i \in \mathcal{S}\}$, where $\mathcal{N}_i$ is the set of sites neighboring i. A single-site or a set of neighboring sites form a clique denoted by c. Here, only up to second-order (pair-site) cliques defined on the 8-neighborhood system are considered.

The underlying image f can consist of blob-like regions or a texture pattern. The different types are due to the different ways that pixels interact with each other (i.e., due to different contextual interactions). Such contextual interactions can be modeled as MRF's or Gibbs distributions of the form $P(f) = Z^{-1} \times e^{-\sum_{c \in \mathcal{C}} V_c(f)}$, where $V_c(f)$ is the potential function for clique c, $\mathcal{C}$ is the set of all cliques, and Z is the normalizing constant. Among various MRF's, the multilevel logistic (MLL) model is a simple, and yet powerful mechanism for encoding a large class of spatial patterns such as textured or nontextured images. In MLL, the pair-site clique potentials take the form: $V_2(f_i, f_{i'}) = \beta_c$ if sites in clique $\{i, i'\} = c$ have the same label or $V_2(f_i, f_{i'}) = -\beta_c$ otherwise, where β_c is a parameter for type c cliques; the single-site potentials are defined by $V_1(f_i) = \alpha_I$, where α_I is the potential for the label $I = f_i$. When the true pixel values are contaminated by identical independently distributed (i.i.d.) Gaussian noise, the observed data, or the likelihood model, is $d_i = f_i + e_i$, where $e_i \sim N(0, \sigma^2)$ is the zero-mean Gaussian distribution with standard deviation σ. With these prior and likelihood models, the energy in the posterior distribution $P(f \mid d) \propto e^{-E(f)}$ is

$$E(f) = \sum_{i \in S}(f_i - d_i)^2/[2\sigma^2] + \sum_{\{i\} \in C} V_1(f_i) + \sum_{\{i,i'\} \in C} V_2(f_i, f_{i'}) \qquad (9.94)$$

Here, we assume that the MRF and noise parameters α, β, and σ are known. The MAP estimate for the restoration or segmentation is defined as $f^* = \arg\min_{f \in \mathcal{L}^m} E(f)$. The minimization of the energy function $E(f)$ is in the discrete space $\mathcal{L}^m$ and hence combinatorial.

The original combinatorial optimization is converted into a constrained optimization in a real space by using the notion of continuous relaxation labeling. Let real value $p_i(I) \in [0, 1]$ represent the strength with which label I is assigned to i, the M-element vector $p_i = [p_i(I) \mid I \in \mathcal{L}]$ the state of the assignment for $i \in S$, and the matrix $p = [p_i(I) \mid i \in S, I \in \mathcal{L}]$ the state of the labeling assignment. The energy with the p variables is given by

$$E(p) = \sum_{i \in S}\sum_{I \in \mathcal{L}} r_i(I)\, p_i(I) + \sum_{i \in S}\sum_{I \in \mathcal{L}}\sum_{i' \in S, i' \neq i}\sum_{I' \in \mathcal{L}} r_{i,i'}(I, I')\, p_i(I)\, p_{i'}(I')$$

$$(9.95)$$

where $I = f_i$ and $I' = f_{i'}$, $r_i(I) = V_1(I \mid d) = (I - d_i)^2/2\sigma + V_1(I)$ is the single-site clique potential function in the posterior distribution $P(f \mid d)$, and $r_{i,i'}(I, I') = V_2(I, I' \mid d) = V_2(I, I')$ is the pair-site clique potential function in $P(f \mid d)$.

With such a representation, the combinatorial minimization is reformulated as the constrained minimization

$$\min_{p} \quad E(p) \qquad (9.96)$$

$$\text{subject to} \quad C_i(p) = \sum_{I} p_i(I) - 1 = 0 \quad i \in S \qquad (9.97)$$

$$p_i(I) \geq 0 \qquad\qquad \forall i \in S, \forall I \in \mathcal{L} \qquad (9.98)$$

The final solution p^* is subject to additional ambiguity constraints: $p_i^*(I) \in \{0, 1\}$.

9.5.2 The ALH Method

The ALH method aims to solve the constrained minimization problem above. It uses the augmented Lagrange technique to satisfy the equality constraints of (9.97) and the Hopfield encoding to impose the inequality constraints of

(9.98). In the Hopfield encoding, internal variables $u_i(I) \in (-\infty, +\infty)$ are introduced and related to $p_i(I)$ via the sigmoid function

$$p_i(I) = \psi_T(u_i(I)) = \frac{1}{1 + e^{-u_i(I)/T}} \tag{9.99}$$

controlled by a temperature parameter $T > 0$. This treatment confines $p_i(I)$ to the range $(0, 1)$ so as to impose the inequality constraints. When $T \to 0^+$, $p_i(I)$ is forced to be 0 or 1 depending on whether $u_i(I)$ is positive or negative. Thereby, the ambiguity constraints $p_i^*(I) \in \{0, 1\}$ are imposed. With the free u variables, the energy function can be considered as $E(u) = E(p(u))$.

The sigmoid function of Hopfield is one of many ways for imposing the inequality constraints. Yuille and Kosowshy (Yuille and Kosowsky 1994) have used a barrier term of the form $T \sum_{i,I} p_i(I) \log p_i(I)$ for imposing $p_i(I) > 0$. However, care must be taken in the numerical implementation of a gradient-descent algorithm involving the barrier term in order to avoid $p_i(I)$ from decreasing to nonpositive values. A truncation operation $\min\{0, p_i(I)\}$ may be used to truncate negative $p_i(I)$ values to 0. But this inevitably causes oscillations and hence non-convergence. We find it easier and more stable to use the sigmoid function in that the u variables can be updated freely.

The augmented Lagrange (AL) function (Powell 1969) takes the form

$$L_\beta(p, \gamma) = E(p) + \sum_{k \in \mathcal{S}} \gamma_k C_k(p) + \frac{\beta}{2} \sum_{k \in \mathcal{S}} [C_k(p)]^2 \tag{9.100}$$

where γ_k are the Lagrange multipliers and $\beta > 0$ is the weight for the quadratic penalty term. It is a function of the $M \times m$ variables of p and the m variables of γ. When $\gamma_i = 0$ for all i, the AL reduces to the penalty method $L_\beta(p) = E(p) + \frac{\beta}{2} \sum_k [C_k(p)]^2$. When $\beta = 0$, it reduces to the standard Lagrange $L(p, \gamma) = E(p) + \sum_k \gamma_k C_k(p)$. With the use of Lagrange multipliers, the weight β for the penalty terms need not be infinitely large and this overcomes the ill-conditioning problem in the penalty method. Smaller β is also beneficial for better quality solution because the original objective function E is more emphasized. The use of quadratic penalty terms "convexifies" the system and hence stabilizes it, and this overcomes the zigzagging problem in the standard Lagrange.

When (p^*, γ^*) is a saddle point for which $L_\beta(p^*, \gamma) \le L_\beta(p^*, \gamma^*) \le L_\beta(p, \gamma^*)$, then p^* is a local minimum of $E(p)$ satisfying (9.97) (Gottfried 1973). The dynamic equations for minimizing L_β are

$$\frac{dp_i(I)}{dt} = -\frac{\partial L_\beta(p, \gamma)}{\partial p_i(I)} = -\frac{\partial E(p)}{\partial p_i(I)} - \sum_k \gamma_k \frac{\partial C_k(p)}{\partial p_i(I)} - \beta \sum_k C_k(p) \frac{\partial C_k(p)}{\partial p_i(I)} \tag{9.101}$$

$$\frac{d\gamma_i}{dt} = +\frac{\partial L_\beta(p, \gamma)}{\partial \gamma_i} = +C_i(p) \tag{9.102}$$

where

$$\frac{\partial C_k(p)}{\partial p_i(I)} = \begin{cases} 1 & \text{if } i = k \\ 0 & \text{otherwise} \end{cases} \qquad (9.103)$$

for $C_i(p)$ defined in (9.97). The system performs energy descent on p but ascent on γ.

In the ALH method, the updating of label assignment is performed on u, rather than on p variables. The first part of Equation (9.101) is replaced by

$$\frac{du_i(I)}{dt} = -\frac{\partial L_\beta(p(u), \gamma)}{\partial p_i(I)} = -[q_i(I) + \gamma_i + \beta C_i(p)] \qquad (9.104)$$

where $q_i(I) = \frac{\partial E}{\partial p_i(I)}$. Because $\frac{\partial p_i(I)}{\partial u_i(I)} = \frac{e^{-u_i(I)/T}}{T(1+e^{-u_i(I)/T})^2}$ is always positive, $\frac{dp_i(I)}{dt} = \frac{\partial p_i(I)}{\partial u_i(I)} \frac{du_i(I)}{dt}$ has the same sign as $\frac{du_i(I)}{dt}$.

Numerically, the ALH algorithm consists of the following three equations, corresponding to (9.104), (9.99) and (9.102):

$$u_i^{(t+1)}(I) \leftarrow u_i^{(t)}(I) - \mu \left\{ q_i^{(t)}(I) + \gamma_i^{(t)} + \beta C_i(p^{(t)}) \right\} \qquad (9.105)$$

$$p_i^{(t+1)}(I) \leftarrow \psi_T(u_i^{(t+1)}(I)) \qquad (9.106)$$

$$\gamma_i^{(t+1)} \leftarrow \gamma_i^{(t)} + \beta C_i(p^{(t)}) \qquad (9.107)$$

In the above, μ is the step size; during the update, β is increased to speed up convergence. In practice, T need not be very low to impose the ambiguity constraints on the final solution p^*. The competitive mechanism in (9.97) will make the winner take all. We set $T = 10^5$ in our experiments and obtained good convergence. In our implementation, the update is performed synchronously in parallel for all i and I.

To summarize, the advantages of the augmented Lagrange-Hopfield (ALH) method are due to the use of the augmented Lagrange method. The ALH method not only overcomes the ill-conditioning problem of the penalty method and the zigzagging problem of the standard Lagrangian method but also improves convergence. With the augmented Lagrangian method, the weights for the penalty terms can be much smaller than those required by the penalty method, and the relative weight for the original objective can be increased. This helps yield lower minimized energy values. The use of ALH in image restoration and segmentation will be demonstrated in Section 10.6.2. The ALH method as a general method has also been used for solving other combinatorial optimization problems. For example, in solving the traveling salesman problem, the ALH finds remarkably better solutions than Hopfield type neural networks (Li 1996a).

Chapter 10

Minimization – Global Methods

The minimal solution is usually defined as the global one or one of them when there are multiple global minima. Finding a global minimum is non-trivial if the energy function contains many local minima. Whereas methods for local minimization are quite mature, with commercial software on the market, the study of global minimization is still young. There are no efficient algorithms that guarantee finding globally minimal solutions as there are for local minimization.

In analytical studies, local and global minima may be studied via convexity analysis. Let $E(f)$ be a real function defined on $\mathbb{F} = \mathbb{R}^n$. $E(f)$ is said to be convex in $\mathbb{F}$ if

$$E(\lambda x + (1 - \lambda)y) \leq \lambda E(x) + (1 - \lambda)E(y) \tag{10.1}$$

for any two points $x, y \in \mathbb{F}$ and any real number $\lambda \in [0, 1]$. It is strictly convex if strict inequality "<" holds in the above. When plotted, a convex $E(f)$ will always lie below the line segment connecting any two points on itself. The neighboring points of a minimum are always above the tangent at the minimum. A local minimum is also a global one if $E(f)$ is convex on $\mathbb{F}$. It is the unique global minimum if $E(f)$ is strictly convex. Therefore, when $E(f)$ is convex, we can find a global minimum by finding a local minimum using the gradient-descent method.

Global minimization requires (1) finding all (a finite number of) local minima and (2) proving that there are no more local minima. Without an efficient algorithm, this amounts to an exhaustive search. In reality, one is always facing one of two choices: (1) to find the exact global minimum with possibly intolerable expense or (2) to find some approximations to it with much less cost.

S.Z. Li, *Markov Random Field Modeling in Image Analysis*,
Advances in Pattern Recognition, DOI: 10.1007/978-1-84800-279-1_10,
© Springer-Verlag London Limited 2009

Two methods may be used to deal with the local minimum problem: random search and annealing. In random search methods, a new configuration does not always make energy descend; occasional energy increases are allowed. The probability with which a configuration is generated is proportional to $e^{-E(f)/T}$, where T is a control parameter called the temperature. Therefore, a lower-energy configuration is generated with a larger probability.

Annealing is incorporated into a local search method to overcome the problem of local minima. It is performed by decreasing the temperature parameter in the Gibbs distribution from a high value to a low value during the iterative minimization. At one extreme (e.g., a high temperature), the energy landscape is convex and smooth and thus the unique minimum can be located easily; the minimum is tracked as the is gradually decreased to a sufficiently low value. This is the so-called continuation method (Wasserstrom 1973). Such a technique can significantly overcome the local minimum problem and improve the quality of the solution. A disadvantage is that they take more time since an annealing algorithm has to decrease the parameter gradually over a range of values and at each value some convergence has to be reached.

There are two types of annealing: deterministic and stochastic. In MRF vision work, the stochastic simulated annealing algorithm (Geman and Geman 1984) and the deterministic graduated nonconvexity (GNC) algorithm of Blake and Zisserman (1987) enjoy a popularity. Other deterministic algorithms include mean field annealing (Peterson and Soderberg 1989; Yuille 1990; Geiger and Girosi 1991) and the Hopfield network approach (Hopfield and Tank 1985; Koch et al. 1986; Yuille 1987).

While stochastic annealing such as simulated annealing is theoretically justified (Geman and Geman 1984), deterministic annealing remains heuristic. There is no guarantee that the minimum at high temperature can always be tracked to the minimum at low temperature.

10.1   Simulated Annealing

Simulated annealing (SA), introduced independently by Cerny (1982), Kirkpatrick, Gellatt, and Vecchi (1982),Cerny (1985), and Kirkpatrick, Gellatt, and Vecchi (1983), is a stochastic algorithm for combinatorial optimization. It simulates the physical annealing procedure in which a physical substance is melted and then slowly cooled down in search of a low energy configuration. Consider a system in which any f in the configuration space $\mathbb{F}$ has probability

$$P_T(f) = [P(f)]^{1/T} \tag{10.2}$$

where $T > 0$ is the temperature parameter. When $T \to \infty$, $P_T(f)$ is a uniform distribution on $\mathbb{F}$; for $T = 1$, $P_T(f) = P(f)$; and as $T \to 0$, $P_T(f)$ is concentrated on the peak(s) of $P(f)$. This gives intuition as to how the samples of f distribute in $\mathbb{F}$.

```
initialize T and f;
repeat
    randomly sample f from N(f) under T;
    decrease T;
until (T → 0);
return f;
```

Figure 10.1: The simulated annealing algorithm.

The SA algorithm is described in Fig. 10.1. SA applies a sampling algorithm, such as the Metropolis algorithm (Metropolis et al. 1953) or Gibbs sampler (Geman and Geman 1984) (see Section 7.1.6), successively at decreasing values of temperature T. Initially, T is set very high and f is set to a random configuration. At a fixed T, the sampling is according to the Gibbs distribution $P_T(f) = e^{-E(f)/T} / \sum_f e^{-E(f)/T}$. After the sampling converges to the equilibrium at current T, T is decreased according to a carefully chosen schedule. This continues until T is close to 0, at which point the system is "frozen" near the minimum of $E(f)$. The cooling schedule, specified by a decrement function and a final value, plays an important part; see below.

Two convergence theorems have been developed (Geman and Geman 1984). The first concerns the convergence of the Metropolis algorithm. It states that if every configuration is visited infinitely often, the distribution of generated configurations is guaranteed to converge to the Boltzmann (i.e., Gibbs) distribution. The second is about SA. It states that if the decreasing sequence of temperatures satisfy

$$\lim_{t \to \infty} T^{(t)} = 0 \tag{10.3}$$

and

$$T^{(t)} \geq \frac{m \times \Delta}{\ln(1 + t)} \tag{10.4}$$

where $\Delta = \max_f E(f) - \min_f E(f)$, then the system converges to the global minimum regardless of the initial configuration $f^{(0)}$. Note that the conditions above are sufficient but not necessary for the convergence.

Unfortunately, the schedule (10.4) is too slow to be of practical use. In practice, heuristic, faster schedules have to be used instead. Geman and Geman (1984) adopt the following

$$T^{(t)} = \frac{C}{\ln(1 + t)} \tag{10.5}$$

where the constant C is set to $C = 3.0$ or $C = 0.4$ for their problem. Kirkpatrick et al. (1983) choose

$$T^{(t)} = \kappa T^{(t-1)} \qquad (10.6)$$

where κ typically takes a value between 0.8 and 0.99. The initial temperature is set high enough that essentially all configuration changes are accepted. At each temperature, enough configurations are tried that either there are $10m$ (10 times the number of sites) accepted configuration transitions or the number of tries exceeds $100m$. The system is frozen and annealing stops if the desired number of acceptances is not achieved at three successive temperatures. In (Kato et al. 1993a), a multi-temperature annealing scheme is proposed for annealing in multiresolution computation; there, the temperature in the Gibbs distribution is related not only to the time but also to the scale.

The two sampling procedures (i.e., the Metropolis algorithm and Gibbs sampler) are proven to be asymptotically equivalent for SA performed on lattices, but this is not generally true for non-lattice structures (Chiang and Chow 1992). The interested reader is referred to (Kirkpatrick et al. 1982; Geman and Geman 1984; Aarts 1989) for discussions devoted on SA.

A performance comparison between SA and GNC for edge-preserving surface reconstruction is given in (Blake 1989). Based on the results of the experiments under controlled conditions, Blake argues that GNC excels SA both in computational efficiency and problem-solving power. An experimental comparison between SA and deterministic RL algorithms for combinatorial optimization will be presented in Section 10.6.1.

10.2 Mean Field Annealing

Mean field annealing (Peterson and Soderberg 1989) provides another approach for solving combinatorial optimization problems using real computation. It can be considered as a special RL algorithm incorporating an annealing process. Let $f = \{f_i(I) \in \{0,1\} | i \in \mathcal{S}, I \in \mathcal{L}\}$ be an unambiguous labeling assignment in $\mathbb{P}^*$ defined in (9.24); each f_i takes the value of one of the vectors $(1,0,\ldots,0)$, $(0,1,\ldots,0)$, and $(0,0,\ldots,1)$. Still denote the energy with the RL representation by $E(f)$. The Gibbs distribution of the energy under temperature $T > 0$ is given as

$$P_T(f) = Z^{-1} e^{-\frac{1}{T}E(f)} \qquad (10.7)$$

where the partition function

$$Z = \sum_{f \in \mathbb{P}^*} e^{-\frac{1}{T}E(f)} \qquad (10.8)$$

sums over $\mathbb{P}^*$. The statistical mean of the distribution is defined as

$$\langle f \rangle_T = \sum_f f_i P_T(f) = \sum_f f_i Z^{-1} e^{-\frac{1}{T} E(f)} \tag{10.9}$$

As the temperature approaches zero, the mean of the Gibbs distribution approaches its mode or the global energy minimum f^*

$$\lim_{T \to 0} \langle f \rangle_T = \lim_{T \to 0} \sum_f f P_T(f) = f^* \tag{10.10}$$

This suggests that instead of minimizing $E(f)$ directly, we could try to evaluate the mean field $\langle f \rangle_T$ at a sufficiently high temperature and then track it down using the continuation method (Wasserstrom 1973) as the temperature is lowered toward zero.

The analysis of the partition function is central in mean field theory because once it is calculated all statistical information about the system can be deduced from it. The mean field trick for calculating the partition function is to (1) rewrite the sum over a discrete space as an integral over a pair of continuous variables p and q, $p = \{p_i(I) \in \mathbb{R} | i \in \mathcal{S}, I \in \mathcal{L}\}$ and $q = \{q_i(I) \in \mathbb{R} | i \in \mathcal{S}, I \in \mathcal{L}\}$ having the same dimensionality as f, and then (2) evaluate its integrand at the saddle point (saddle-point approximation). As will be seen later, p and q correspond to a labeling assignment as defined in (9.20) and a gradient as defined in (9.29), respectively. In the following, f, p, and q are considered $m \times M$ matrices. To do step (1), we begin by rewriting any function $g(f)$ as the integral

$$g(f) = \int_{\mathbb{D}_R} g(p) \delta(f - p) dp \tag{10.11}$$

where δ is the Dirac delta function and $\mathbb{D}_R$ is the $|\mathcal{S}| \times |\mathcal{L}|$ dimensional real space. The definition of the Dirac delta function

$$\delta(q) = C \int_{\mathbb{D}_I} e^{p^T q} dp dq \tag{10.12}$$

is used in the above, where T denotes transpose, $\mathbb{D}_I$ is the $|\mathcal{S}| \times |\mathcal{L}|$-dimensional imaginary space, and C is a normalizing constant. Therefore,

$$g(f) = C \int_{\mathbb{D}_R} \int_{\mathbb{D}_I} g(p) e^{(f^T q - p^T q)} dp dq \tag{10.13}$$

Using the formula above, we can rewrite the partition function, which is the sum of the function $e^{-\frac{1}{T} E(p)}$, as

$$
\begin{aligned}
Z &= \sum_{f \in \mathbb{P}^*} C \int_{\mathbb{D}_R} \int_{\mathbb{D}_I} e^{-\frac{1}{T} E(p)} e^{(f^T q - p^T q)} dp dq \tag{10.14} \\
&= C \int_{\mathbb{D}_R} \int_{\mathbb{D}_I} e^{-\frac{1}{T} E(p)} e^{(-p^T q)} \left[\sum_{f \in \mathbb{P}^*} e^{(f^T q)} \right] dp dq
\end{aligned}
$$

The summation inside the integrand can be written as

$$\sum_{f \in \mathbb{P}^*} e^{(f^\mathrm{T} q)} = \prod_{i \in \mathcal{S}} \sum_{I \in \mathcal{L}} e^{q_i(I)} \tag{10.15}$$

Therefore,

$$Z = C \int_{\mathbb{D}_R} \int_{\mathbb{D}_I} e^{-\frac{1}{T} E_{eff}(p,q)} \mathrm{d}p \mathrm{d}q \tag{10.16}$$

where

$$E_{eff}(p,q) = E(p) + T p^\mathrm{T} q - T \sum_{i \in \mathcal{S}} \ln \sum_{I \in \mathcal{L}} \left(e^{q_i(I)} \right) \tag{10.17}$$

is called the *effective energy*.

The integral expression of Z above is exact but too complicated for precise calculation. The calculation may be approximated based on the following heuristic argument: The double integral is dominated by saddle points in the integration intervals, and therefore the integral is approximated by

$$Z \approx e^{-\frac{1}{T} E_{eff}(p^*,q^*)} \tag{10.18}$$

where (p^*, q^*) is a saddle point of E_{eff}. The saddle points are among the roots of the equations

$$\nabla_p E_{eff}(p,q) = 0 \quad \text{and} \quad \nabla_q E_{eff}(p,q) = 0 \tag{10.19}$$

where ∇_p is the gradient w.r.t. p. This yields the mean field theory equations

$$q_i(I) = -\frac{1}{T} \frac{\partial E(p)}{\partial p_i(I)} \quad \text{and} \quad p_i(I) = \frac{e^{q_i(I)}}{\sum_J e^{q_i(J)}} \tag{10.20}$$

The p matrix represents a labeling assignment in the feasibility space defined by (9.19). The mean field theory equations for combinatorial minimization have been derived in many places, for example, in (Yuille 1990; Simic 1990; Elfadel 1993; Haykin 1994; Yang and Kittler 1994).

Note that in terms of the RL representation, the energy can be written as $E(f) = Const - G(f)$, and so q can be computed as

$$q_i(I) \;=\; \tfrac{1}{T} \left[r_i(I) + 2 \sum_{i' \in \mathcal{S}} \sum_{I' \in \mathcal{L}} r_{i,i'}(I, I') \, p_{i'}(I') \right] \tag{10.21}$$

(see 9.29). Now we obtain the fixed-point iteration

$$p_i^{(t+1)}(I) \leftarrow \frac{e^{q_i^{(t)}(I)}}{\sum_J e^{q_i^{(t)}(J)}} \tag{10.22}$$

In iterative computation, the temperature T in q is decreased toward 0^+ as the iteration preceeds. The unambiguity of p is achieved when $T \to 0^+$.

in (Peterson and Soderberg 1989), the convergence is judged by a quantity
defined as

$$S = \frac{1}{m} \sum_I p_i^2(I) \tag{10.23}$$

also called the "saturation" of f. The iteration (10.22) is considered converged if $S > 0.99$. Analog network implementation schemes for mean field
algorithms are investigated in (Elfadel 1993).

It is worth pointing out that the idea of mean field annealing for global op-
timization was proposed earlier by Pinkus (1968) as the integral limit method.
There, the global minimum is expressed in closed form as a limit of an integral.
Assume that $\mathbb{F}$ is the closure of a bounded domain in $\mathbb{R}^m$. Suppose that $E(f)$
is continuous on $\mathbb{F}$ and has a unique global minimum $f^* = \{f_1^*, \ldots, f_m^*\} \in \mathbb{F}$.
Pinkus shows that the integral

$$\bar{f}_i = \frac{\int_{\mathbb{F}} f_i e^{-\frac{1}{T}E(f)} df}{\int_{\mathbb{F}} e^{-\frac{1}{T}E(f)} df} \qquad i = 1, \ldots, m \tag{10.24}$$

approaches the global minimum f^* as $T \to 0^+$. Obviously, the integral above
is the continuous version of the mean field defined in (10.9), and the idea of the
integral limit is similar to that of mean field annealing. Theoretical analysis
shows that this approximation has rather good asymptotic properties, but
its realization may need some heuristic trials. Computationally, the values
f_i^* can be obtained analytically only in exceptional cases, and a numerical
method has to be used in general.

10.3 Graduated Nonconvexity

Graduated nonconvexity (GNC) (Blake and Zisserman 1987) is a deterministic annealing method for approximating a global solution for nonconvex
minimization of unconstrained, continuous problems such as (9.5). It finds
good solutions with much less cost than stochastic simulated annealing. Energies of the form

$$E(f) = \sum_{i=1}^{m} \chi_i (f_i - d_i)^2 + \lambda \sum_{i=1}^{m} \sum_{i' \in \mathcal{N}_i} g_\gamma(f_i - f_{i'}) \tag{10.25}$$

will be considered in the subsequent discussion of GNC. When g is a nonconvex function, a gradient-based method such as (9.7) finds only a local
minimum.

10.3.1 GNC Algorithm

The idea of GNC is the following. Initially, γ is set to a sufficiently large
value $\gamma^{(0)}$ such that $E(f|\gamma^{(0)})$ is strictly convex. It is easy to find the unique

minimum of $E(f|\gamma^{(0)})$ regardless of the initial f by using the gradient-descent method. The minimum $f^*_{\gamma^{(0)}}$ found under $\gamma^{(0)}$ is then used as the initial value for the next phase of minimization under a lower value $\gamma^{(1)} < \gamma^{(0)}$ to obtain the next minimum $f^*_{\gamma^{(1)}}$. As γ is lowered, $E(f|\gamma)$ becomes nonconvex and local minima appear. However, if we track the sequence of minima as γ decreases from the high value $\gamma^{(0)}$ to the target value γ, we may approximate the global minimum f^*_γ under the target value γ (i.e., the global minimum f^* of the target energy $E(f)$).

The first term on the right-hand side of (10.25) is quadratic and hence strictly convex w.r.t. f. The second term may or may not be convex, depending on f and g. Its nonconvexity could be partly compensated by the convexity of the closeness term. If $g(\eta) = g_q(\eta) = \eta^2$, then the second term is strictly convex and so is $E(f)$. If $g(\eta) = g_\alpha(\eta) = \min\{\eta^2, \alpha\}$ (see (5.10)), the second term is nonconvex and so is $E(f)$. However, if the function g and the parameters involved are chosen to satisfy

$$g''(f_i - f_{i-1}) \geq -c^* \qquad \forall i \qquad\qquad (10.26)$$

where $c^* > 0$ is some constant, then the convexity of $E(f)$ is guaranteed. The value of c^* depends on the type of model and whether the data d are complete. Provided d_i are available for all i, the value is $c^* = 1/2$ for the string, $1/4$ for the membrane, $1/8$ for the rod, and $1/32$ for the plate (Blake and Zisserman 1987). If d_i are missing altogether, it requires $g''(f_i - f_{i-1}) > 0$, so that the second term is convex by itself, to ensure the convexity of $E(f)$. The practical situation is generally better than this worst case because the data cannot be missing altogether.

Therefore, to construct an initially convex g_γ, we can choose an initial parameter $\gamma = \gamma^{(0)}$ for which $g''_{\gamma^{(0)}}(f_i^{(0)} - f_{i-1}^{(0)}) \leq 0$ for all i. This is equivalent to choosing a $\gamma^{(0)}$ such that $f_i^{(0)} - f_{i-1}^{(0)} \in B_{\gamma^{(0)}}$ where B_γ is the band. For APFs, 1, 2 and 3, the $\gamma^{(0)}$ must be larger than $2v$, $3v$ and v, respectively, where $v = \max_i[f_i^{(0)} - f_{i-1}^{(0)}]^2$.

The GNC algorithm is outlined in Fig. 10.2. Given d, λ, and a target value γ_{target} for γ, the algorithm aims to construct a sequence $\{\gamma^{(t)}\}$ ($\gamma^{(\infty)} \rightarrow \gamma_{target}$) and thus $\{f_{\gamma^{(t)}}^{(t)}\}$ to approach the global minimum $f^* = \lim_{t\to\infty} f_{\gamma^{(t)}}^{(t)}$ for which $E(f^*) = \min$. In the algorithm, ϵ is a constant for judging the convergence, and κ is a factor for decreasing $\gamma^{(t)}$ toward γ_{target}, $0 < \kappa < 1$. The choice of κ controls the balance between the quality of the solution and the computational time. In principle, $\gamma^{(t)}$ should vary continuously to keep good track of the global minimum. In discrete computation, we choose $0.9 \leq \kappa \leq 0.99$. A more rapid decrease of $\gamma^{(t)}$ (with smaller κ) is likely to lead the system to an unfavorable local minimum. Witkin, Terzopoulos, and Kass (1987) present a more sophisticated scheme for decreasing γ by relating the step to the energy change: $\gamma^{(t+1)} = \gamma^{(t)} - c_1 e^{-c_2|\nabla E|}$, where c_1 and c_2 are constants. This seems reasonable.

Choose a convex $\gamma^{(0)}$; Set $f^{(0)} \leftarrow d$ and $t = 0$;
Do 1) Update $f^{(t)}$ using (9.7);
   2) Set $t \leftarrow t + 1$;
   3) If $(f^{(t)} - f^{(t-1)} < \epsilon)$
        set $\gamma^{(t)} \leftarrow \max\{\gamma_{target}, \kappa\gamma^{(t-1)}\}$;
Until $(f^{(t)} - f^{(t-1)} < \epsilon)$ and $(\gamma^{(t)} = \gamma_{target})$;
Set $f^* \leftarrow f^{(t)}$

Figure 10.2: A GNC algorithm for finding the DA solution.

To approximate the truncated quadratic function $\lambda g_\alpha(\eta) = \lambda \min\{\eta^2, \alpha\}$, Blake and Zisserman (1987) construct the function

$$\lambda g_{\alpha,\lambda}^{(p)}(\eta) = \begin{cases} \lambda\eta^2 & |\eta| < q \\ \alpha - c(|\eta| - r)^2/2 & q \le |\eta| < r \\ \alpha & \text{otherwise} \end{cases} \qquad (10.27)$$

where $p \in [0, 1]$ is a continuation parameter, $c = c^*/p$, $r^2 = \alpha(2/c + 1/\lambda)$, and $q = \alpha/(\lambda r)$. The corresponding interaction function is

$$\lambda h_\alpha^{(p)}(\eta) = \begin{cases} \lambda\eta & |\eta| < q \\ -c(|\eta| - r)\text{sign}(\eta) & q \le |\eta| < r \\ 0 & \text{otherwise} \end{cases} \qquad (10.28)$$

The continuation parameter p is decreased from 1 to 0. When $p = 1$, the corresponding $E_\alpha^{(p=1)}$ is convex, where E_α is the energy with g_α. When $p = 0$, $E_\alpha^{(p=0)} = E_\alpha$ equals the original energy. A performance comparison between GNC and SA for reconstruction is given in (Blake 1989).

However, the complexity of the convexity treatment above can be avoided. Instead of modifying the definition of g_α, we can simply modify the parameter α in it. To achieve graduated nonconvexity, we can use α as the control parameter and decrease it from a sufficiently high value, $\alpha^{(0)} > (\max|d_i - d_{i-1}|)^2$, toward the target value. Let f be initialized so that $\max|f_i^{(0)} - f_{i-1}^{(0)}| \le \max|d_i - d_{i-1}|$. Then the energy is convex at such $\alpha^{(0)}$. An assumption is that if so initialized, $\max|f_i^{(t)} - f_{i-1}^{(t)}| < \max|d_i - d_{i-1}|$ always holds at any t during the iteration. Parameter values for the convexity in this way are related to the band B defined in (5.29) for the DA (discontinuity-adaptive) model. In a similar way, parameter values for the convexity can be derived for the LP (line process) approximation models given by Koch, Marroquin, and Yuille (1986), Yuille (1987), and Geiger and Girosi (1991).

The computation can be performed using an analog network. Let f_i be the potential of neural cell i. Let $C_i = C = 1/2\mu$ be the membrane capacitance

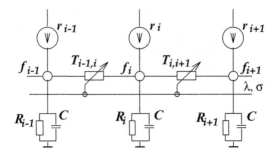

Figure 10.3: Schematic diagram of the analog network circuit.

and $R_i = 1/\chi_i$ the membrane resistance. Let

$$T_{i,i'} = T_{i',i} = \lambda h_\gamma(f_i - f_{i'}) \tag{10.29}$$

be the conductance or synaptic efficacy between neurons i and i', where $|i - i'| = 1$ and $h_\gamma \in \mathbb{H}_\gamma$. If the exponential $g_{1\gamma}$ is used, then

$$T_{i,i'} = T_{i',i} = \lambda \exp\{-[f_i - f_{i'}]^2/\gamma\} \tag{10.30}$$

Let d_i be the external current input to i, with $d_i = 0$ when $\chi_i = 0$. Now (9.7) can be written as

$$C\frac{\partial f_i}{\partial t} = -\frac{1}{R_i}f_i + d_i + T_{i-1,i}[f_{i-1} - f_i] + T_{i+1,i}[f_{i+1} - f_i] \tag{10.31}$$

The above is the dynamic equation at neuron i of the network. The diagram of the network circuit is shown in Fig. 10.3. The synaptic current from i to i' is

$$I_{i,i'} = \lambda g_\gamma'(f_i - f_{i'}) = \lambda[f_i - f_{i'}]h_\gamma(f_i - f_{i'}) \tag{10.32}$$

If the exponential $g_{1\gamma}$ is used, then

$$I_{i,i'} = \lambda[f_i - f_{i'}]\exp\{-[f_i - f_{i'}]^2/\gamma\} \tag{10.33}$$

A plot of current $I_{i,i'}$ versus potential difference $f_i - f_{i'}$ was shown at the bottom of Fig. 5.2. The voltage-controlled nonlinear synaptic conductance $T_{i,i'}$, characterized by the h function in (5.27), realizes the adaptive continuity control; the corresponding nonlinear current $I_{i,i'}$ realizes the adaptive smoothing. The current $I_{i,i'}$ diminishes asymptotically to zero as the potential difference between neurons i and i' reaches far beyond the band B_γ.

Comparing the DA and LP models, the former is more suitable for VLSI implementation than the latter. This is because the continuous shape of the adaptive conductance $T_{i,i'} = \lambda h_\gamma(f_i - f_{i'})$ in the DA is easier to implement using analog circuits than the piecewise λh_α in the LP. This advantage of the DA model is also reflected in the work by Harris, Koch, Staats, and

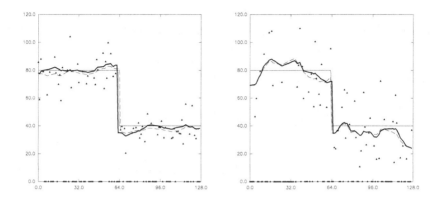

Figure 10.4: Stability of the DA solution under disturbances in parameters.

Lou (1990) and Lumsdaine, Waytt, and Elfadel (1990). There, the piecewise current-voltage characteristic $I = \Delta V h_\alpha(\Delta V)$ is implemented by a resistive fuse. Interestingly, the resistive fuse is of *finite gain* and thus does not give rise to a sharp switch-off as described by the piecewise h_α. The actual current-voltage characteristic of the resistive fuse is more like $I = \Delta V h_\gamma(\Delta V)$ than $I = \Delta V h_\alpha(\Delta V)$. It seems to satisfy all the requirements in (5.27) except for the C^1 continuity. The definition of $\mathbb{H}_\gamma$ offers guidelines for the DA circuit design.

Figure 10.4 shows the behavior of the analog DA network under component defects such as manufacturing inadequacy, quality changes, etc. The defects are simulated by adding $\pm 25\%$ evenly distributed random noise into R, C, and T in (10.31). The data d are shown in triangles with a 50% missing rate; the locations of the missing data, for which $\chi_i = 0$, are indicated by triangles at the bottom. The noise in the data is white Gaussian with standard deviation $\sigma = 10$ (left) and $\sigma = 20$ (right). The interaction function is chosen to be $h_{2\gamma}(\eta) = \frac{1}{[1+\eta^2/\gamma]^2}$. Solutions obtained with simulated component defects are shown as dashed lines in comparison with those obtained without such noise, shown as thicker solid lines. The ideal signal is shown as thinner solid lines. As can be seen, there is only a little difference between the solutions obtained with and without such noise. This demonstrates not only the stability of the network circuit but also the error-tolerance property of the DA model.

10.3.2   Annealing Labeling for MAP-MRF Matching

Here we describe a special GNC algorithm, called annealing labeling (Li et al. 1994), which may be used to improve the optimization of the posterior energy (see 4.18) formulated for the MAP matching.

We first do an energy-gain conversion (Section 9.3.2). Substituting (4.11), (4.12), (4.16), and (4.17) into (4.18), we have

$$
\begin{aligned}
U(f|d) = \ & \{v_{10}N_1 + v_{20}N_2\} + \sum_{i \in \mathcal{S}:f_i \neq 0} V_1(d_1(i)|f_i) \\
& + \sum_{i \in \mathcal{S}:f_i \neq 0} \sum_{i' \in \mathcal{S}-\{i\}:f_{i'} \neq 0} V_2(d_2(i,i')|f_i,f_{i'})
\end{aligned}
\tag{10.34}
$$

where $N_1 = \#\{f_i = 0|i \in \mathcal{S}\}$ is the number of NULL labels in f and $N_2 = \#\{f_i = 0 \text{ or } f_{i'} = 0|i \in \mathcal{S}, i' \in \mathcal{N}_i\}$ is the number of label pairs at least one of which is NULL . The constants v_{10} and v_{20} control the number of sites assigned the NULL label. The smaller they are, the more sites will be assigned the NULL . The corresponding gain function is

$$
\begin{aligned}
G(f) = \ & \{g_{10}N_1 + g_{20}N_2\} + \\
& \sum_{i \in \mathcal{S}:f_i \neq 0} r_i(f_i) + \sum_{i \in \mathcal{S}:f_i \neq 0} \sum_{i' \in \mathcal{S}-\{i\}:f_{i'} \neq 0} r_{i,i'}(f_i,f_{i'})
\end{aligned}
\tag{10.35}
$$

In the above, $g_{10} = Const_1 - v_{10}$, $g_{20} = Const_2 - v_{20}$, and $r_i(f_i)$ and $r_{i,i'}(f_i,f_{i'})$ are related to $V_1(d_1(i)|f_i)$ and $V_2(d_2(i,i')|f_i,f_{i'})$ by (9.26) and (9.27). These determine the compatibilities in RL. The gain is to be maximized.

The parameters g_{10} and g_{20} affect not only the NULL labels but also the local behavior of the maximization algorithm. When both are zero, there will be no NULL labels. The larger their values, the more sites will be assigned the NULL and the more serious the problem of local maxima becomes. In other words, we found that a labeling f in which $f_i = 0$ for some i, is likely a local optimum and that the labeling f in which $f_i = 0$ for all i, is a deep local optimum. This has motivated us to incorporate a heuristic annealing procedure into the labeling process, in a way similar to the graduated non-convexity (GNC) algorithm (Blake and Zisserman 1987), to overcome the local optimum problem.

Introduce a temperature parameter into the gain

$$
\begin{aligned}
G(f|T) = \ & \{g_{10}N_1 + g_{20}N_2\}/T + \sum_{i \in \mathcal{S}:f_i \neq 0} r_i(f_i) \\
& + \sum_{i \in \mathcal{S}:f_i \neq 0} \sum_{i' \in \mathcal{S}-\{i\}:f_{i'} \neq 0} r_{i,i'}(f_i,f_{i'})
\end{aligned}
\tag{10.36}
$$

with $G(f) = \lim_{T \to 1} G(f|T)$. T is initially set to a very high value $T^{(0)} \to \infty$ and gradually lowered toward $T^{(\infty)} = 1$. The maximum f^* obtained at $T^{(t-1)}$ is used as the initial value for the next new phase of maximization at $T^{(t)}$. In this way, the f^* are tracked from high T to low T. Because g_{10} and g_{20} are weighted by $\frac{1}{T}$, the optimum f^* obtained at high T is less affected by the local optimum problem; when it is tracked down, a better quality solution may be obtained than one found by a nonannealing algorithm. The improvement of annealing labeling ICM, a procedure incorporating annealing labeling into ICM, over the simple ICM will be shown shortly.

Note that the T in the annealing labeling acts on only the prior part, whereas in SA, T acts on both the prior and likelihood parts. So the present annealing procedure is more like GNC (Blake and Zisserman 1987) than SA (Geman and Geman 1984).

10.4 Graph Cuts

Graph cuts is a class of algorithms that uses max-flow algorithms to solve discrete energy minimization problems. It was first proposed to obtain the global minimum of a two-label MRF model (Ising model) by Greig, Porteous, and Seheult (1989). It was then extended to solve convex multilabel problems (Roy and Cox 1998) and approximate the global solution for more general multilabel MRF problems (Boykov et al. 2001). Graph cuts has now been widely used in image analysis problems such as image segmentation, restoration, super-resolution, and stereo. This section introduces basic concepts of graph cuts. The reader is referred to the Website of Zabih () for more about the theories and applications.

10.4.1 Max-Flow

Max-flow (or min-cut or *s-t* cut) algorithms play a key role in graph cuts. There are two types of max-flow algorithms: augmenting paths algorithms (Ford-Fulkerson style) and push-relabel algorithms (Goldberg-Tarjan style). A comparison of these algorithms for energy minimization in image analysis can be found in (Boykov and Kolmogorov 2004).

Let $\mathcal{G} = (\mathcal{V}, \mathcal{E})$ be a digraph (directed graph) with nonnegative weight on each edge, where $\mathcal{V}$ is the set of vertices and $\mathcal{E}$ the set of edges. In max-flow, $\mathcal{V}$ contains two special vertices, the source s and sink t. A cut $\mathcal{E}_c$ of $\mathcal{G}$ is a subset of $\mathcal{E}$, satisfying: (1) the subgraph excluding $\mathcal{E}_c$, $\tilde{\mathcal{G}} = (\mathcal{V}, \mathcal{E} - \mathcal{E}_c)$, is disconnected; (2) adding to this subgraph any edge in $\mathcal{E}_c$, giving $\tilde{\mathcal{G}} = (\mathcal{V}, (\mathcal{E} - \mathcal{E}_c) \cup \{e\})$ ($\forall e \in \mathcal{E}_c$), the subgraph is connected. The cost of a cut $\mathcal{E}_c$ is the sum of its edge weights. The min-cut problem is to find the cut that has the minimum cost. According to the Ford-Fulkerson theorem, this problem is equivalent to find the maximum flowing from s to t.

Fig. 10.5 shows an example of max-flow. On each edge there is a pair of number (x, y), where x is the maximum flux permitted by this edge and y is the maximum flux in practice. The edges are cut by the dashed line and this forms the min-cut. The flux of this cut is equivalent to the maximum flux from the source s to the terminal t, which in this digraph is 11. The dashed line divides the nodes of the digraph into two parts, S and T, with all the nodes in S connected to the source s, and all the nodes in T connected to the terminal t. In a two-label MRF problem, this can be explained as all the nodes in S having the same label ℓ_1 and all the nodes in T having the label ℓ_2.

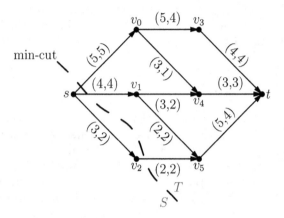

Figure 10.5: An example of max-flow.

10.4.2   Two-Label Graph Cuts

The simplest case of a labeling problem is binary labeling, where $\mathcal{L} = \{\ell_1, \ell_2\}$. If the prior of an MRF is defined as an Ising model, the exact optimum solution can be obtained by finding the max-flow on a special defined graph constructed according to the energy function; in other words, the minimum cut or the maximum flow of this graph is the same as the global minimum of the energy function (Greig et al. 1989).

A related concept for the max-flow is the submodularity. Let $\mathcal{S}$ be a finite set and $g : 2^{\mathcal{S}} \to \mathbb{R}$ be a real-valued function defined on the set of all subsets of $\mathcal{S}$. U is called submodular if for arbitrary $X, Y \subset \mathcal{S}$

$$g(X) + g(Y) \geq g(X \cup Y) + g(X \cap Y) \tag{10.37}$$

When the energy function is submodular, the following algorithm produces a graph that represents the energy function.

Introduce two auxiliary vertices s and t (source and sink), and let $\mathcal{G} = (\mathcal{V}, \mathcal{E})$ be the digraph to be constructed, with $\mathcal{V} = \mathcal{S} \cup \{s, t\}$. The following describes how to construct the edge set $\mathcal{E}$. This can be done in two parts.

For the closeness terms $V_1(i, f_i)$, $\forall i \in \mathcal{S}$, according to the definition of submodularity, all functions with only one variable are submodular. Therefore, an energy function with the V_1 term only is submodular. If $V_1(i, 1) > V_1(i, 0)$, then add an edge $< s, i >$ with the weight $w_{s,i} = V_1(i, 1) - V_1(i, 0)$; otherwise, add an edge $< i, t >$ with the weight $w_{i,t} = V_1(i, 0) - V_1(i, 1)$. This can be done using a procedure $edge(i, \ell)$: if $\ell > 0$, then add an edge $< s, i >$ with the weight ℓ. Otherwise, add an edge $< i, t >$ with the weight $-\ell$; see Fig. 10.6.

The smoothness term V_2 is submodular if it satisfies the following condition:

$$V_2(i, i', 0, 1) + V_2(i, i', 1, 0) \geq V_2(i, i', 0, 0) + V_2(i, i', 1, 1) \quad \forall i, i' \in \mathcal{S} \quad (10.38)$$

The digraph for representing a smoothness term $V_2(i, i', f_i, f_{i'})$ can be constructed as follows (see also Fig. 10.6):

- $edge(i, V_2(i, i', 1, 0) - V_2(i, i', 0, 0))$;

- $edge(i', V_2(i, i', 1, 1) - V_2(i, i', 1, 0))$;

- add an edge $< i, i' >$ with the weight $w_{i,i'} = V_2(i, i', 0, 1) + V_2(i, i', 1, 0) - V_2(i, i', 0, 0) - V_2(i, i', 1, 1)$.

A digraph is thus constructed for a submodular energy function. A max-flow algorithm can then be applied to this graph to find the optimum. The necessary condition for finding the exact optimum of binary labeling in polynomial time is that the energy function be submodular (Kolmogorov and Zabih 2004). If the energy function can be expressed as $V_{C_1}(f) + V_{C_2}(f) + V_{C_3}(f)$, this is also the sufficient condition (Kolmogorov and Zabih 2004). Methods exist for constructing digraphs for nonsubmodular energy functions; see (Kolmogorov and Rother 2007).

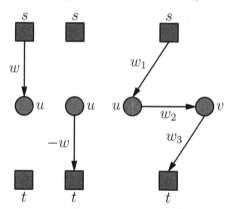

Figure 10.6: Construction of edges of a digraph for the closeness term (left and middle) and the smoothness term (right) of a submodular energy function.

10.4.3 Multilabel Graph Cuts

Minimization of a convex energy function with multiple discrete labels can be solved exactly. Assuming that the discrete label set is ordered (such as

$1 < 2 < \ldots < M$ as the depths for stereo vision (Roy and Cox 1998)) and the smoothness term is convex, then the original multilabel problem can be transformed into a two-label one for a graph with an augmented set of nodes (sites) and solved using a two-label graph cut method (Roy and Cox 1998). This finds a global solution.

Minimization of a more general energy function with multiple discrete labels is an NP-hard problem. Numerous algorithms have been proposed to approximate the global solution with a lower complexity. Such an approximation could be done using graph cuts by transforming the original problem to an equivalent problem of multiway cuts of a special graph constructed according to the primal minimization problem (Boykov et al. 1998). The two algorithms presented in (Boykov et al. 2001), α–β swap and α-expansion, use methods of iteratively solving multiple two-label graph cuts and achieve better computational performance. It is shown that the α-expansion algorithm has a linear time complexity w.r.t. the number of labels $(O(M))$, whereas α–β has an $O(M^2)$ complexity. The α-expansion can be further improved to achieve a logarithmical time complexity (Lempitsky et al. 2007). The following introduces the α-expansion and α–β swap algorithms.

α-expansion

The α-expansion algorithm (Boykov et al. 2001) assumes that the smoothness prior term V_2 is a metric, that is

- $V_2(f_i, f_{i'}) \geq 0, V_2(f_i, f_{i'}) = 0 \Leftrightarrow f_i = f_{i'}$,

- $V_2(f_i, f_{i'}) = V_2(f_{i'}, f_i)$,

- $V_2(f_i, f_{i'}) \leq V_2(f_i, f_{i''}) + V_2(f_{i''}, f_{i'})$.

such that the submodularity condition (10.37), which is weaker than the aforementioned convexity condition, is satisfied. Each iteration considers an $\alpha \in \mathcal{L}$ value and the two-label set of $\{\alpha, \text{non-}\alpha\}$, and solves the two-label graph cut problem. The algorithm decides whether or not a site $i \in \mathcal{S}, f_i \neq \alpha$ needs to change from non-α to α. The iteration continues until the energy does not decrease any more; see Fig. 10.7. The energy of the approximate solution is upper-bounded by cU^*, where U^* is the global minimum of the original energy function and $c \geq 1$ is a constant that depends on the original energy function.

α–β Swap

The α–β Swap (Boykov et al. 2001) is an iterative graph cut algorithm applicable to situations when the smoothness term V_2 is semimetric, i.e.,

- $V_2(f_i, f_{i'}) \geq 0, V_2(f_i, f_{i'}) = 0 \Leftrightarrow f_i = f_{i'}$;

- $V_2(f_i, f_{i'}) = V_2(f_{i'}, f_i)$;

```
Begin
    1. Initialize f with an arbitrary labeling;
    2. Flag: = false;
    3. For each label α ∈ L;
        3.1 Find f̂ = arg min E(f') among f' within one α expansion of f;
        3.2 If E(f̂) < E(f), set f := f̂ and Flag: = true;
    4. If Flag = true goto step 2
    5. Return f;
End
```

Figure 10.7: α-expansion Algorithm

It chooses two labels $\alpha, \beta \in \mathcal{L}$ in each iteration. Let $\mathcal{S}_\alpha = \{i \in \mathcal{S} \mid f_i = \alpha\}$ and $\mathcal{S}_\beta = \{i \in \mathcal{S} \mid f_i = \beta\}$. A two-label graph cut algorithm is used to determine two subsets, $\hat{\mathcal{S}}_\alpha \subset \mathcal{S}_\alpha$ and $\hat{\mathcal{S}}_\beta \subset \mathcal{S}_\beta$, that need to swap the labels. The iteration continues until the energy function stops decreasing; see Fig. 10.8.

```
Begin
    1. Initialize f with an arbitrary labeling;
    2. Flag: = false;
    3. For each pair {α, β} ∈ L;
        3.1 Find f̂ = arg min E(f') among f' within one α − β swap of f;
        3.2 If E(f̂) < E(f), set f := f̂ and Flag: = true;
    4. If Flag = true goto step 2
    5. Return f;
End
```

Figure 10.8: α–β swap algorithm

10.5   Genetic Algorithms

Genetic algorithms (GAs) (Holland 1975; Goldberg 1989) are another class of heuristic procedures for global optimization. Inspired by the principle of natural evolution in the biological world, the procedures simulate the evolutionary process in which a population of individuals who have the highest goodness-of-fit values survives.

---

generate initial population $P = \{f^1, \ldots, f^n\}$;
compute $E(f)\ \forall f \in P$;
repeat
      select two individuals from P;
      recombine the individuals to produce two offsprings;
      compute $E(f)$ of the offsprings;
      update P with the offsprings;
until (converged);
return $f = \arg\min_{f \in P} E(f)$;

Figure 10.9: A standard genetic algorithm.

10.5.1 Standard GA

Figure 10.9 illustrates a standard genetic algorithm for unconstrained, combinatorial optimization.[1] Initially, a number of (say, $n = 50$) individuals are *generated*, yielding a population P. An individual is determined by his chromosome f, which is a string of m genes (labels) f_i $(i = 1, \ldots, m)$.

At a point of the evolution, two individuals are randomly *selected* for mating. The selection is done according to a scheme that favors the fitter individuals (with lower E value); the lower the value of $E(f)$, the more likely f will be selected. For example, f may be selected with a probability proportional to $1 - \frac{E(f)}{\sum_{f \in P} E(f)}$ or a monotonically increasing function of it, assuming $E(f)$ is nonnegative. There are numerous selection schemes (Goldberg 1989).

Recombination takes place between the two individuals selected. The mechanisms of *crossover* and *mutation* are typically used for this. Figure 10.10 illustrates two such basic operations in which a gene can take any alphanumeric value. Crossover takes two individuals, cuts their chromosome strings at some randomly chosen position, and recombines the opposite segments to create two offsprings. Crossover is not always invoked; it is applied with a probability p_c typically between 0.6 and 1. If it is not applied, the offsprings are simply the duplicates of the selected individuals. Mutation is applied to each offspring after the crossover. It randomly alters a randomly chosen gene with a small probability p_m, typically 0.001. As is usual in GA practice, there are many crossover and mutation operators.

The offsprings are then added to P, and the two least fit individuals (i.e., those with the highest $E(f)$ values) are removed from the population P. As the evolution continues in this way, the fitness of the best individual as

---

[1]Special coding of solutions into chromosomes needs to be done when GA is used for (piecewise) continuous optimization problems.

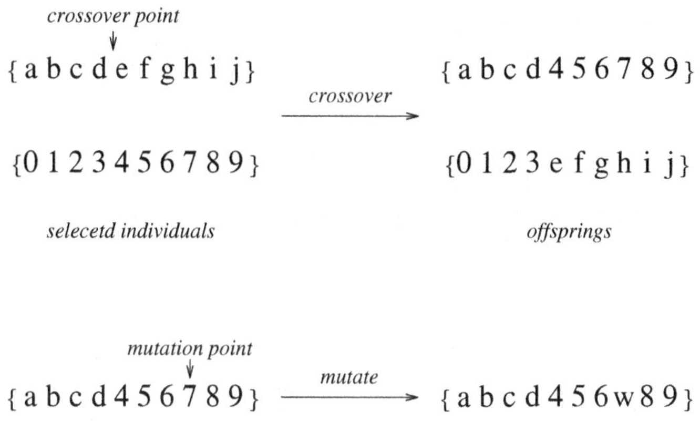

crossover point
▼
{a b c d e f g h i j}   {a b c d 4 5 6 7 8 9}

                          crossover
                        ──────────────→

{0 1 2 3 4 5 6 7 8 9}   {0 1 2 3 e f g h i j}

selecetd individuals   *offsprings*

mutation point
▼   *mutate*
{a b c d 4 5 6 7 8 9}   ──────────→   {a b c d 4 5 6 w 8 9}

Figure 10.10: The two basic operations in recombination: crossover and mutation.

well as the average fitness increases toward the global optimum. *Convergence* is judged by uniformity: A gene is said to have converged when most (say 95%) of the population share the same value. The population is said to have converged when all the genes have converged. The convergence of GAs is obtained, but no theoretical proof is given.

Impressive empirical results for solving real and combinatorial, unconstrained and constrained optimization problems, such as the traveling salesman problem, and neural network optimization and scheduling are reported (see (Goldberg 1989) for a review). Applications in computer vision are also seen (Bhanu et al. 1989; Ankenbrandt et al. 1990; Hill and Taylor 1992). Currently, there is no theory that explains why GAs work. However, some hypotheses exist. Among these are the schema theorem (Holland 1975) and building block hypothesis (Goldberg 1989).

10.5.2 Hybrid GA: Comb Algorithm

Combining a local search with a GA yields a hybrid GA, also called a memetic algorithm (Moscato 1989; Radcliffe and Surry 1994). Here, a new random search method, called the Comb method, is described for combinatorial optimization. Assume that an energy function has been given that is formulated based on the MRF theory for image restoration and segmentation. The Comb method maintains a number of best local minima found so far, as a population based method. It uses the common structure of the local minima to infer the structure of the global minimum. In every iteration, it derives one or two new initial configurations based on the Common structure (common labels) of the Best local minima (hence "Comb"): If two local minima have the same label (pixel value) in a pixel location, the label is copied to the corresponding location in the new configuration; otherwise, a label randomly chosen from either local minimum is set to it. The configuration thus derived contains about the same percentage of common labels as the two local minima (assuming the two have about the same percentage of common labels). But the configuration derived is no longer a local minimum, and thus further improvement is possible. The new local minimum then updates the best existing ones. This process is repeated until some termination conditions are satisfied.

The resulting Comb algorithm is equivalent to a GA hybridized with steepest descent, in which the Comb initialization therein works like a uniform crossover operator. There have been various interpretations for the crossover operation. The idea of encouraging common structures in the Comb initialization provides a new perspective for interpreting the crossover operation in GA.

Experimental results in both image restoration and segmentation are provided to compare the Comb method with the ICM, HCF (Chou et al. 1993) and SA. The results show that the Comb yields solutions of better quality than the ICM and HCF and comparable to SA.

The Comb method maintains a number N of best local minima found so far, denoted $F = \{f^{[1]}, \ldots, f^{[N]}\}$, as a population based method. In every iteration, it derives a new initial configuration from F and performs steepest descent using the initial configurations derived. If the local minimum found is better than an existing one in F, it replaces it.

Ideally, we desire that all local minima in F converge towards the global minimum $f^{[global]}$, in which case, there must be

$$f_i^{[n]} = f_i^{[global]} \qquad 1 \le n \le N \qquad (10.39)$$

for all $i \in \mathcal{S}$. We call $f_i^{[global]}$ the *minimal label* at i. To achieve (10.39), all the labels at i, $\{f_i^{[n]} | \forall n\}$, should finally converge to the minimal label $f_i^{[global]}$. The Comb is performed with this objective.

The following heuristic is the basis for deriving new initial configurations. Although $f^{[1]}, \ldots, f^{[N]}$ are local minima, they share some structure with the global minimum $f^{[global]}$. More specifically, some local minima $f^{[n]}$ have the minimal label $f_i^{[n]} = f_i^{[global]}$ for some $i \in \mathcal{S}$. Figure 10.11 shows the (approximate) global minimum for an MAP-MRF restoration problem and some local minima found by using the multi-start method with initially random configurations. A statistic over a number of $N = 10$ local minima is made to see how many minimal labels they have. Table 10.1 shows the statistic in terms of the percentile of the sites (pixels) $i \in \mathcal{S}$ at which at least k local minima $f^{[n]}$ have the minimal label $f_i^{[n]} = f_i^{[global]}$.

The Comb initialization is aimed at deriving configurations having a substantial number of minimal labels so as to improve F toward the objective

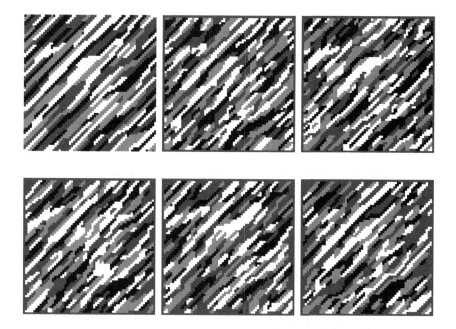

Figure 10.11: The global minimum (upper left) and five local minima. The local minima share some structure with the global minimum.

of (10.39). Although a configuration with a larger number of minimal labels does not necessarily have a lower-energy value, we hope that it provides a good basis to start with (i.e., it can serve as a good initial configuration).

Table 10.1: Percentile (rounded up to integers) of the sites (pixels) $i \in \mathcal{S}$ at which *at least* k local minima $f^{[n]}$ have the same label $f_i^{[n]} = f_i^{[global]}$ as the global minimum $f^{[global]}$.

k	0	1	2	3	4	5	6	7	8	9	10
%	100	98	95	87	75	60	43	28	16	7	2

The Comb algorithm is described in Fig. 10.12. The initialization at the beginning of the Comb algorithm is done according to a uniform distribution like the multi-start method. This is followed by iterations of four steps. First, two local minima in F, $f^{[a]}$ and $f^{[b]}$, $(a \neq b)$, are randomly selected according to a uniform distribution. Second, a new initial configuration $f^{[0]}$ is derived from $f^{[a]}$ and $f^{[b]}$ using the standard Comb initialization, which will be

comb_initialization($f^{[a]}$, $f^{[b]}$, F)
begin

        for (each $i \in \mathcal{S}$) do:

           if ($f_i^{[a]} == f_i^{[b]}$ && $rand[0,1] < 1 - \tau$)

                then $f_i^{[0]} = f_i^{[a]}$;

           else

                $f_i^{[0]} = rand(\mathcal{L})$;

end

Comb_Algorithm
begin

        initialize the set $F = \{f^{[1]}, \ldots, f^{[N]}\}$;

        do {

           random_selection($f^{[a]}$, $f^{[b]}$, F);

           comb_initialization($f^{[0]}$, $f^{[a]}$, $f^{[b]}$);

           steepest_descent(f^*, $f^{[0]}$);

           update(F, f^*);

        } until (termination condition satisfied);

        return(arg $\min_{f \in F} E(f)$);

end

Figure 10.12: The Comb algorithm.

explained shortly. Then, steepest descent is applied to $f^{[0]}$ to produce a local minimum f^*. Finally, the set F is updated by f^*: If $E(f^*) < \max_{f \in F} E(f)$, then the configuration arg $\max\{f|f \in F\}$, which has the highest energy value, higher than $E(f^*)$, is replaced by f^*. The termination condition may be that all configurations in F are the same or that a certain number of local minima have been performed. The algorithm returns the best local minimum in F (i.e., the one having the lowest energy).

The central part of the Comb method is the derivation of new initial configurations. The Comb is aimed at deriving $f^{[0]}$ in such a way that $f^{[0]}$ contains as many minimal labels as possible. Because the minimal labels are not known a priori, the Comb attempts to use common structure, or *common labels*, of $f^{[a]}$ and $f^{[b]}$ to infer the minimal labels. We say that f_i^{comm} is a common label of $f^{[a]}$ and $f^{[b]}$ if $f_i^{comm} = f_i^{[a]} = f_i^{[b]}$. The Comb makes a hypothesis that f_i^{comm} is a minimal label if $f_i^{[a]} = f_i^{[b]}$. The Comb initialization schemes are illustrated as follows:

1. The *basic Comb* initialization. For each $i \in \mathcal{S}$, if $f_i^{[a]}$ and $f_i^{[b]}$ are identical, then set $f_i^{[0]} = f_i^{[a]}$; otherwise set $f_i^{[0]} = rand(\mathcal{L})$, which is a label randomly drawn from $\mathcal{L}$. The use of the common label in the initial configuration encourages the enlargement of common structure in the local minimum to be found subsequently.

2. The *standard Comb* initialization. The basic Comb initialization is accepted with a probability $1 - \tau$, where $0 < \tau < 1$. The probabilistic acceptance of common labels diversifies the search and prevents F from converging to a local minimum too soon. The standard Comb initialization is shown in the upper part of Fig. 10.12, where $rand[0, 1]$ stands for an evenly distributed random number in $[0, 1]$.

Then, how many minimal labels are there in $f^{[0]}$ as the result of copying common labels(i.e., as the result of inferring minimal labels using common labels)? In supervised tests where the (near) global minimum is known, we find that the percentage of minimal labels in $f^{[0]}$ is usually only slightly (about 1.0–2.0%) lower than those in $f^{[a]}$ and $f^{[b]}$. That is, the number of minimal labels retained in $f^{[0]}$ is about the same as those in $f^{[a]}$ and $f^{[b]}$. Given this and that $f^{[0]}$ is no longer a local minimum like $f^{[a]}$ and $f^{[b]}$, there is room to improve $f^{[0]}$ using a subsequent local minimization. This makes it possible to yield a better local minimum from $f^{[0]}$.

There are two parameters in the Comb algorithm, N and τ. The solution quality increases (i.e., the minimized energy value decreases) as the size of F, N, increases from 2 to 10, but remains about the same (probabilistically) for greater N values; and a larger N leads to more computational load. Therefore, we choose $N = 10$. Empirically, when $\tau = 0$, the algorithm converges sooner or later to a unique configuration, and choosing a smaller N makes such a convergence quicker. But $\tau = 0$ often gives a premature solution. The value of $\tau = 0.01$ is empirically a good choice.

The Comb algorithm corresponds to a hybrid GA as described in Fig. 10.13. The standard Comb initialization is effectively the same as a crossover operation followed by a mutation operation, the major and minor operations in genetic algorithms (GA) (Goldberg 1989). More exactly,

*   the basic Comb corresponds to uniform crossover and

*   the probability acceptance in the standard Comb corresponds to mutation.

In GA, two offspring, $f_i^{[01]}$ and $f_i^{[02]}$, are produced as the result of crossover. In the uniform crossover, either of the following two settings are accepted with equal probability:

(i) $f^{[01]} = f_i^{[a]}$ and $f^{[02]} = f_i^{[b]}$,

(ii) $f^{[01]} = f_i^{[b]}$ and $f^{[02]} = f_i^{[a]}$.

So, if $f_i^{[a]} = f_i^{[b]}$, there must be $f_i^{[01]} = f_i^{[02]} = f_i^{[a]}$, just as in the Comb initialization. This encourages common labels because common labels are copied to the new initial configurations; in contrast, noncommon labels are subject to swap. The discussion above is about the uniform crossover. It should be noted that even the simplest one-point crossover also works in a way that encourages common labels.

The above suggests that the essence of both the Comb and GA is captured by using common structures of local minima. This is supported by the fact that the original Comb and the GA-like Comb yield comparable results: In the GA-like Comb algorithm (Fig. 10.13), when $f_i^{[a]} \neq f_i^{[b]}$, $f_i^{[01]}$ and $f_i^{[02]}$ inherit the values of $f_i^{[a]}$ and $f_i^{[b]}$, as does a crossover operator. However, setting $f_i^{[01]}$ and $f_i^{[02]}$ to a random label $rand(\mathcal{L})$ (i.e., not necessarily inheriting $f_i^{[a]}$ and $f_i^{[b]}$) leads to comparable results as long as the common labels are copied to $f_j^{[0]}$ when $f_i^{[a]} = f_i^{[b]}$. Moreover, whether to derive one initial configuration $f^{[0]}$ or two initial configurations $f^{[01]}$ and $f^{[02]}$ does not matter; both schemes yield comparable results. In summary, the Comb and the GA-like Comb produce comparable results, and this suggests that retaining common labels is important and provides an interpretation for the crossover operation in GA.

The Comb is better than the multi-start method. Running a steepest descent algorithm a number of times using the Comb initialization gets a better solution than running it the same number of times using the independent initialization of multi-start. The Comb has a much higher efficiency in descending to good local minima because it makes use of the best local minima.

To summarize, the Comb attempts to derive good initial configurations from the best local minima found so far in order to achieve better solutions. To do so, it uses the common structure of the local minima to infer label values in the global minimum. An initial configuration thus derived has about the same number of minimal labels as the two local minima from which it is derived. However, it is no longer a local minimum and thus its quality can be improved by a subsequent local minimization. This makes it possible to yield a better local minimum and thus increments the solution quality step by step. The comparison shows that the Comb produces better results than the ICM and HCF though at higher computational cost, and results comparable to the SA at lower cost (Section 10.6.3). This suggests that the Comb can provide a good alternative to the well-known global minimizer SA. Further, the Comb algorithm is applicable, in principle, to many optimization problems of vision and pattern recognition.

GA_comb_initialization($f^{[a]}$, $f^{[b]}$, F)
begin

      for each $i \in \mathcal{S}$

         /* uniform crossover */

         if ($f_i^{[a]} == f_i^{[b]}$)

              then $f_i^{[01]} = f_i^{[02]} = f_i^{[a]}$;

         else if ($rand[0,1] < 0.5$)

             $f_i^{[01]} = f_i^{[a]}$ and $f_i^{[02]} = f_i^{[b]}$;

         else

             $f_i^{[01]} = f_i^{[b]}$ and $f_i^{[02]} = f_i^{[a]}$;

         /* mutation */

         if ($rand[0,1] < \tau$)

             $f_i^{[01]} = rand(\mathcal{L})$;

         if ($rand[0,1] < \tau$)

             $f_i^{[02]} = rand(\mathcal{L})$;

end

GA_Comb_Algorithm
begin

      initialize $F = \{f^{[1]}, \ldots, f^{[N]}\}$;

      do {

         random_selection($f^{[a]}$, $f^{[b]}$, F);

         GA_comb_initialization($f^{[01]}$, $f^{[02]}$, $f^{[a]}$, $f^{[b]}$);

         steepest_descent(f^{*1}, $f^{[01]}$);

         steepest_descent(f^{*2}, $f^{[02]}$);

         update(F, f^{*1}, f^{*2});

      } until (termination condition satisfied);

      return($\arg\min_{f \in F} E(f)$);

end

Figure 10.13: A GA-like Comb algorithm.

10.6 Experimental Comparisons

10.6.1 Comparing Various Relaxation Labeling Algorithms

In the following experiments, we compare several algorithms for combinatorial minimization and for constrained minimization:

1. ICM of Besag (1986),

2. RL of Hummel and Zucker (1983),

3. RL of Rosenfeld, Hummel, and Zucker (1976),

4. MF annealing of Peterson and Soderberg (1989),

5. SA of Geman and Geman (1984),

6. Annealing ICM (Section 10.3.2).

The comparison is in terms of (1) the solution quality measured by the maximized gain $G(f^*)$, (2) the computational cost measured by the number of iterations required and (3) the need for heuristics to tune the algorithm.

The schedules for the annealing algorithms are as follows. There are two annealing schedules for MFA. In the first schedule, which is given in (Peterson and Soderberg 1989), T is decreased according to $T^{(t+1)} \leftarrow 0.99T^{(t)}$; we set the initial temperature as $T^{(0)} = 10^9 \times M \times m$. The second is also in the form $T^{(t+1)} \leftarrow \kappa^{(t+1)}T^{(t)}$, but $\kappa^{(t+1)}$ is chosen on an ad hoc basis of trial and error to get the best result, which is

$$\kappa^{(t+1)} = \begin{cases} 0.9 & \text{if } S < 0.9 \ rmand \ \kappa^{(t)} \leq 0.9 \\ 0.95 & \text{if } S < 0.95 \ rmand \ \kappa^{(t)} \leq 0.95 \\ 0.99 & \text{otherwise} \end{cases} \qquad (10.40)$$

where S is the saturation defined in (10.23). We refer to these two schedules as MFA-1 and MFA-2, respectively. For the SA, it is $T^{(t+1)} \leftarrow 0.99T^{(t)}$ but with the initial temperature set to $10^9 \times M \times m$; annealing stops if the desired number of acceptances is not achieved at 100 successive temperatures (Kirkpatrick et al. 1983). These are purposely tuned for the best result possible. The schedule for the annealing ICM is quite tolerant; it is chosen as $T^{(t+1)} \leftarrow 0.9T^{(t)}$ with $T^{(0)} = 10$.

The initial labeling is assigned as follows. First, we set $p_i^{(0)}(I) = 1 + 0.001 * rnd$ $(\forall i, I)$ as the start point common to all the algorithms compared, where rnd is a random number evenly distributed between 0 and 1. For the continuous algorithms (2, 3, and 4), we normalized $p^{(0)}$ to satisfy (9.19). For the discrete algorithms (1, 5, and 6), i is initially labeled according to maximal selection, $f_i^{(0)} = I^* = \arg\max_I p_i^{(0)}(I)$. The convergence of the continuous algorithms is judged by checking whether the saturation, S, is larger than 0.99.

The test bed is the MRF matching of weighted graphs. Here, a node corresponds to a point in the $X - Y$ plane. Two random graphs, each containing $m = M$ nodes, are randomly generated as follows. First, a number of $\lfloor \frac{2}{3}m \rfloor$ nodes, where $\lfloor \cdot \rfloor$ is the "floor" operation, are generated using random numbers uniformly distributed within a box of size 100×100. They are for the first graph. Their counterparts in the other graph are generated by adding

Gaussian noise of $N(0, \sigma^2)$ to the x and y coordinates of each of the nodes in the first graph. This gives $\lfloor \frac{2}{3}m \rfloor$ deviated locations of the nodes. Then, the rest of the $\lceil \frac{1}{3}m \rceil$ nodes in each of the two graphs are generated independently by using random numbers uniformly distributed in the box, where $\lceil \cdot \rceil$ is the "ceiling" operation. This simulates outlier nodes.

The steps above generate two weighted graphs, $\mathcal{G} = (\mathcal{S}, d)$ and $\mathcal{G}' = (\mathcal{L}, D)$, where $\mathcal{S} = \{1, \ldots, m\}$ and $\mathcal{L} = \{1, \ldots, M\}$ index the sets of nodes and $d = [d(i, i')]_{i,i' \in \mathcal{S}}$ and $D = [D(I, I')]_{I, I' \in \mathcal{L}}$ are the distances between nodes in the first and second graphs, respectively. The weights in d and D, which reflect bilateral relations between nodes, are the basis for matching, whereas unary properties of nodes are not used in the test. We augment $\mathcal{L}$ by a special node, called the NULL node and indexed by 0, into $\mathcal{L}' = \{0, 1, \ldots, M\}$. The purpose is to cater for the matching of outlier nodes. Refer also to Section 4.1 for the graph representation. The matching is to assign a node from $\mathcal{L}'$ to each of the nodes in $\mathcal{S}$ so that some goodness (cost) of matching is maximized (minimized).

The posterior energy formulated in Section 4.2 is used to measure the cost of a matching $f = \{f_1, \ldots, f_m\}$. To be suitable for real, as opposed to combinatorial, computation, the problem is then reformulated in terms of relaxation labeling (see Section 9.3.2). From the two weighted graphs, the compatibility matrix $[r_{i,i'}(I, I')]$ is defined as

$$r_{i,i'}(I, I') = \begin{cases} Const_2 - [d(i, i') - D(I, I')]^2 & \text{if } I' \neq 0 \text{ and } I' \neq 0 \\ Const_2 - v_{20} & \text{otherwise} \end{cases}$$

(10.41)

where $v_{20} > 0$ is a constant. The gain function is then

$$G(f) = \sum_{i \in \mathcal{S}} \sum_{I \in \mathcal{L}'} \sum_{i' \in \mathcal{S}, i' \neq i} \sum_{I' \in \mathcal{L}'} r_{i,i'}(I, I')\, p_i(I)\, p_{i'}(I')$$

(10.42)

To reduce storage, we used only one byte (8 bits) to represent the compatibility coefficients $r_{i,i'}(I, I')$. Positive values are truncated to an integer between 0 and 255 while negative values are truncated to zero. In this case, we set $Const_2 = 255$. Although the precision of the compatibility coefficients is low, good results are still obtained; this demonstrates the error-tolerant aspect of the RL minimization approach.

The purpose of the MRF and RL formulations is to provide the compatibility matrix. Our ultimate objective here is more abstract: to compare the ability of the algorithms to solve the minimization problem expressed, given the posterior probability function or the corresponding compatibility matrix.

The following results are obtained after 200 runs. Figure 10.14 illustrates the solution quality in terms of the maximized gain as a function of the noise level σ. SA provides the best result when carefully tuned, followed by the Hummel-Zucker algorithm, whereas the quality of the ICM solution is the poorest. The annealing procedure (see Section 10.3.2) significantly improves the quality of the ICM solution.

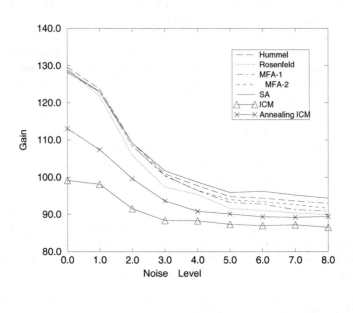

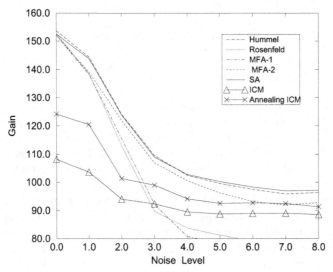

Figure 10.14: The maximized gain $G(f^*)$ (after dividing by m^2) for $m = 10$ (top) and $m = 20$ (bottom); the higher, the better.

The performance of MFT annealing is after the Hummel-Zucker algorithm when $m = M = 10$. However, for "harder" problems (e.g., when $m = M = 20$) the results produced by using the simple schedule MFA-1 deteriorate quickly as the noise level σ increases, dropping to even lower than that of ICM. For large m and M and high σ^2 values, the algorithm is often trapped in the all-NULL solution, which is a significant local optimum. We also found that a lower schedule (i.e., smaller κ in $T \leftarrow \kappa T$) does not necessarily yield better solutions.

Figure 10.15 demonstrates the cost in terms of the number of iterations as a function of the noise level σ. ICM converges very fast, after just a few iterations. In contrast, the number for SA with the specified heuristic schedule is two to four orders higher than all the deterministic algorithms. Figure 10.16 shows the efficiency measured by the maximized-gain/iteration-number ratio. The ordering by the efficiency is roughly consistent with the inverse ordering by the iteration number.

Besides the solution quality and the cost, an important factor that must be taken into account is the need for, and difficulties in, tuning the annealing schedule. It is well known that the schedule is critical to the success of SA, and it is an area of study (Aarts 1989). We add that the schedule is also critical to the mean field annealing. How to choose an optimal schedule depends not only on the type of the problem, but also on the size. Finding a good heuristic schedule based on trial and error can be tedious.

The comparison leads to a conclusion in favor of the Hummel-Zucker algorithm. It yields good-quality solutions quite comparable with the time-consuming simulated annealing; yet, it is much more efficient than SA, thus balancing well between quality and cost. Furthermore, it avoids the cumbersome needs for the heuristic tuning of annealing schedules in annealing algorithms. MFT annealing would also be a good choice if the rapid deterioration of solutions to "harder" problems could be remedied. An experimental result (Li 1995c) shows that the Lagrange-Hopfield algorithm described in Section 9.5 yields solutions of quality similar to those produced by the Hummel-Zucker algorithm.

An earlier experimental comparison of several RL algorithms in terms of the number of iterations (cost) and the number of errors (quality) was done by Price (1985). Price concluded that the algorithm of Faugeras and Price (1981) is the best, that of Hummel and Zucker (1983) is about as good, that by Peleg (1980) converges too fast to yield a good result, and that of Rosenfeld, Hummel, and Zucker (1976) performs only adequately. The goodness of interpretation in (Price 1985) relates not only to algorithms themselves but also to how the problem is formulated (e.g., the definition of compatibilities), which is a combination of the two issues, problem formulation and computational algorithm. Here, we measure the solution quality by the quantitative gain $G(f^*)$ and regard RL as just a mechanism of minimization rather than

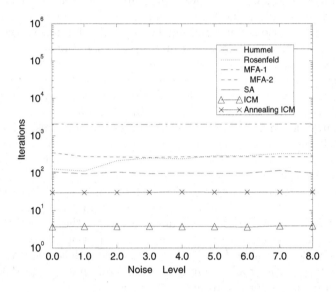

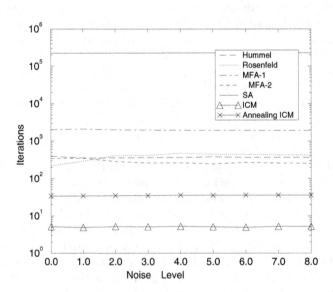

Figure 10.15: The number of iterations for $m = 10$ (top) and $m = 20$ (bottom); the lower, the better.

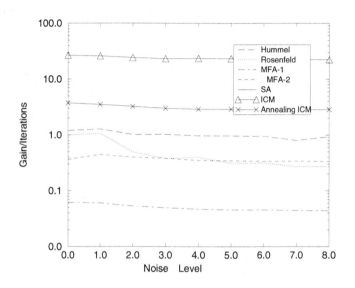

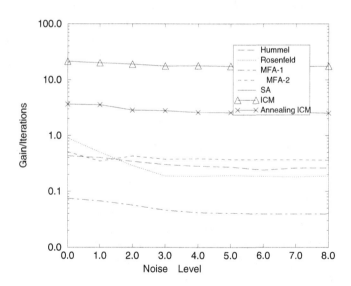

Figure 10.16: The efficiency measured by the maximized-gain/iteration-number ratio for $m = 10$ (top) and $m = 20$ (bottom); the higher, the better. The efficiency of SA is near zero.

one of interpretation. In our comparison, our interest is in how well an algorithm can minimize an energy regardless of the problem formulation.

10.6.2 Comparing the ALH Algorithm with Others

In the following, two experiments are presented, one for MAP-MRF image restoration and the other for segmentation, to compare the performance of the following algorithms: (1) ALH of this chapter, (2) ICM (Besag 1986), (3) HCF (Chou et al. 1993) (the parallel version), (4) MFA (Peterson and Soderberg 1989; Yuille and Kosowsky 1994) (see Section 10.2), and (5) SA with the Metropolis sampler (Kirkpatrick et al. 1983) implemented based on a procedure given in (Press et al. 1988). The comparison is in terms of (i) the solution quality measured by the minimized energy values and (ii) the convergence rate measured by the number of iterations. In calculating the energy, $E(f)$ of (9.94) is used where, for the continuous algorithms of ALH and MFA, f is obtained from p by a maximum selection (winner-take-all) operation.

For ALH, $\mu = 100$ and $T = 10^5$ are fixed, and β is increased from 1 to 100 according to $\beta \leftarrow 1.01\beta$. The convergence criterion is $\|u^{(t)} - u^{(t-1)}\|_\infty < 0.0001$. The schedule for SA is $T^{(t+1)} \leftarrow 0.999T^{(t)}$ (with $T^{(0)} = 10^4$) and for MFA is $T^{(t+1)} \leftarrow 0.99T^{(t)}$ (with $T^{(0)} = 10^3$).

The first set of experiments is for MAP-MRF restoration performed on three synthetic images of $M = 4$ gray levels, and Figs. 10.17–10.19 show the results. In each figure, (a) is the true image with $M = 4$ gray levels, the label set $\mathcal{L} = \{1, 2, 3, 4\}$, and the pixel gray values also in $\{1, 2, 3, 4\}$. The clique potential parameters α_I and $(\beta_1, \cdots, \beta_4)$ for generating the three images are shown in Table 10.2. (b) is the observed image in which every pixel takes a real value that is the true pixel value plus zero-mean i.i.d. Gaussian noise with standard deviation $\sigma = 1$. (c) is the maximum likelihood estimate that was used as the initial labeling. (d) to (h) are the solutions found by the algorithms compared.

Table 10.3 shows the minimized energy values, the error rates, and the iteration numbers required. It can be seen from the table that objectively ALH performs the best out of the three deterministic algorithms in terms of both the minimized energy values. Overall, the solution quality is ranked as "ICM < HCF < MFA < ALH < SA", which is in agreement with a subjective evaluation of the results. We also implemented a parallel ICM using codings but the results were not as good as for the serial ICM.

The second experiment compares the algorithms in performing MAP-MRF segmentation on the Lena image of size 256×240 into a tri-level segmentation map. The results are illustrated in Fig. 10.20. The input image (a) is the original Lena image corrupted by the i.i.d. Gaussian noise with standard deviation 10. The observation model is a Gaussian distribution of standard deviation 10 with mean values 40, 125, and 200 for the three-level segmentation. An isometric MRF prior is used, with the four β parameters

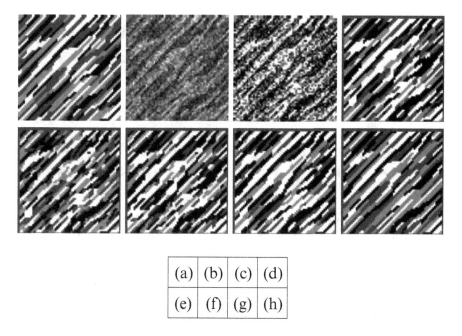

(a)	(b)	(c)	(d)
(e)	(f)	(g)	(h)

Figure 10.17: Restoration of image no. 1. (a) Original image. (b) Observed noisy image. (c) Maximum likelihood estimate. (d) ALH solution. (e) ICM solution. (f) HCF solution. (g) MFA solution. (h) SA solution.

Table 10.2: The MRF parameters (α and β) and noise parameter σ for generating the three images.

	σ	α_I	β_1	β_2	β_3	β_4
Image No. 1	1	0	-1	-1	-1	1
Image No. 2	1	0	-2	-2	1	1
Image No. 3	1	0	1	1	1	1

being $(-1, -1, -1, -1)$. (b) is the maximum likelihood estimate that was used as the initial segmentation. (c)–(f) are the segmentation results of ALH, ICM, MFA, and SA, respectively. Table 10.4 illustrates the minimized energy (i.e., maximized posterior probability) values and the iteration numbers numerically. The solution quality is ranked as "HCF < ICM < SA < ALH < MFA".

According to the iteration numbers, the ALH method takes much fewer iterations than SA, about 0.2%–4.4%, to converge. Although it takes more iterations than the other deterministic algorithms, it is not as slow as it might seem. That the ALH converged after thousands of iterations was due to the stringent convergence criterion $\|u^{(t)} - u^{(t-1)}\|_\infty < 0.0001$. However,

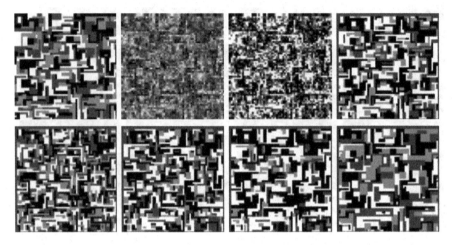

Figure 10.18: Restoration of image no.2 (refer to Fig. 10.17 for legend).

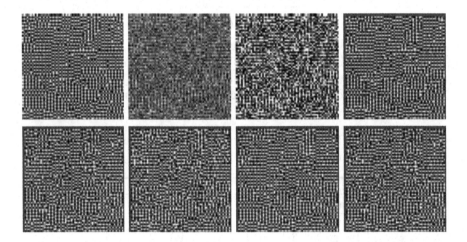

Figure 10.19: Restoration of image no.3 (refer to Fig. 10.17 for legend).

looking at the energy evolution curve in Fig. 10.21 (SA is not included there because it is far from the convergence within a thousand iterations), we see that the ALH reaches a solution better than ICM and HCF after only a dozen iterations and becomes nearly convergent after a hundred of iterations. The MFA needs five hundred iterations to converge.

10.6.3   Comparing the Comb Algorithm with Others

In the following, two experiments are presented, one for MAP-MRF image restoration and the other for segmentation, to compare the performance of

Table 10.3: The minimized energy values (top block), error rates (middle block), and iteration numbers (bottom block) for images 1–3.

	ALH	ICM	HCF	MFA	SA
Image No. 1	−11049	−10003	−10049	−10824	−11988
Image No. 2	−10191	−8675	−9650	−10073	−11396
Image No. 3	−26974	−25881	−26629	−26716	−27526
Image No. 1	0.268	0.349	0.367	0.291	0.164
Image No. 2	0.360	0.438	0.414	0.381	0.306
Image No. 3	0.212	0.327	0.273	0.181	0.125
Image No. 1	3654	7	31	553	83034
Image No. 2	1456	6	29	553	68804
Image No. 3	2789	7	36	560	92721

Table 10.4: Numerical comparison of the algorithms on the segmentation of the Lena image.

	ALH	ICM	HCF	MFA	SA
Min. Energy	−180333	−171806	−176167	−180617	−173301
Iterations	1255	7	38	545	593916

the following algorithms: (1) the Comb algorithm; (2) the ICM (Besag 1986); (3) the HCF (Chou et al. 1993) (the parallel version); and (4) the SA with the Metropolis sampler (Kirkpatrick et al. 1983). For the Comb, the parameters are $N = 10$ and $\tau = 0.01$. The implementation of SA is based on a procedure given in (Press et al. 1988). The schedules for SA are set to $T^{(t+1)} \leftarrow 0.999T^{(t)}$ with $T^{(0)} = 10^4$. The initial configurations for ICM, HCF, and SA are taken as the ML estimate, whereas those in F for the Comb are entirely random. The termination condition for Comb is that all configurations in F are the same or that 10000 new local minima have been generated.

The first set of experiments is for MAP-MRF restoration performed on three synthetic images of $M = 4$ gray levels, shown in Figs. 10.22–10.24. The original has the label set $\mathcal{L} = \{1, 2, 3, 4\}$ and the pixel gray values also in $\{1, 2, 3, 4\}$. Table 10.5 gives the clique potential parameters α_I and $\beta_1, \cdots,$ β_4 for generating the three types of textures and the standard deviation σ of the Gaussian noise.

The second experiment compares the algorithms in performing MAP-MRF segmentation on the Lena image of size 256×240 into a tri-level segmentation map. The results are illustrated in Fig. 10.25. The input image is the original Lena image corrupted by the i.i.d. Gaussian noise with standard

Figure 10.20: Segmentation of the Lena image. Top raw: Input image, maximum likelihood segmentation, ALH solution. Bottom raw: ICM solution, MFA solution, SA solution.

Table 10.5: The MRF parameters (α and β) and noise parameter σ for generating the three images.

Image	σ	α_I	β_1	β_2	β_3	β_4
No. 1	1	0	−1	−1	−1	1
No. 2	1	0	−2	−2	1	1
No. 3	1	0	1	1	1	1

deviation 10. The observation model is assumed to be the Gaussian distribution superimposed on the mean values of 40, 125 and 200 for the three-level segmentation. An isometric MRF prior is used, with the four β parameters being $(-1, -1, -1, -1)$.

Table 10.6 compares the quality of restoration and segmentation solutions in terms of the minimized energy values. We can see that the Comb outperforms the ICM and the HCF and is comparable to SA. A subjective evaluation of the resulting images would also agree to the objective numerical comparison. The quality of the Comb solutions is generally also better than that produced by using a continuous augmented Lagrange method developed previously (Li 1998b).

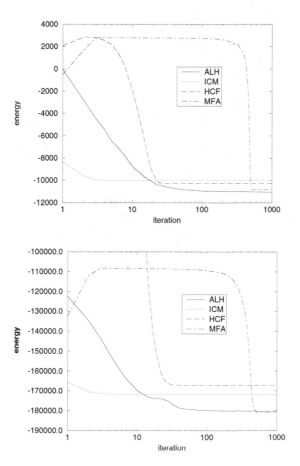

Figure 10.21: The solution evolution curves for image no.1 (top) and the Lena image (bottom).

The Comb as a random search method needs many iterations to converge, the number increasing as τ decreases. All the Comb solutions with $\tau = 0.01$ are obtained when the limit of generating 10,000 local minima is reached. This is about 1000 times more than the fast-converging ICM and HCF. Nonetheless, the Comb takes about 1/20 of the computational effort needed by the SA.

The Comb algorithm does not rely on initial configurations at the beginning of the algorithm to achieve better solutions; the maximum likelihood estimate can lead to a better solution for algorithms that operate on a single

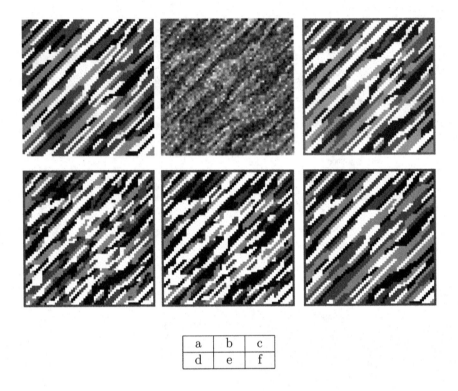

a	b	c
d	e	f

Figure 10.22: Restoration of image 1. (a) Original clean image. (b) Observed noisy image (input data). (c) Comb solution. (d) ICM solution. (e) HCF solution. (f) SA solution.

configuration, such as the ICM, HCF, and SA, but not necessarily for the Comb.

10.7   Accelerating Computation

The MRF configuration space for image and vision analysis is generally large, and the search for an energy minimum is computationally intensive. When the global solution is required, the computation is further increased by complications incurred by the problem of local minima and can well become intractable. Unrealistic computational demand has been criticism of the MAP-MRF framework. Efforts have been made to design efficient algorithms.

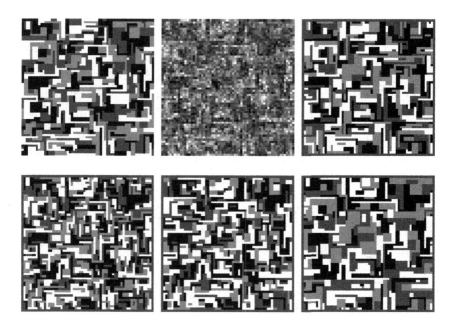

Figure 10.23: Restoration of image 2. Legends same as Fig. 10.22.

Table 10.6: The minimized energy values for the restoration of images 1–3 and the segmentation of the Lena image.

	Comb	ICM	HCF	SA
No. 1	−12057	−10003	−10269	−11988
No. 2	−10944	−8675	−9650	−11396
No. 3	−27511	−25881	−26629	−27526
Lena	−175647	−171806	−167167	−173301

10.7.1 Multiresolution Methods

Multiresolution methods provide a means for improving the convergence of iterative relaxation procedures (Hackbusch 1985). It was shown by Terzopoulos (1986a) that multiresolution relaxation can be used to efficiently solve a number of low-Level vision problems. This class of techniques has been used for MRF computation by many authors (Konrad and Dubois 1988b; Barnard 1989; Bouman and Liu 1991; Kato et al. 1993b; Bouman and Shapiro 1994). Gidas (1989) proposed a method that uses renormalization group theory, MRF's, and the Metropolis algorithm for global optimization. How to

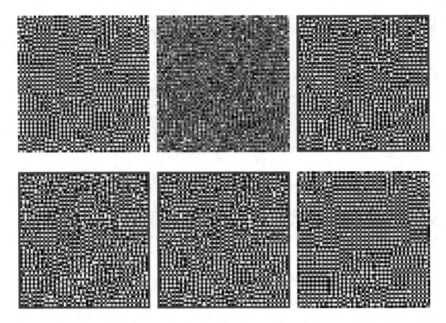

Figure 10.24: Restoration of image 3. Legends same as Fig. 10.22.

preserve the inter-relationships between objects in the scene using the renor-malization group transformation has been studied in (Geiger and Kogler Jr. 1993; Hurn and Gennison 1993; Petrou 1993).

An important issue in multiresolution computation of MRF's is how to preserve the Markovianity and define consistent model descriptions at different resolutions. In general, the local Markovian property is not preserved at the coarse levels after a sub-sampling. In (Jeng 1992), two theorems are given for a periodic sub-sampling of MRF's. One gives necessary and sufficient conditions for preserving the Markovianity and the other states that there is at least one sub-sampling scheme by which the Markovianity is preserved. A multiresolution treatment is presented by Lakshmanan and Derin (1993) in which (possibly) non-Markov Gaussian fields are approximated by linear Gaussian MRF's. In (Heitz and Bouthemy 1994), a consistent set of parameters are determined for objective functions at different resolutions; a nonlinear multiresolution relaxation algorithm that has fast convergence towards quasi-optimal solutions, is developed. A general transformation model is considered in (Perez and Heitz 1994) as the "restriction" of an MRF, defined on a finite arbitrary non-directed graph, to a subset of its original sites; several results are derived for the preservation of the Markovianity which may be useful for designing consistent and tractable multiresolution relaxation algorithms.

a	b	c
d	e	f

Figure 10.25: Segmentation of the Lena image. (a) The noisy Lena image. (b) ML configuration. (c) Comb solution. (d) ICM solution. (e) HCF solution. (f) SA solution.

10.7.2 Use of Heuristics

Apart from theory-supported methods, heuristics are often combined into the search. Solution candidates are quickly located using some efficient means. They are then evaluated by the derived energy function. Because the number of such candidates is usually small, the energy values can be compared exhaustively to give the minimal solution.

Not to miss the true global solution is crucial to successfully applying heuristics. One has to balance between efficiency and the danger of missing the true solution. More restrictive heuristics reduce the number of hypotheses generated and increase the efficiency, but they also cause a greater chance of missing the true minimum.

The *bounded noise model* (Baird 1985; Breuel 1992) is a simple heuristic for approximating noise distributions to quickly reduce the search space. By checking whether an error is within the allowed bounds, numerical constraints are converted into symbolic ones so that a yes-or-no decision can be made to prune the search space. This strategy has been used in many methods where numerical constraints have to be converted into symbolic ones. For example, in maximal cliques (Ambler et al. 1973), dynamic programming (Fischler and Elschlager 1973), and constraint search (Faugeras and Hebert 1986; Grimson and Lozano-Prez 1987), symbolic compatibilities are determined; in Hough transform (Hough 1962; Duda and Hart 1972) and geometric hashing (Lamdan and Wolfson 1988), numerical values are quantized to vote accumulators.

Hypothesis-verification is another approach for efficient search. Hypotheses are generated, which may correspond to peaks in Hough transform space or geometric hashing space, leaves of an interpretation tree (Grimson and Lozano-Prez 1987), random samples in random sampling (Fischler and Bolles 1981; Roth and Levine 1993), or result from minimal sets of image-model correspondences in the alignment method (Huttenlocher and Ullman 1987). Because the number of hypothesesgenerated is much smaller than the number of points in the original solution space, costs incurred by hypothesis evaluation is much reduced. In (Lowe 1992), matching and measurement errors are used to determine the probability of correctness for individual labels. Techniques presented therein may be used to speed verification. All these have potential applications in MRF's to tackle the problem of low computational efficiency.

References

Aarts, E. H. L. (1989). *Simulated annealing and Boltzmann machines : a stochastic approach to combinatorial optimization and neural computing*. Wiley.

Abend, K., T. J. Harley, and L. N. Kanal (1965). "Classification of binary random patterns". *IEEE Transactions on Information Theory 11*(4), 538–544.

Agarwala, A., M. Dontcheva, M. Agrawala, S. Drucker, A. Colburn, B. Curless, D. Salesin, and M. Cohen (2004). "Interactive digital photomontage". *ACM Trans. Graph. 23*(3), 294–302.

Akaike, H. (1974). "A new look at the statistical model indentification". *IEEE Transactions on Automatic Control 19*, 716–722.

Allain, M., J. Idier, and Y. Goussard (2006). "On global and local convergence of half-quadratic algorithms". *Image Processing, IEEE Transactions on 15*(5), 1130–1142.

Aloimonos, J. and D. Shulman (1989). *Integration of Visual Modules*. London, UK: Academic Press.

Ambler, A. P., H. G. Barrow, C. M. Brown, R. M. Burstall, and R. J. Popplestone (1973). "A versatile computer-controlled assembly system". In *Proceedings of International Joint Conference on Artificial Intelligence*, pp. 298–307.

Amini, A., S. Tehrani, and T. Weymouth (1988). "Using dynamic programming for minimizing the energy of active contours in the presence of hard constraints". In *Proceedings of IEEE International Conference on Computer Vision*, pp. 95–99.

Amini, A. A., T. E. Weymouth, and R. C. Jain (1990). "Using dynamic programming for solving variational problems in vision". *IEEE Transactions on Pattern Analysis and Machine Intelligence 12*, 855–867.

Ankenbrandt, C., B. Buckles, and F. Petry (1990). "Scene recognition using genetic algorithms with semantic nets". *Pattern Recognition Letters 11*, 285–293.

Arrow, K. J., L. Hurwicz, and H. Uzawa (1958). *Studies in Linear and Nonlinear Programming.* Stanford University Press.

Ayache, N. and O. D. Faugeras (1986). "HYPER: A new approach for the representation and positioning of two-dimensional objects". *IEEE Transactions on Pattern Analysis and Machine Intelligence 8*(1), 44–54.

Baddeley, A. J. and M. N. M. van Lieshout (1992). "Object recognition using Markov spatial processes". In *Proceedings of International Conference on Pattern Recognition*, Volume B, pp. 136–139.

Baddeley, A. J. and M. N. M. van Lieshout (1993). "Stochastic geometry in high level vision". In K. V. Mardia and G. K. Kanji (Eds.), *Statistics and Images: 1.*

Baird, H. S. (1985). *Model-based image matching using location.* Cambridge, Mass: MIT Press.

Ballard, D. H. (1981). "Generalizing the Hough transform to detect arbitrary shapes". *Pattern Recognition 13*(2), 111–122.

Ballard, D. H. and C. M. Brown (1982). *Computer Vision.* Prentice-Hall.

Barker, S. A. and P. J. W. Rayner (1997). "Unsupervised image segmentation using Markov random field models". In *Proceedings of International Workshop on Energy Minimization Methods in Computer Vision and Pattern Recognition*, Venice, Italy.

Barnard, S. (1989). "Stochastic stereo matching over scale". *International Journal of Computer Vision 3*, 17–32.

Barnard, S. T. (1987). "Stereo matching by hierarchical, microcanonical annealing". In *Proceedings of International Joint Conference on Artificial Intelligence*, pp. 832–835.

Barrow, H. G. and J. M. Tenenbaum (1981a). "Computational vision". *Proceedings of the IEEE 69*(5), 572–595.

Barrow, H. G. and J. M. Tenenbaum (1981b). "Interpreting line drawings as three dimensional surfaces". *Artificial Intelligence 17*, 75–117.

Battiti, R., E. Amaldi, and C. Koch (1991). "Computing optical flow across multiple scales: An adaptive coarse-to-fine strategy". *International Journal of Computer Vision 6*, 133–145.

Bellman, R. E. and S. E. Dreyfus (1962). *Applied Dynamic programming.* Princeton University Press.

Ben-Arie, J. and A. Z. Meiri (1987). 3d objects recognition by optimal matching search of multinary relations graphs. *Computer Vision, Graphics and Image Processing 37*, 345–361.

Bertero, M., T. A. Poggio, and V. Torre (1988). "Ill-posed problems in early vision". *Proceedings of the IEEE 76*(8), 869–889.

Besag, J. (1974). "Spatial interaction and the statistical analysis of lattice systems" (with discussions). *Journal of the Royal Statistical Society, Series B 36*, 192–236.

Besag, J. (1975). "Statistical analysis of non-lattice data". *The Statistician 24*(3), 179–195.

Besag, J. (1977). "Efficiency of pseudo-likelihood estimation for simple Gaussian fields". *Biometrika 64*, 616–618.

Besag, J. (1986). "On the statistical analysis of dirty pictures" (with discussions). *Journal of the Royal Statistical Society, Series B 48*, 259–302.

Besag, J. (1989). "Towards Bayesian image analysis". *Journal of Applied Statistics 16*(3), 395–406.

Besag, J. and P. J. Green (1993). "Spatial statistics and beyesian computation. *Journal of the Royal Statistical Society, Series B 55*(1), 25–37.

Besag, J., P. J. Green, D. Higdon, and K. Mengersen (1995). "Bayesian computation and stochastic systems". *Statistical Science 10*, 3–66.

Besl, P. J., J. B. Birch, and L. T. Watson (1988). "Robust window operators". In *Proceedings of Second International Conference on Computer Vision*, Florida, pp. 591–600.

Besl, P. J. and R. C. Jain (1985). "Three-Dimensional object recognition". *Computing Surveys 17*(1), 75–145.

Bhanu, B. (1984). "Representation and shape matching of 3D objects". *IEEE Transactions on Pattern Analysis and Machine Intelligence 6*(3), 340–350.

Bhanu, B. and O. D. Faugeras (1984). "Shape matching of two-dimensional objects". *IEEE Transactions on Pattern Analysis and Machine Intelligence 6*(2), 137–155.

Bhanu, B., S. Lee, and J. Ming (1989). "Adaptive image segmentation using a genetic algorithm". In *Proceedings of the Image Understanding Workshop*, pp. 1043–1055.

Bienenstock, E. (1988). "Neural-like graph-matching techniques for image-processing". In D. Z. Anderson (Ed.), *Neural Information Processing Systems*, pp. 211–235. Addison-Wesley.

Bilbro, G. and W. E. Snyder (1989). "Range image restoration using mean field annealing". In *Advances in neural information processing systems*, Volume 1, pp. 594–601. Morgan Kaufmann Publishers.

Black, M. and A. Rangarajan (1994). "The outlier process: Unifying line processes and robust statistics". In *Proceedings of IEEE Computer Society Conference on Computer Vision and Pattern Recognition*.

Black, M. J. and P. Anandan (1990). "A model for the detection of motion over time. In *Proceedings of IEEE International Conference on Computer Vision*, pp. 33–37.

Black, M. J. and P. Anandan (1993). "A framework for the robust estimation of optical flow. In *Proceedings of IEEE International Conference on Computer Vision*, pp. 231–236.

Blake, A. (1983). "The least disturbance principle and weak constraints". *Pattern Recognition Letters 1*, 393–399.

Blake, A. (1989). "Comparison of the efficiency of deterministic and stochastic algorithms for visual reconstruction". *IEEE Transactions on Pattern Analysis and Machine Intelligence 11*(1), 2–12.

Blake, A., B. Bascle, M. Isard, and J. MacCormick (1998). "Statistical models of visual shape and motion". *Phil. Trans. R. Soc. Lond. A 356*, 1283–1302.

Blake, A. and A. Zisserman (1987). *Visual Reconstruction*. Cambridge, MA: MIT Press.

Blomgren, P. (1998). *"Total Variation Methods for Restoration of Vector Valued Images"*. Ph.d, UCLA.

Blomgren, P. and T. F. Chan (1998). "Color TV: total variation methods for restoration of vector-valued images". *Image Processing, IEEE Transactions on 7*(3), 304–309.

Bolles, R. C. and P. Horaud (1986). "3DPO: A three-dimensional part orientation system". *International Journal of Robotic Research 5*(3), 3–26.

Boser, B. E., I. M. Guyon, and V. N. Vapnik (1992). "A training algorithm for optimal margin classifiers". In *Proceedings of the Fifth Annual ACM Workshop on Computational Learning Theory*, Pittsburgh, Pennsylvania.

Bouhlel, N., S. Sevestre-Ghalila, and C. Graffigne (06a). "New Markov random field model based on Nakagami distribution for modelling ultrasound RF envelope". In *14th European Signal Processing Conference*, Florence, Italy.

Bouhlel, N., S. Sevestre-Ghalila, M. JAIDANE, and C. Graffigne (06b). "Ultrasound backscatter characterization by using Markov random field model". In *IEEE International Conference on Acoustics, Speech and Signal Processing 2006.*, Volume 2.

Bouhlel, N., S. Sevestre-Ghalila, H. Rajhi, and R. Hamza (2004). "New Markov random field model based on K-distribution for textured ultrasound image". In *Medical Imaging 2004: Ultrasonic Imaging and Signal Processing. SPIE International Symposium*, Volume 5373, pp. 363–372.

Boult, T. E. (1987). "What is regular in regularization?". In *Proceedings of First International Conference on Computer Vision*, pp. 457–462.

Bouman, C. and B. Liu (1991). "Multiple resolution segmemtation of texture segmenation". *IEEE Transactions on Pattern Analysis and Machine Intelligence 13*(2), 99–113.

Bouman, C. and K. Sauer (1993). "A generalized Gaussian image model for edge preserving MAP estimation". *IEEE Transactions on Image Processing 2*(3), 296–310.

Bouman, C. and M. Shapiro (1994). "A multiscale random field model for Bayesian segmenation". *IEEE Transactions on Image Processing 3*(2), 162–177.

Bouthemy, P. (1989). "A maximum likelihood framework for determining moving edges". *IEEE Transactions on Pattern Analysis and Machine Intelligence 11*, 499–511.

Bowyer, A. (1981). "Computing Dirichlet tessellations". *Computer Journal 24*, 162–166.

Boyer, K. L., M. J. Mirza, and G. Ganguly (1994). "The robust sequential estimator: A general approach and its application to surface organization in range data". *IEEE Transactions on Pattern Analysis and Machine Intelligence 16*(10), 987–1001.

Boykov, Y. and V. Kolmogorov (2004). "An experimental comparison of min-cut/max-flow algorithms for energy minimization in vision". *IEEE Transactions on Pattern Analysis and Machine Intelligence 26*(9), 1124–1137.

Boykov, Y., O. Veksler, and R. Zabih (1998). "Markov random fields with efficient approximations". In *IEEE Conference on Computer Vision and Pattern Recognition*, pp. 648–655.

Boykov, Y., O. Veksler, and R. Zabih (2001). "Fast approximate energy minimization via graph cuts". *IEEE Transactions on Pattern Analysis and Machine Intelligence 23*(11), 1222–1239.

Bozma, H. I. and J. S. Duncan (1988). "Admissibility of constraint functions in relaxation labeling". In *Proceedings of Second International Conference on Computer Vision*, Florida, pp. 328–332.

Breuel, T. M. (1992). "Fast recognition using adaptive subdivision of transformation space". In *Proceedings of IEEE Computer Society Conference on Computer Vision and Pattern Recognition*, pp. 445–451.

Breuel, T. M. (1993). "Higher-order statistics in visual object recognition". Memo #93-02, IDIAP, Martigny, Switzerland.

Canny, J. (1986). "A computational approach to edge detection". *IEEE Transactions on Pattern Analysis and Machine Intelligence 6*(6), 679–698.

Carter, J. (2001). *"Dual methods for total variation-based image restoration"*. Ph.d, UCLA.

Cerny, V. (1982). "A thermodynamical approach to the traveling salesman problem: an efficient simulation algorithm". Preprint, Institute of Physics and Biophysics, Comenius University, Bratislava. also appear as "Thermodynamical approach to the traveling salesman problem: an efficient simulation algorithm", *Journal of Optimization Theory and Applications*, vol.45, pp.41-51, 1985.

Cerny, V. (1985). "Thermodynamical approach to the traveling salesman problem: an efficient simulation algorithm". *Journal of Optimization Theory and Applications 45*, 41–51.

Cesari, L. (1983). *Optimization – Theory and Applications : Problems with Ordinary Differential Equations*. New York: Springer-Verlag.

Chalmond, B. (1989). "An iterative Gibbsian technique for reconstruction of m-ary images". *Pattern Recognition 22*(6), 747–761.

Chambolle, A. (2004). "An algorithm for total variation minimization and applications". *J. Math. Imaging Vis. 20*(1-2), 89–7.

Chan, T. F. and S. Esedoglu (2005). "Aspects of total variation regularized L1 function approximation". *SIAM Journal of Applied Mathematics 65*, 1817–1837.

Chan, T. F., G. H. Golub, and P. Mulet (1996). "A nonlinear primal-dual method for total variation-based image restoration". In *ICAOS '96 (Paris, 1996),*, Volume 219, pp. 241–252. Springer-Verlag.

Chan, T. F. and J. J. Shen (2005). *Image Processing and Analysis - Variational, PDE, wavelet, and stochastic methods.* SIAM.

Chandler, D. (1987). *Introduction to modern statistical mechanics.* Oxford University Press.

Charbonnier, P., B.-F. Laure, G. Aubert, and M. Barlaud (1997). "Deterministic edge-preserving regularization in computed imaging". *Image Processing, IEEE Transactions on 6*(2), 298–311.

Chekuri, C., S. Khanna, J. Naor, and L. Zosin (2001). "Approximation algorithms for the metric labeling problem via a new linear programming formulation". In *SODA '01: Proceedings of the twelfth annual ACM-SIAM symposium on Discrete algorithms*, pp. 109–118.

Chellappa, R. (1985). "Two-dimensional discrete gaussian Markov random field models for image processing". In L. N. Kanal and A. Rosenfeld (Eds.), *Progress in Pattern Recognition 2*, pp. 79–112.

Chellappa, R. and A. Jain (1993). *Markov Random Fields: Theory and Applications.* Academic Press.

Chellappa, R. and R. L. Kashyap (1982). "Digital image restoration using spatial interaction models". *IEEE Transactions on Acoustic, Speech and Signal Processing 30*, 461–472.

Chen, C. C. (1988). *Markov Random Fields in Image Analysis.* Ph. D. thesis, Michigan State University.

Cheng, J. K. and T. S. Huang (1984). "Image registration by matching relational structures". *Pattern Recognition 17*(1), 149–159.

Cheng, Y. (1995). Mean shift, mode seeking and clustering. *IEEE Transactions on Pattern Analysis and Machine Intelligence 17*(8), 790–799.

Chiang, T.-S. and Y. Chow (1992). "A comparison of simulated annealing of Gibbs sampler and Metropolis algorithms". In P. Barone, A. Frigessi, and M. Piccioni (Eds.), *Proceedings of the special year on image analysis held in Rome, Italy, 1990.* Springer-Verlag:Berlin.

Chou, P. B. and C. M. Brown (1990). "The theory and practice of Bayesian image labeling". *International Journal of Computer Vision 4*, 185–210.

Chou, P. B., P. R. Cooper, M. J. Swain, C. M. Brown, and L. E. Wixson (1993). "Probabilistic network inference for cooperative high and low level vision". In R. Chellappa and A. Jain (Eds.), *Markov Random Fields: Theory and Applications*, Boston, pp. 211–243. Academic Press.

Chow, C. K. (1962). "A recognition method using neighbor dependence". *IRE Transactions on Electronic Computer" 11*, 683–690.

Clark, J. J. and A. L. Yuille (1990). *Data Fusion for Sensory Information Processing Systems*. Norwell, MA: Kluwer Academic Publishers.

Cohen, F. S. and D. B. Cooper (1987). "Simple parallel hierarchical and relaxation algorithms for segmenting noncasual Markovian random fields". *IEEE Transactions on Pattern Analysis and Machine Intelligence 9*(2), 195–218.

Cohen, F. S. and Z. Fan (1992). "Maximum likelihood unsupervised textured image segmentation". *CVGIP: Graphics Model and Image Processing 54*, 239–251.

Comaniciu, D. and P. Meer (1997). Robust analysis of feature space: Color image segmentation. In *Proceedings of IEEE Computer Society Conference on Computer Vision Pattern Recognition*, San Juan, Puerto Rico.

Comaniciu, D. and P. Meer (1999). Mean shift analysis and applications. In *IEEE International Conference on Computer Vision (ICCV'99)*, Kerkyra, Greece.

Cooper, D. B., J. Subrahmonia, Y. P. Hung, and B. Cernuschi-Frias (1993). "The use of Markov random fields in estimating amd recognizing object in 3D space". In R. Chellappa and A. Jain (Eds.), *Marokov Random Fields: Theory and Applications*, Boston, pp. 335–367. Academic Press.

Cooper, P. R. (1990). "Parallel structure recognition with uncertainty: Coupled segmentation and matching". In *Proceedings of IEEE International Conference on Computer Vision*, pp. 287–290.

Cootes, T. F., C. J. Taylor, D. H. Cooper, and J. Graham (1995). "Active shape models: Their training and application". *CVGIP: Image Understanding 61*, 38–59.

Courant, R. and D. Hilbert (1953). *Methods of Mathematical Physics*, Volume 1. New York: Interscience Publishers Inc.

Craven and G. Wahba (1979). "Smoothing noisy data with spline functions: estimating the correct degree of smoothing by the methods of generalized cross-validation". *Numerische Mathematik 31*, 377–403.

Cross, G. C. and A. K. Jain (1983). "Markov random field texture models". *IEEE Transactions on Pattern Analysis and Machine Intelligence 5*(1), 25–39.

Dahlquist, G. and A. Bjorck (1974). *Numerical Methods*. Englewood, NJ: N. Anderson (trans.). Prentice-Hall.

Darbon, J. and M. Sigelle (2004). "Exact optimization of discrete constrained total variation minimization problems". In *In LNCS series vol. 3322, editor, Tenth International Workshop on Combinatorial Image Analysis (IWCIA 2004)*, pp. 540–549.

Darbon, J. and M. Sigelle (2005). "A fast and exact algorithm for total variation minimization". In *In LNCS series, editor, 2nd Iberian Conference on Pattern Recognition and Image Analysis (IbPria)*.

Darrell, T., S. Sclaroff, and A. Pentland (1990). "Segmentation by minimal description". In *Proceedings of IEEE International Conference on Computer Vision*, pp. 112–116.

Dass, S. C., A. K. Jain, and X. Lu (2002). "Face detection and synthesis using Markov random field models". *Pattern Recognition, 2002. Proceedings. 16th International Conference on 4*, 201–204.

Davis, L. S. (1979). "Shape matching using relaxation techniques". *IEEE Transactions on Pattern Analysis and Machine Intelligence 1*(1), 60–72.

Dempster, A. P., N. M. Laird, and D. B. Bubin (1977). "Maximum likelihood from imcomplete data via EM algorithm". *Journal of the Royal Statistical Society, Series B 39*, 1–38.

Dengler, J. (1991). "Estimation of discontinuous displacement vector fields with the minimum description length criterion". In *Proceedings of IEEE Computer Society Conference on Computer Vision and Pattern Recognition*, pp. 276–282.

Derin, H. and W. S. Cole (1986). "Segmentation of textured images using using Gibbs random fields". *Computer Vision, Graphics and Image Processing 35*, 72–98.

Derin, H. and H. Elliott (1987). "Modeling and segmentation of noisy and textured images using Gibbs random fields". *IEEE Transactions on Pattern Analysis and Machine Intelligence 9*(1), 39–55.

Derin, H., H. Elliott, R. Cristi, and D. Geman (1984). "Bayes smoothing algorithms for segmentation of binary images modeled by Markov random fields". *IEEE Transactions on Pattern Analysis and Machine Intelligence 6*(6), 707–720.

Derin, H. and P. A. Kelly (1989). "Discrete-index Markov-type random fields". *Proceedings of the IEEE* 77(10), 1485–1510.

Dubes, R. C. and A. K. Jain (1989). "Random field models in image analysis". *Journal of Applied Statistics* 16(2), 131–164.

Dubes, R. C., A. K. Jain, S. G. Nadabar, and C. C. Chen (1990). "MRF model-based algorithms for image segmentation". In *Proceedings of International Conference on Pattern Recognition*, Volume B, pp. 808–814.

Duda, R. O. and P. E. Hart (1972). "Use of Hough transform to detect lines and curves in picture". *Communications of the ACM* 15(1), 11–15.

Duda, R. O. and P. E. Hart (1973). *Pattern Classification and Scene Analysis*. Wiley.

Elfadel, I. M. (1993). *From Random Fields to Networks*. Ph. D. thesis, Department of Mechanical Engineering.

Elliott, H., H. Derin, R. Cristi, and D. Geman (1984). "Application of the Gibbs distribution to image segmentation". In *Proceedings of the International Conference on Acoustic, Speech and Signal Processing*, San Diego, pp. 32.5.1–32.5.4.

Fan, T. J., G. Medioni, and R. Nevatia (1989). "Recognizing 3D objects using surface descriptions". *IEEE Transactions on Pattern Analysis and Machine Intelligence* 11(11), 1140–1157.

Faugeras, O. D. and M. Berthod (1981). "Improving consistency and reducing ambiguity in stochastic labeling: An optimization approach". *IEEE Transactions on Pattern Analysis and Machine Intelligence* 3, 412–423.

Faugeras, O. D. and M. Hebert (1986). "The representation, recognition and locating of 3D objects". *International Journal of Robotic Research* 5(3), 27–52.

Faugeras, O. D. and K. Price (1981). "Semantic description of arrial images using stochastic labeling". *IEEE Transactions on Pattern Analysis and Machine Intelligence* 3, 638–642.

Fischler, M. and R. C. Bolles (1981). "Random sample consensus: A paradigm for model fitting with applications to image analysis and automated cartography". *Communications of the ACM* 24(6), 381–395.

Fischler, M. and R. Elschlager (1973). "The representation and matching of pictorial structures". *IEEE Transactions on Computers* C-22, 67–92.

Fletcher, R. (1987). *Practical Methods of Optimization*. Wiley.

Frey, B. J. (1997). *Bayesian Networks for Pattern Classification, Data Compression and Channel Coding.* MIT Press.

Friedland, N. S. and A. Rosenfeld (1992). "Compact object recognition using energy-function based optimization". *IEEE Transactions on Pattern Analysis and Machine Intelligence 14*, 770–777.

Fu, K. S. and T. S. Yu (1980). *Statistical pattern classification using contextual information.* Pattern Recognition & Image Processing Series. Research Studies Press.

Fukunaga, K. (1990). *Introduction to statistical pattern recognition* (2 ed.). Boston: Academic Press.

Galatsanos, N. P. and A. K. Katsaggelos (1992). "Methods for choosing the regularization parameter and estimating the noise variance in image restoration and their relations". *IEEE Transactions on Image Processing 1*(3), 322–336.

Gamble, E. and T. Poggio (1987). "Visual integration and detection of discontinuities". *A. I. Lab. Memo* No. 970, MIT.

Geiger, D. and F. Girosi (1989). "Parallel and deterministic algorithms from MRF's: surface reconstruction and integration". *A. I. Lab. Memo* No. 1114, MIT, Cambridge, MA.

Geiger, D. and F. Girosi (1991). "Parallel and deterministic algorithms from MRF's: surface reconstruction". *IEEE Transactions on Pattern Analysis and Machine Intelligence 13*(5), 401–412.

Geiger, D. and J. E. Kogler Jr. (1993). "Scaling images and image features via the renormalization group". In *Proceedings of IEEE Computer Society Conference on Computer Vision and Pattern Recognition*, pp. 47–53.

Geiger, D. and T. Poggio (1987). "An optimal scale for edge detection". In *Proc. International Joint Conference on AI.*

Geman, D., S. Geman, C. Graffigne, and P. Dong (1990). "Boundary detection by constrained optimization". *IEEE Transactions on Pattern Analysis and Machine Intelligence 12*(7), 609–628.

Geman, D. and B. Gidas (1991). *"Image analysis and computer vision"*, Chapter 2, pp. 9–36. National Academy Press.

Geman, D. and G. Reynolds (1992). "Constrained restoration and the recovery of discontinuities". *IEEE Transactions on Pattern Analysis and Machine Intelligence 14*(3), 767–783.

Geman, S. and D. Geman (1984). "Stochastic relaxation, Gibbs distribution and the Bayesian restoration of images". *IEEE Transactions on Pattern Analysis and Machine Intelligence 6*(6), 721–741.

Geman, S. and C. Graffigne (1987). "Markov random field image models and their applications to computer vision". In A. M. Gleason (Ed.), *Proceedings of the International Congress of Mathematicians: Berkeley, August 3-11, 1986*, pp. 1496–1517.

Geman, S. and D. McClure (1985). "Bayesian image analysis: An application to single photon emission tomography". In *Proceedings of the Statistical Computing Section*, Washington, DC, pp. 12–18.

Geman, S., D. McClure, and D. Geman (1992). "A nonlinear filter for film restoration and other problems in image processing". *CVGIP: Graphics Model and Image Processing 54*, 281–289.

Ghahraman, D. E., A. K. C. Wong, and T. Au (1980). "Graph optimal monomorphism algorithms". *IEEE Transactions on Systems, Man and Cybernetics SMC-10*(4), 181–188.

Gidas, B. (1989). "A renormalization group approach to image processing problems". *IEEE Transactions on Pattern Analysis and Machine Intelligence 11*, 164–180.

Gilks et al., W. R. (1993). "Modeling complexity: Applications of Gibbs sampling in medicine". *Journal of the Royal Statistical Society, Series B 55*(1), 39–52.

Gimel'farb, G. L. (1996). "Texture modelling by multiple pairwise pixel interactions". *IEEE Transactions on Pattern Analysis and Machine Intelligence 18*(11), 1110–1114.

Goemans, M. X. and D. P. Williamson (1995). "Improved approximation algorithms for maximum cut and satisfiability problems using semidefinite programming". *J. ACM 42*(6), 1115–1145.

Goldberg, D. E. (1989). *Genetic Algorithms in Search, Optimization, and Machine Learning*. Addison-Wesley.

Golub, G. H., M. Heath, and G. Wahba (1979). "Generalized cross-validation as a method for choosing a good ridge parameter". *Technometrics 21*, 215–223.

Gottfried, B. S. (1973). *Introduction to Optimization Theory*. Prentice-Hall.

Green, P. (1990). "Bayesian reconstructions from emission tomography data using a modified EM algorithm". *IEEE Transactions on Medical Imaging 9*(1), 84–93.

Green, P. (1995). "Reversible jump Markov chain Monte Carlo computation and Bayesian model determination. *Biometrika 82*, 711–732.

Greig, D. M., B. T. Porteous, and A. H. Seheult (1989). "Exact maximum a posteriori estimation for binary images". *Journal of the Royal Statistical Society. Series B (Methodological) 51*(2), 271–279.

Grenander, U. (1976). *Pattern synthesis*, Volume 1. New York: Springer-Verlag.

Grenander, U. (1983). *Tutorials in Pattern synthesis*. Brown University, Division of Applied Mathematics.

Grenander, U., Y. Chow, and D. M. Keenan (1991). *Hands: a pattern theoretic study of biological shapes*. New York: Springer-Verlag.

Grenander, U. and M. Miller (1994). "Representation of knowledge in complex systems". *Journal of the Royal Society, B 56*(3), 549–603.

Griffeath, D. (1976). Introduction to random fields. In J. G. Kemeny, J. L. Snell, and A. W. Knapp (Eds.), *Denumerable Markov Chains* (2nd ed.)., Chapter 12, pp. 425–458. New York: Springer-Verlag.

Grimmett, G. (1982). *Probability and Random Processes*. Oxford University Press.

Grimson, W. E. L. (1981). *From Images to Surfaces: A Computational Study of the Human Early Visual System*. Cambridge, MA: MIT Press.

Grimson, W. E. L. (1990). *Object Recognition by Computer – The Role of Geometric Constraints*. Cambridge, MA: MIT Press.

Grimson, W. E. L. and T. Lozano-Prez (1987). "Localizing overlapping parts by searching the interpretation tree". *IEEE Transactions on Pattern Analysis and Machine Intelligence 9*(4), 469–482.

Grimson, W. E. L. and T. Pavlidis (1985). "Discontinuity detection for visual surface reconstruction". *Computer Vision, Graphics and Image Processing 30*, 316–330.

Gurelli, M. I. and L. Onural (1994). "On a parameter estimation method for Gibbs-Markov random fields". *IEEE Transactions on Pattern Analysis and Machine Intelligence 4*(16), 424–430.

Hackbusch, W. (1985). *Multi-Grid methods and applications*. Berlin: Springer-Verlag.

Hammersley, J. M. and P. Clifford (1971). "Markov field on finite graphs and lattices". unpublished.

Hammersley, J. M. and D. C. Handscomb (1964). *Monte Carlo Methods*. New York: Wiley.

Hansen, F. R. and H. Elliott (1982). "Image segmentation using simple Markov random field models". *Computer Graphics Image Processing 20*, 101–132.

Hanson, A. R. and E. M. Riseman (1978). *Computer Vision Systems*, Chapter "Segmentation of natural scenes", pp. 129–163. Academic Press.

Haralick, R. M. (1983). "Decision making in context". *IEEE Transactions on Pattern Analysis and Machine Intelligence 5*(4), 417–428.

Haralick, R. M., H. Joo, C. Lee, X. Zhuang, V. Vaidya, and M. Kim (1989). "Pose estimation from corresponding point data". *IEEE Transactions on Systems, Man and Cybernetics 19*, 1426–1446.

Haralick, R. M. and L. G. Shapiro (1992). *Computer and Robot Vision*. Reading, MA: Addison-Wesley.

Harris, J. G. (1987). "A new approach to surface reconstruction: The coupled depth/slope model". In *Proceedings of First International Conference on Computer Vision*, London, England, pp. 277–283.

Harris, J. G., C. Koch, E. Staats, and J. Lou (1990). "Analog hardware for detecting discontinuities in early vision". *International Journal of Computer Vision 4*, 211–223.

Hassner, M. and J. Slansky (1980). "The use of Markov random field as models of texture". *Computer Graphics Image Processing 12*, 357–370.

Hastings, W. K. (1970). "Monte Carlo sampling methods using Markov chains and their applications". *Biometrika 57*, 97–109.

Haykin, S. (1994). *Neural networks : a comprehensive foundation*. New York: Macmillan.

Hebert, T. and R. Leahy (1992). "Statistic-based MAP image restoration from Poisson data using Gibbs priors". *IEEE Transactions on Signal Processing 40*(9), 2290–2303.

Heitz, F. and P. Bouthemy (1993). "Multimodal estimation of discontinuous optical flow using Markov random fields". *IEEE Transactions on Pattern Analysis and Machine Intelligence 15*(12), 1217–1232.

Heitz, F. and P. Bouthemy (1994). "Multiscale minimization of global energy functions in visual recovery problems". *CVGIP: Image Understanding 59*(1), 125–134.

Herault, L. and R. Horaud (1993). "Figure-ground discrimination: A combinatorial optimization approach. *IEEE Transactions on Pattern Analysis and Machine Intelligence 15*, 899–914.

Hestenes, M. R. (1969). "Multipler and gradient methods". *Journal of Optimization Theory and Applications 4*, 303–320.

Hildreth, E. C. (1984). *The Measurement of Visual Motion*. Cambridge, MA: MIT Press.

Hill, A. and C. Taylor (1992). "Model-based image interpretation using genetic algorithms". *Image and Vision Computing 10*, 295–300.

Hinton, G. E. (1978). *Relaxation and Its Role in Vision*. Ph. D. thesis, University of Edinburgh, Edinburgh.

Holland, J. (1975). *Adaptation in Natural and Artificial Systems*. University of Michigan Press.

Hopfield, J. J. (1984). "Neurons with graded response have collective computational properties like those of two state neurons". *Proceedings of National Academic Science, USA 81*, 3088–3092.

Hopfield, J. J. and D. W. Tank (1985). "'Neural' computation of decisions optimization problems". *Biological Cybernetics 52*, 141–152.

Horn, B. K. P. (1988). "Parallel networks for machine vision". *A. I. Memo.* No. 1071, MIT.

Horn, B. K. P. and B. G. Schunck (1981). "Determining optical flow". *Artificial Intelligence 17*, 185–203.

Hough, P. V. C. (1962). "A method and means for recognizing complex patterns". U.S. Patent No. 3,069,654.

Hu, R. and M. M. Fahmy (1987). "Texture segmentation based on a hierarchical Markov random field model". *Signal Processing 26*, 285–385.

Huang, R., V. Pavlovic, and D. N. Metaxas (2004). "A hybrid face recognition method using Markov random fields". *Pattern Recognition, 2004. ICPR 2004. Proceedings of the 17th International Conference on 3*, 157–160.

Huber, P. (1981). *Robust Statistics*. Wiley.

Hummel, R. A. and S. W. Zucker (1983). "On the foundations of relaxation labeling process". *IEEE Transactions on Pattern Analysis and Machine Intelligence 5*(3), 267–286.

Hung, Y. P., D. B. Cooper, and B. Cernuschi-Frias (1991). "Asymtotic Baysian surface estimation using an image sequence". *International Journal of Computer Vision 6*(2), 105–132.

Hurn, M. and C. Gennison (1993). "Multiple-site method for image estimation". See Mardia and Kanji (1994), pp. 155–186.

Huttenlocher, D. P. and S. Ullman (1987). "Object recognition using alignment". In *Proceedings of IEEE International Conference on Computer Vision*, pp. 102–111.

Ikeuchi, K. and B. K. P. Horn (1981). "Numerical shape from shading and occluding boundaries". *Artificial Intelligence 17*, 141–184.

Illingworth, J. and J. Kittler (1988). "A survey of Hough transform". *Computer Vision, Graphics and Image Processing 43*, 221–238.

Jacobus, C. J., R. T. Chien, and J. M. Selander (1980). "Motion detection and analysis of matching graphs of intermediate level primitives". *IEEE Transactions on Pattern Analysis and Machine Intelligence 2*(6), 495–510.

Jain, A. K. (1981). "Advances in mathematical models in image processing". *Proceedings of IEEE 69*, 502–528.

Jain, A. K. and R. C. Dubes (1988). *Algorithms for Clustering Analysis*. Englewood Cliff, NJ: Printice-Hall.

Jain, A. K. and V. Moreau (1987). "Bootstrap technique in cluster analysis". *Pattern Recognition 20*(5), 547–568.

Jain, A. K. and S. G. Nadabar (1990). "MRF model-based segmentation of range images". In *Proceedings of IEEE International Conference on Computer Vision*, pp. 667–671.

Jain, A. K., Y. Zhong, and M.-P. Dubuisson-Jolly (1998). "Deformable template models: A review". *Signal Processing 71*, 109–129.

Jain, A. K., Y. Zhong, and S. Lakshmanan (1996). "Object matching using deformable templates". *IEEE Transactions on Pattern Analysis and Machine Intelligence 18*(3), 267–278.

Jakeman, E. and P. Pusey (1976). "A model for non-Rayleigh sea echo". *Antennas and Propagation, IEEE Transactions on 24*(6), 806–814.

Jamison, T. A. and R. J. Schalkoff (1988). "Image labeling: a neural network approach". *Image and Vision Computing 6*(4), 203–214.

Jaynes, E. (1982). "On the rationale of maximum-entropy methods". *Proceedings of the IEEE 70*(9), 939–952.

Jeng, F. C. (1992). "Subsampling of Markov random fields". *Journal of Visual Communication and Image Representation 3*, 225–229.

Jeng, F. C. and J. W. Wood (1990). "Simulated annealing in compound Gauss-Markov random fields". *IEEE Transactions on Information Theory 36*, 94–107.

Jeng, F. C. and J. W. Wood (1991). "Compound Gauss-Markov random fields for image estimation". *IEEE Transactions on Signal Processing 39*, 683–679.

Jensen, F. V. (2001). *Bayesian Networks and Decision Graphs*. Springer.

Jolion, J. M., P. Meer, and S. Bataouche (1991). "Robust clustering with applications in computer vision". *IEEE Transactions on Pattern Analysis and Machine Intelligence 13*, 791–802.

Jordan, M. I. (1998). *Learning in Graphical Models*. MIT Press.

Julesz, B. (1962). Visual pattern discrimination. *IRE Transactions of Information Theory IT-8*, 84–92.

Kanatani, K. (1993). *Geometric computation for machine vision*. New York: Oxford University Press.

Kashyap, R. L. (1988). "Optimal choice of AR and MA parts in autoregressive moving average models". *IEEE Transactions on Pattern Analysis and Machine Intelligence 4*, 99–104.

Kashyap, R. L., R. Chellappa, and A. Khotanzad (1982). "Texture classification using features derived from random process models". *Pattern Recognition Letters 1*, 43–50.

Kashyap, R. L. and K. N. Eom (1988). "Robust image modeling techniques with their applications". *IEEE Transactions on Acoustic, Speech and Signal Processing 36*(8), 1313–1325.

Kass, M., A. Witkin, and D. Terzopoulos (1987). "Snakes: Active contour models". In *Proceedings of IEEE International Conference on Computer Vision*, pp. 259–268.

Kass, M., A. Witkin, and D. Terzopoulos (1988). "Snakes: Active contour models". *International Journal of Computer Vision 1*, 321–331.

Kato, Z., M. Berthod, and J. Zerubia (1993a). "A hierarchical markov random field model and multi-temperature annealing for parallel image classification". RR 1938, INRIA, Sophia-Antipolis Cedex, France.

Kato, Z., M. Berthod, and J. Zerubia (1993b). "Multiscale Markov random field models for parallel image classification". In *Proceedings of IEEE International Conference on Computer Vision*, pp. 253–257.

Keeler, K. (1991). "Map representations and coding based priors for segmentation". In *Proceedings of IEEE Computer Society Conference on Computer Vision and Pattern Recognition*, pp. 420–425.

Kelly, P. A., H. Derin, and K. D. Hartt (1988). "Adaptive segmentation of speckled images using a hierarchical random field model". *IEEE Transactions on Acoustic, Speech and Signal Processing 36*, 1628–1641.

Kim, I. Y. and H. S. Yang (1992). "Efficient image understanding based on the Markov random field model and error backpropagation network". In *Proceedings of International Conference on Pattern Recognition*, Volume A, pp. 441–444.

Kim, S. and M. Kojima (2001). "Second order cone programming relaxations of quadratic optimization problems". *Optimization Methods and Software 15*(3-4), 201–224.

Kindermann, R. and J. L. Snell (1980). *Markov Random Fields and Their Applications*. Providence, R.I.: American Mathematical Society.

Kirkpatrick, S., C. D. Gellatt, and M. P. Vecchi (1982). "Optimization by simulated annealing". Research report rc, IBM. also appear as "Optimization by simulated annealing", *Science*, Vol.220, pp.671-680, 1983.

Kirkpatrick, S., C. D. Gellatt, and M. P. Vecchi (1983). "Optimization by simulated annealing". *Science 220*, 671–680.

Kittler, J., W. J. Christmas, and M. Petrou (1993). "Probabilistic relaxation for matching problem in computer vision". In *Proceedings of Forth International Conference on Computer Vision*, Gemany, pp. 666–673.

Koch, C. (1988). "Computing motion in the presence of discontinuities: algorithm and analog networks". In R. Eckmiller and C. c. d. Malsburg (Eds.), *Neural Computers*, Volume F41 of *NATO ASI Series*, pp. 101–110. Springer-Verlag.

Koch, C., J. Marroquin, and A. Yuille (1986). "Analog 'neuronal' networks in early vision". *Proceedings of National Academic Science, USA 83*, 4263–4267.

Koenderink, J. J. (1990). *Solid Shape*. MIT Press.

Kolmogorov, V. and C. Rother (2007). "Minimizing nonsubmodular functions with graph cutsA review". *IEEE Transactions on Pattern Analysis and Machine Intelligence 29*(7), 1274–1279.

Kolmogorov, V. and M. Wainwright (2005). "On the optimality of tree-reweighted max-product message-passing". In *Proceedings of the 21th Annual Conference on Uncertainty in Artificial Intelligence (UAI-05)*, pp. 316–32. AUAI Press.

Kolmogorov, V. and R. Zabih (2002). "Multi-camera sence reconstruction via graph cuts". In *ECCV*, Volume 3, pp. 82–96.

Kolmogorov, V. and R. Zabih (2004). "What energy functions can be minimized via graph cuts?". *IEEE Transactions on Pattern Analysis and Machine Intelligence 26*(2), 147–159.

Konrad, J. and E. Dubois (1988a). "Estimation of image motion fields: Bayesian formulation and stochastic solution". In *Proceedings of the International Conference on Acoustic, Speech and Signal Processing*, pp. 354–362.

Konrad, J. and E. Dubois (1988b). "Multigrid bayesian estimation of image motion fields using stochastic relaxation". In *Proceedings of IEEE International Conference on Computer Vision*, pp. 354–362.

Konrad, J. and E. Dubois (1992). "Bayesian estimation of motion vector fields". *IEEE Transactions on Pattern Analysis and Machine Intelligence 14*, 910–927.

Korte, B. and J. Vygen (2006). *Combinatorial Optimization: Theory and Algorithms* (3rd ed.). Springer-Verlag.

Koster, A. M. C. A., S. P. M. van Hoesel, and A. W. J. Kolen (1998). "The partial constraint satisfaction problem: Facets and lifting theorems". *Operations Research Letters 23*(3-5), 89–97.

Kumar, M. P., P. Torr, and A. Zisserman (2006). "Solving Markov random fields using second order cone programming relaxations". *Computer Vision and Pattern Recognition, 2006 IEEE Computer Society Conference on 1*, 1045–1052.

Kumar, R. and A. Hanson (1989). "Robust estimation of camera location and orientation from noisy data having outliers". In *Proc Workshop on Interpretation of Three-Dimensional Scenes*, pp. 52–60.

Kumar, S. (2005). *"Models for Learning Spatial Interactions in Natural Images for Context-Based Classification"*. Ph.d, CMU.

Kumar, S. and M. Hebert (2003). "Discriminative Random Fields: A discriminative framework for contextual interaction in classification". In *IEEE International Conference on Computer Vision(ICCV 03)*.

Kumar, S. and M. Hebert (2006). "Discriminative random fields". *International Journal of Computer Vision(IJCV) 68*(2), 179–201.

Lafferty, J., A. McCallum, and F. Pereira (2001). "Conditional Random Fields: Probabilistic models for segmenting and labeling sequence data". In *Proc. 18th International Conf. on Machine Learning*, pp. 282–289.

Lai, K. F. and R. T. Chin (1995). "Deformable contour: modeling and extraction". *IEEE Transactions on Pattern Analysis and Machine Intelligence 17*, 1084–1090.

Lakshmanan, S. and H. Derin (1989). "Simultaneous parameter estimation and segmentation of gibbs random fields using simulated annealing". *IEEE Transactions on Pattern Analysis and Machine Intelligence 11*, 799–813.

Lakshmanan, S. and H. Derin (1993). "Gaussian Markov random fields at multiple resolutions". In R. Chellappa and A. Jain (Eds.), *Markov Random Fields: Theory and Applications*, Boston, pp. 131–157. Academic Press.

Lamdan, Y. and H. Wolfson (1988). "Geometric hashing: a general and efficient model-based recognition scheme". In *ICCV88*, pp. 238–249.

Lange, K. (1990). "Convergence of EM image reconstruction algorithm with Gibbs smoothing". *IEEE Transactions on Medical Imaging 9*(4), 439–446.

Lanitis, A., C. J. Taylor, and T. F. Cootes (1997). "Automatic interpretation and coding of face images using flexible models". *IEEE Transactions on Pattern Analysis and Machine Intelligence 19*, 743–756.

Lauritzen, S. L. (1996). *Graphical Models*. Clarendon Press-Oxford.

Leclerc, Y. G. (1989). "Constructing simple stable descriptions for image partitioning". *International Journal of Computer Vision 3*, 73–102.

Leclerc, Y. G. and M. A. Fischler (1992). "An optimization-based approach to the interpretation of single line drawings as 3D wire frames". *International Journal of Computer Vision 9*, 113–136.

Lee, D. and T. Pavlidis (1987). "One dimensional regularization with discontinuities". In *Proc. 1st International Conference on Computer Vision*, London, England, pp. 572–577.

Lempitsky, V., C. Rother, and A. Blake (2007). "LogCut-efficient graph cut optimization for Markov random fields". In *ICCV07*.

Li, S. Z. (1990a). "Invariant surface segmentation through energy minimization with discontinuities". *International Journal of Computer Vision 5*(2), 161–194.

Li, S. Z. (1990b). "Reconstruction without discontinuities". In *Proceedings of Third International Conference on Computer Vision*, Osaka, Japan, pp. 709–712.

Li, S. Z. (1991). *Towards 3D Vision from Range Images: An Optimisation Framework and Parallel Distributed Networks*. Ph. D. thesis, University of Surrey, Guilford, Surrey, UK.

Li, S. Z. (1992a). "Matching: invariant to translations, rotations and scale changes". *Pattern Recognition 25*(6), 583–594.

Li, S. Z. (1992b). "Object recognition from range data prior to segmentation". *Image and Vision Computing 10*(8), 566–576.

Li, S. Z. (1992c). "Towards 3D vision from range images: An optimization framework and parallel networks". *CVGIP: Image Understanding 55*(3), 231–260.

Li, S. Z. (1993). "Similarity invariants for 3D space curve matching". In *Proceedings of the First Asian Conference on Computer Vision*, Osaka, Japan, pp. 454–457.

Li, S. Z. (1994a). "A Markov random field model for object matching under contextual constraints". In *Proceedings of IEEE Computer Society Conference on Computer Vision and Pattern Recognition*, Seattle, Washington, pp. 866–869.

Li, S. Z. (1994b). "Markov random field models in computer vision". In *Proceedings of the European Conference on Computer Vision*, Volume B, Stockholm, Sweden, pp. 361–370.

Li, S. Z. (1995a). "Discontinuity-adaptive MRF prior and robust statistics: A comparative study". *Image and Vision Computing 13*(4), 227–233.

Li, S. Z. (1995b). "On discontinuity-adaptive smoothness priors in computer vision". *IEEE Transactions on Pattern Analysis and Machine Intelligence 17*(6), 576–586.

Li, S. Z. (1995c). "Relaxation labeling using Lagrange multipliers and Hopfield network". In *Proceedings of IEEE International Conference on Image Processing*, Volume 1, Washington, D.C., pp. 266–269.

Li, S. Z. (1996a). "Improving convergence and solution quality of Hopfield-type neural network with augmented Lagrange multipliers". *IEEE Transactions on Neural Networks 7*(6), 1507–1516.

Li, S. Z. (1996b). "Robustizing robust M-estimation using deterministic annealing". *Pattern Recognition 29*(1), 159–166.

Li, S. Z. (1997a). "Invariant representation, recognition and pose estimation of 3d space curved under similarity transformations". *Pattern Recognition 30*(3), 447–458.

Li, S. Z. (1997b). "Parameter estimation for optimal object recognition: Theory and application". *International Journal of Computer Vision 21*(3), 207–222.

Li, S. Z. (1998a). "Bayesian object matching". *Journal of Applied Statistics* *25*(3), 425–443.

Li, S. Z. (1998b). "MAP image restoration and segmentation by constrained optimization". *IEEE Transactions on Image Processing* *7*(12), 1730–1735.

Li, S. Z. (2000). "Roof-edge preserving image smoothing based on MRF". *IEEE Transactions on Image Processing* *9*(6), 1134–1138.

Li, S. Z., Y. H. Huang, and J. Fu (1995). "Convex MRF potential functions". In *Proceedings of IEEE International Conference on Image Processing*, Volume 2, Washington, D.C., pp. 296–299.

Li, S. Z., J. Kittler, and M. Petrou (1993). "Automatic registration of aerial photographs and digital maps". *Optical Engineering* *32*(6), 1213–1221.

Li, S. Z., H. Wang, and M. Petrou (1994). "Relaxation labeling of Markov random fields". In *Proceedings of International Conference on Pattern Recognition*, Volume 1, Jerusalem, Israel, pp. 488–492.

Lin, W.-C., F.-Y. Liao, and C.-K. Tsao (1991). "A hierarchical multiple-view approach to three-dimensional object recognition". *IEEE Transactions on Neural Networks* *2*(1), 84–92.

Liu, J. and L. Wang (1999). "MRMRF texture classification and MCMC parameter estimation". *The proceedings of Visual Interface'99* *20*, 171–182.

Liu, S. C. and J. G. Harris (1989). "Generalized smoothing networks in early vision". In *Proceedings of the IEEE Computer Society Conference on Computer Vision and Pattern Recognition*, pp. 184–191.

Lowe, D. G. (1985). *Perceptual Organization and Visual Recognition*. Kluwer.

Lowe, D. G. (1992). "Robust model-based motion tracking through the integration of search and estimation". *International Journal of Computer Vision* *8*, 113–122.

Lumsdaine, A., J. Waytt, and I. Elfadel (1990). "Nonlinear analog networks for image smoothing and segmentation". In *Proc. IEEE International Symposium on Circuits and Systems*, New Orleans, LA, pp. 987–991.

Mackay, D. J. C. and R. M. Neal (1995). "Good error correcting codes based on very sparse matrices". In *Cryptograph and Coding-LNCS 1025*.

Manjunath, B. S. and R. Chellappa (1991). "Unsupervised texture segmentation using Markov random field models". *IEEE Transactions on Pattern Analysis and Machine Intelligence* *13*, 478–482.

Manjunath, B. S., T. Simchony, and R. Chellappa (1990). "Stochastic and deterministic networks for texture segmentation". *IEEE Transactions on Acoustic, Speech and Signal Processing 38*, 1030–1049.

Mardia, K. V. (1989). *Journal of Applied Statistics*, Volume 16(2). Special Issue on Statistic Image Analysis. An extended edition appears in K.V. Mardia and G.K. Kanji (ed) (1993), *Statistics and Images: 1*, Carfax.

Mardia, K. V., T. J. Hainsworth, and J. F. Haddon (1992). "Deformable templates in image sequences". In *Proceedings of International Conference on Pattern Recognition*, Volume B, pp. 132–135.

Mardia, K. V. and G. K. Kanji (1993,1994). *Statistics and Images: 1 & 2*. Advances in Applied Statistics. Carfax.

Mardia, K. V., J. T. Kent, and A. N. Walder (1991). "Statistical shape models in image analysis". In *Proceedings of 23rd Symposium Interface*, pp. 550–575.

Marr, D. (1982). *Vision*. San Francisco: W. H. Freeman and Co.

Marr, D. and T. Poggio (1979). "A computational theory of human stereo vision". *Proceedings Royal Society London B*(204), 301–328.

Marroquin, J. L. (1985). "Probabilistic solution of inverse problems". *A. I. Lab. Tech. Report* No. 860, MIT, Cambridge, MA.

Marroquin, J. L., S. Mitter, and T. Poggio (1987). "Probabilistic solution of ill-posed problems in computational vision". *Journal of the American Statistical Association 82*(397), 76–89.

Mayhew, J. E. W. and J. P. Frisby (1981). "Towards a computational and psychophysical theory of stereropsis". *Artificial Intelligence 17*, 349–385.

McMillin, B. M. and L. M. Ni (1989). "A reliable parallel algorithm for relaxation labeling". In P. M. Dew, R. A. Earnshaw, and T. R. Heywood (Eds.), *Parallel Processing for Computer Vision and Display*, pp. 190–209. Addison-Wesley.

Meer, P., D. Mintz, A. Rosenfeld, and D. Kim (1991). "Robust regression methods for computer vision: A review". *International Journal of Computer Vision 6*, 59–70.

Metropolis, N., A. W. Rosenbluth, M. N. Rosenbluth, and E. Teller (1953). "Equations of state calculations by fast computational machine". *Journal of Chemical Physics 21*, 1087–1092.

Meyer, Y. (2001). *"Oscillating Patterns in Image Processing and Nonlinear Evolution Equations: The Fifteenth Dean Jacqueline B. Lewis Memorial Lectures"*. American Mathematical Society.

Modestino, J. W. and J. Zhang (1989). "A Markov random field model-based approach to image interpretation". In *Proceedings of the IEEE Computer Society Conference on Computer Vision and Pattern Recognition*, pp. 458–465.

Moghaddam, B. and A. Pentland (1997). "Probabilistic visual learning for object representation". *IEEE Transactions on Pattern Analysis and Machine Intelligence 7*, 696–710.

Mohammed, J., R. Hummel, and S. Zucker (1983). "A feasible direction operator for relaxation method". *IEEE Transactions on Pattern Analysis and Machine Intelligence 5*(3), 330–332.

Mohan, R. and R. Nevatia (1989). "Using perceptual organization to extract 3-d structures". *IEEE Transactions on Pattern Analysis and Machine Intelligence 11*, 1121–1139.

Moscato, P. (1989). "On evolution, search, optimization, genetic algorithms and martial arts: Towards memetic algorithms". C3P Report 826, Caltech Concurrent Computation Program.

Moussouris, J. (1974). "Gibbs and Markov systems with constraints". *Journal of statistical physics 10*, 11–33.

Mumford, D. (1996). "Pattern theory: a unified perspective". In D. Knill and W. Richard (Eds.), *Perception as Bayesian Inference*, pp. 25–62. Cambridge University Press.

Mumford, D. and J. Shah (1985). "Boundary detection by minimizing functionals: I". In *Proceedings of the IEEE Computer Society Conference on Computer Vision and Pattern Recognition*, San Francisco, CA, pp. 22–26.

Mundy, J. L. and A. Zisserman (1992). *Geometric Invariants in Computer Vision*. Cambridge, MA: MIT Press.

Muramatsu, M. and T. Suzuki (2003). "A new second-order cone programming relaxation for max-cut problems". *Journal of Operations Research of Japan 43*, 164–177.

Murphy, K. P., Y. Weiss, and M. I. Jordan (1999). "Loopy belief propagation for approximate inference: An empirical study". In *Uncertainty in Artificial Intelligence*, pp. 467–475.

Murray, D. and B. Buxton (1987). "Scene segmentation from visual motion using global optimization". *IEEE Transactions on Pattern Analysis and Machine Intelligence 8*, 220–228.

Myers, R. H. (1990). *Classical and Modern Regression with Applications*. PWS-Kent Publishing Company.

Nadabar, S. G. and A. K. Jain (1992). "Parameter estimation in MRF line process models". In *Proceedings of IEEE Computer Society Conference on Computer Vision and Pattern Recognition*, pp. 528–533.

Nadabar, S. G. and A. K. Jain (1995). "Fusion of range and intensity images on a connection machine". *Pattern Recognition 28*(1), 11–26.

Nagel, H. H. (1983). "Displacement vectors derived from second-order intensity variations in image sequences". *Computer Vision, Graphics and Image Processing 21*, 85–117.

Nagel, H. H. and W. Enkelmann (1986). "An investigation of smoothness constraints for the estimation of displacement vector fields from image sequences". *IEEE Transactions on Pattern Analysis and Machine Intelligence 8*, 565–593.

Nasrabadi, N., W. Li, and C. Y. Choo (1990). "Object recognition by a Hopfield neural network". In *Proceedings of Third International Conference on Computer Vision*, Osaka, Japan, pp. 325–328.

Neal, R. M. (1993). "Probabilistic inference using markov chain monte carlo methods". CRG-TR 93-1, Dept. of Computer Science, University of Toronto.

Ng, A. Y. and M. I. Jordan (2002). "On discriminative vs. generative classifiers: A comparison of logistic regression and naive bayes". In *Advances in Neural Information Processing Systems(NIPS)*.

Nguyen, H. T., Q. Ji, and A. W. Smeulders (2007). "Spatio-temporal context for robust multitarget tracking". *Transactions on Pattern Analysis and Machine Intelligence 29*(1), 52–64.

Nikolova, M. and M. K. NG (2005). "Analysis of half-quadratic minimization methods for signal and image recovery". *SIAM Journal on Scientific Computing 27*(3), 937–966.

Nordstrom, N. (1990). "Biased anisotropic diffusion – a unified regularization and diffusion approach to edge detection". In *Proceedings of the European Conference on Computer Vision*, pp. 18–27.

O'Leary, D. P. and S. Peleg (1983). "Analysis of relaxation processes: the two-node two label case". *IEEE Transactions on Systems, Man and Cybernetics SMC-13*(4), 618–623.

Oshima, M. and Y. Shirai (1983). "Object recognition using three-dimensional information". *IEEE Transactions on Pattern Analysis and Machine Intelligence 5*(4), 353–361.

Paget, R. (2004). "Strong Markov random field model". *Transactions on Pattern Analysis and Machine Intelligence 26*(3), 408–413.

Pappas, T. N. (1992). "An adaptive clustering algorithm for image segmentation". *IEEE Transactions on Signal Processing 40*(4), 901–914.

Paquin, R. and E. Dubios (1983). "A spatial-temporal gradient method for estimating the displacement field in time-varying imagery". *Computer Vision, Graphics and Image Processing 21*, 205–221.

Parisi, G. (1988). *Statistical field theory*. Addison-Wesley.

Park, B.-G., K.-M. Lee, and S.-U. Lee (2005). "Face recognition using face-ARG matching". *Transactions on Pattern Analysis and Machine Intelligence 27*(12), 1982–1988.

Pavlidis, T. (1986). "A critical survey of image analysis methods". In *ICPR*, pp. 502–511.

Pavlidis, T. (1992). "Why progress in machine vision is so slow". *Pattern Recognition Letters 13*, 221–225.

Pearl, J. (1988). *Probabilistic Reasoning in Intelligent Systems: Networks of Plausible Inference*. San Francisco: Morgan Kaufmann.

Peleg, S. (1980). "A new probability relaxation scheme". *IEEE Transactions on Pattern Analysis and Machine Intelligence 8*, 362–369.

Peleg, S. and R. A. Rosenfeld (1978). "Determining compatibility coefficients for curve enhancement relaxation processes". *IEEE Transactions on Systems, Man and Cybernetics 8*, 548–554.

Pelillo, M. and M. Refice (1994). "Learning compatibility coefficients for relaxation labeling processes". *IEEE Transactions on Pattern Analysis and Machine Intelligence 16*(9), 933–945.

Pentland, A. P. (1990). "Automatic extraction of deformable part models". *International Journal of Computer Vision 4*, 107–126.

Pérez, P., M. Gangnet, and A. Blake (2003). "Poisson image editing". *ACM Trans. Graph. 22*(3), 313–318.

Perez, P. and F. Heitz (1994). "Restriction of a Markov random field on a graph and multiresolution image analysis". RR 2170, INRIA, Sophia-Antipolis Cedex, France.

Perona, P. and J. Malik (1990). "Scale-space and edge detection using anisotropic diffusion". *IEEE Transactions on Pattern Analysis and Machine Intelligence 12*(7), 629–639.

Peterson, C. and B. Soderberg (1989). "A new method for mapping optimization problems onto neural networks". *International Journal of Neural Systems 1*(1), 3–22.

Petrou, M. (1993). "Accelerated optimization in image processing via the renormalization group transform". In *Proceedings of IMA conference on Complex Stochastic Systems and Engineering*, Leeds.

Pinkus, M. (1968). "A closed form solution of certain programming problems". *Operations Research 16*, 690–694.

Platt, J. C. and A. H. Barr (1988). "Constrained differential optimization". In *Proceedings of NIPS conference*.

Poggio, T. and S. Edelman (1990). "A network that learn to recognize three-dimensional objects". *Nature 343*, 263–266.

Poggio, T., V. Torre, and C. Koch (85a). "Computational vision and regularization theory". *Nature 317*, 314–319.

Poggio, T., H. Voorhees, and A. Yuille (85b). "Regularizing edge detection". *A. I. Lab. Memo* No. 773, MIT, Cambridge, MA.

Pope, A. R. and D. G. Lowe (1993). "Learning object recognition models from images". In *Proceedings of IEEE International Conference on Computer Vision*, pp. 296–301.

Powell, M. J. D. (1969). "A method of nonlinear constraints in minimization problems". In R. Fletcher (Ed.), *Optimization*, London. Academic Press.

Press, W. H., S. A. Teukolsky, W. T. Vetterling, and B. P. Flannery (1988). *Numerical recipes in C* (2 ed.). Cambridge University Press.

Price, K. E. (1985). "Relaxation matching techniques – A comparison". *IEEE Transactions on Pattern Analysis and Machine Intelligence 7*(5), 617–623.

Pritch, Y., A. Rav-Acha, A. Gutman, and S. Peleg (2007). "Webcam synopsis: Peeking around the world". In *Computer Vision, 2007. ICCV 2007. IEEE 11th International Conference on*, pp. 1–8.

Qian, W. and D. M. Titterington (1989). "On the use of Gibbs Markov chain models in the analysis of image based on second-order pairwise interactive distributions". *Journal of Applied Statistics 16*(2), 267–281.

Qian, W. and D. M. Titterington (1992). "Stochastic relaxation and EM algorithm for Markov random fields". *Journal of statistical computation and simulation 40*, 55–69.

Radcliffe, N. J. and P. D. Surry (1994). Formal memetic algorithms. In T. Fogarty (Ed.), *Evolutionary Computing: AISB Workshop*, Lecture Notes in Computer Science, pp. 1–14. Springer-Verlag.

Radig, B. (1984). "Image sequence analysis using relational structures". *Pattern Recognition 17*(1), 161–167.

Rangarajan, A. and R. Chellappa (1990). "Generalized graduated non-convexity algorithm for maximum a posteriori image estimation". In *Proceedings of International Conference on Pattern Recognition*, pp. 127–133.

Rav-Acha, A., Y. Pritch, and S. Peleg (2006). "Making a long video short: Dynamic video synopsis". *Computer Vision and Pattern Recognition, 2006 IEEE Computer Society Conference on 1*, 435–441.

Ravikumar, P. and J. Lafferty (2006). "Quadratic programming relaxations for metric labeling and Markov random field MAP estimation". In *ICML '06: Proceedings of the 23rd international conference on Machine learning*, pp. 737–744.

Redner, R. A. and H. F. Walker (1984). "Mixture densities, maximum likelihood and the EM algorithm". *SIAM Review 26*, 195–239.

Reeves, S. J. (1992). "A cross-validation framework for solving image resporation problems". *Journal of Visual Communication and Image Representation 3*(4), 433–445.

Richardson, S. and P. J. Green (1997). "On Bayesian analysis of mixtures with an unknown number of components". *Journal of Royal Stat. Soc., B 59*, 731–792.

Ripley, B. D. (1981). *Spatial Statistics*. New York: Wiley.

Rissanen, J. (1978). "Modeling by shortest data description". *Automatica 14*, 465–471.

Rissanen, J. (1983). "A universal prior for integers and estimation by minimal discription length". *Annals of Statistics 11*(2), 416–431.

Roberts, L. G. (1965). "Machine perception of three-dimensional solids". In e. a. J. T. Tippett (Ed.), *Optical and Electro-Optical Information Processing*. Cambridge, MA: MIT Press.

Rockafellar, R. (1970). *Convex Analysis*. Princeton Press.

Rosen, S. A. (1960). "The gradient projection method for nonlinear programming — part i: Linear constraints". *Journal of the Society for Industrial and Applied Mathematics 8*(1), 181–217.

Rosenblatt, F. (1962). *Principles of Neurodynamics*. New York: Spartan.

Rosenfeld, A. (1993,1994). "Some thoughts about image modeling". SeeMardia and Kanji (1994), pp. 19–22.

Rosenfeld, A., R. Hummel, and S. Zucker (1976). "Scene labeling by relaxation operations". *IEEE Transactions on Systems, Man and Cybernetics 6*, 420–433.

Rosenfeld, A. and A. C. Kak (1976). *Digital Image Processing*. New York: Academic Press.

Roth, G. and M. D. Levine (1993). "Extracting geometric primitives". *CVGIP: Image Understanding 58*, 1–22.

Rother, C., L. Bordeaux, Y. Hamadi, and A. Blake (2006). "AutoCollage". *ACM Trans. Graph. 25*(3), 847–852.

Rousseeuw, P. J. (1984). "Least meadian of squares regression". *Journal of the American Statistical Association 79*(388), 871–880.

Roy, S. and I. J. Cox (1998). "A maximum-flow formulation of the n-camera stereo correspondence problem". In *ICCV98*, pp. 492–502.

Rubinstein, Y. D. and T. Hastie (1997). "Discriminative vs informative learning". In *Third Int. Conf. on Knowledge Discovery and Data Mining*, pp. 49–53.

Rudin, L. I., S. Osher, and E. Fatemi (1992). "Nonlinear total variation based noise removal algorithms". *Physica D: Nonlinear Phenomena 60*, 259–268.

Sakamoto, Y., M. Ishiguro, and G. Kitagawa (1987). *Akaike information criterion statistics*. D. Reidel Publishing Company.

Sapiro, G. and D. L. Ringach (1996). "Anisotropic diffusion of multivalued images with applications to color filtering". *Image Processing, IEEE Transactions on 5*(11), 1582–1586.

Schellewald, C. and C. Schnorr (2003). "Subgraph matching with semidefinite programming". In *IWCIA*.

Schlesinger, M. (1976). "Syntactic analysis of two-dimensional visual signals in the presence of noise". *Cybernetics and Systems Analysis 12*(4), 612–628.

Schultz, R. R. and R. L. Stevenson (1994). "A Bayesian approach to image expansion for improved definition". *IEEE Transactions on Image Processing 3*(3), –.

Schwartz, G. (1987). "Estimating the dimension of a model". *Annals of Statistics 6*, 461–464.

Sclove, S. L. (1983). "Application of the conditional population mixture model to image segmentation". *IEEE Transactions on Pattern Analysis and Machine Intelligence 5*, 428–433.

Sclove, S. L. (1987). "Application of model-selection criteria to some problems in multivariate analysis". *Psychmetrika 52*, 333–343.

Shahraray, B. and D. Anderson (1989). "Optimal estimation of contour properties by cross-validated regularization". *IEEE Transactions on Pattern Analysis and Machine Intelligence 11*, 600–610.

Shapiro, L. G. and R. M. Haralick (1981). "Structural description and inexact matching". *IEEE Transactions on Pattern Analysis and Machine Intelligence 3*, 504–519.

Shulman, D. and J. Herve (1989). "Regularization of discontinuous flow fields". In *Proc. Workshop on Visual Motion*, pp. 81–86.

Siegel, A. F. (1982). "Robust regression using repeated meadians". *Biometrika 69*(1), 242–244.

Silverman, J. F. and D. B. Cooper (1988). "Bayesian clustering for unsupervised estimation of surface and texture models". *IEEE Transactions on Pattern Analysis and Machine Intelligence 10*, 482–495.

Simic, P. D. (1990). "Statisrical mechanics as the underlying theory of 'elastic' and 'neural' optimization". *Network 1*, 89–103.

Sinha, S. N. and M. Pollefeys (2005). "Multi-view reconstruction using photo-consistency and exact silhouette constraints: a maximum-flow formulation". *Computer Vision, 2005. ICCV 2005. Tenth IEEE International Conference on 1*, 349–356.

Smith, A. F. M. and G. O. Robert (1993). "Beyesian computation via the gibbs sampler and related Markov chain Monte Carlo methods". *Journal of the Royal Statistical Society, Series B 55*(1), 3–23.

Smyth, P., D. Heckerman, and M. I. Jordan (1997). "Probabilistic independence networks for hidden Markov probability models". *Neural Computation 9*(2), 227–269.

Snyder, M. A. (1991). "On the mathematical foundations of smoothness constraints for the determination of optical flow and for surface reconstruction". *IEEE Transactions on Pattern Analysis and Machine Intelligence 13*, 1105–1114.

Staib, L. and J. Duncan (1992). "Boundary finding with parametrically deformable medels". *IEEE Transactions on Pattern Analysis and Machine Intelligence 14*, 1061–1075.

Stein, A. and M. Werman (1992). "Robust statistics in shape fitting". In *Proceedings of IEEE Computer Society Conference on Computer Vision and Pattern Recognition*, pp. 540–546.

Stevenson, R. and E. Delp (1990). "Fitting curves with discontinuities". In *Proceedings of International Workshop on Robust Computer Vision*, Seattle, WA, pp. 127–136.

Stevenson, R. L., B. E. Schmitz, and E. J. Delp (1994). "Discontinuity preserving regularization of inverse visual problems". *IEEE Transactions on Systems, Man and Cybernetics 24*(3), 455–469.

Stockman, G. (1987). "Object recognition and localization via pose clustering". *Computer Vision, Graphics and Image Processing 40*, 361–387.

Stockman, G. C. and A. K. Agrawala (1977). "Equivalence of Hough curve detection to template matching". *Communications of the ACM 20*, 820–822.

Storvik, G. (1994). "A Bayesian approach to dynamic contours through stochastic sampling and simulated annealing". *IEEE Transactions on Pattern Analysis and Machine Intelligence 16*(10), 976–986.

Strauss, D. J. (1977). "Clustering on colored lattice". *Journal of Applied Probability 14*, 135–143.

Sun, J., W. Zhang, X. Tang, and H.-Y. Shum (2006). "Background cut". In *ECCV*, pp. 628–641.

Szeliski, R. (1989). *Bayesian modeling of uncertainty in low-level vision*. Kluwer.

Tan, H. L., S. B. Gelfand, and E. Delp (1992). "A cost minimization approach to edge detection using simulated annealing". *IEEE Transactions on Pattern Analysis and Machine Intelligence 14*, 3–18.

Tang, B., G. Sapiro, and V. Caselles (2001). "Color image enhancement via chromaticity diffusion". *Image Processing, IEEE Transactions on 10*(5), 701–707.

Terzopolous, D., J. Platt, A. Barr, and K. Fleischer (1987). "Elastically deformable models". *Comput. Graphic 21*(4), 205–214.

Terzopolous, D., A. Witkin, and M. Kass (1988). "Constraints on deformable models: Recovering 3d shape and nonrigid motion". *AI 36*, 91–123.

Terzopoulos, D. (1986a). "Image analysis using multigrid relaxation methods". *IEEE Transactions on Pattern Analysis and Machine Intelligence 8*, 129–139.

Terzopoulos, D. T. (1983a). "Multilevel computational process for visual surface reconstruction". *Computer Vision, Graphics and Image Processing 24*, 52–96.

Terzopoulos, D. T. (1983b). "The role of constraints and discontinuities in visible surface reconstruction". In *Proc. 8th International Joint Conference on AI*, Karlsruhe, W. Germany, pp. 1073–1077.

Terzopoulos, D. T. (1986b). "Regularization of inverse visual problems involving discontinuities". *IEEE Transactions on Pattern Analysis and Machine Intelligence 8*(4), 413–424.

Therrien, C. W. (1989). *Decision, estimation, and classification: an introduction to pattern recognition and related topics.* New York: Wiley.

Thompson, A. M., J. C. Brown, J. W. Kay, and D. M. Titterington (1991). "A study of methods of choosing the smoothing parameter in image restoration by regularization". *IEEE Transactions on Pattern Analysis and Machine Intelligence 13*, 326–339.

Tikhonov, A. N. and V. A. Arsenin (1977). *Solutions of Ill-posed Problems.* Washington: Winston & Sons.

Titterington, D. M. and N. H. Anderson (1994). "Boltzmann machine". In F. P. Kelly (Ed.), *Probability, Statistics and Optimisation*, pp. 255–279. John Wiley & Sons.

Torre, V. and T. Poggio (1986). "On edge detection". *IEEE Transactions on Pattern Analysis and Machine Intelligence 8*(2), 147–163.

Tukey, J. W. (1977). *Explortary Data Analysis.* Reading, MA: Addison-Wesley.

Turk, M. A. and A. P. Pentland (1991). "Eigenfaces for recognition". *Journal of Cognitive Neuroscience 3*(1), 71–86.

Ullman, S. (1979). *The Interpolation of Visual Motion.* Cambridge, MA: MIT Press.

Umeyama, S. (1991). "Least squares estimation of transformation parameters between two point patterns". *IEEE Transactions on Pattern Analysis and Machine Intelligence 13*, 376–380.

Vapnik, V. (1982). *Estimation of Dependences Based on Empirical Data.* New York: Springer-Verlag.

Vese, L. A. and S. J. Osher (2003). "Modeling textures with total variation minimization and oscillating patterns in image processing". *J. Sci. Comput. 19*(1-3), 553–572.

Vogiatzis, G., P. Torr, and R. Cipolla (2005). "Multi-view stereo via volumetric graph-cuts". *Computer Vision and Pattern Recognition, 2005. CVPR 2005. IEEE Computer Society Conference on 2*, 391–398.

Wacholder, E., J. Han, and R. C. Mann (1989). "A neural network algorithm for the multiple traveling salesman problem". *Biological Cybernetics 61*, 11–19.

Wahba, G. (1980). "Spline bases, regularization, and generalized cross-validation for solving approximation problems with large quantities of noisy data". In E. Cheney (Ed.), *Approximation Theory III*, Volume 2. Academic Press.

Wahba, G. (1990). *Spline Models for Observation Data*. Society for Industrial and Applied Mathematics.

Wainwright, M. J. (2002). *"Stochastic processes on graphs with cycles: geometric and variational approaches"*. Phd, Massachusetts Institute of Technology. Supervisor-Alan S. Willsky and Supervisor-Tommi S. Jaakkola.

Wang, H. and M. Brady (1991). "Corner detection for 3D vision using array processors". In *BARNAIMAGE 91*, Barcelona. Springer-Verlag.

Wang, H. and S. Z. Li (1994). "Solving the bas-relief ambiguity". In *ICIP (2)*, Austion, pp. 760–764.

Wasserstrom, E. (1973). Numerical solutions by the continuition method". *SIAM Review 15*, 89–119.

Watson, D. F. (1981). "Computing the n-dimentional Delaunay tessellation with application to Voronoi polytopes". *Computer Journal 24*, 167–172.

Weiss, I. (1990). "Shape reconstruction on a varying mesh". *IEEE Transactions on Pattern Analysis and Machine Intelligence 12*, 345–362.

Weiss, Y. (1997). "Belief propagation and revision in networks with loops". Technical Report AI Momo 1616 & CBCL Memo 155, MIT.

Wells, W. M. (1991). "MAP model matching". In *Proceedings of IEEE Computer Society Conference on Computer Vision and Pattern Recognition*, pp. 486–492.

Weng, J., N. Ahuja, and T. S. Huang (1992). "Matching two perspective views". *IEEE Transactions on Pattern Analysis and Machine Intelligence 14*, 806–825.

Weng, J. J., N. Ahuja, and T. S. Huang (1993). "Learning recognition and segmentation of 3-d objects from 2-d images". In *Proceedings of IEEE International Conference on Computer Vision*, pp. 121–128.

Werner, T. (2007). "A linear programming approach to Max-Sum problem: A review". *Transactions on Pattern Analysis and Machine Intelligence 29*(7), 1165–1179.

Wierschin, T. and S. Fuchs (2002). "Quadratic minimization for labeling problems". Technical report, Dresden University of Technology.

Witkin, A., D. T. Terzopoulos, and M. Kass (1987). "Signal matching through scale space". *International Journal of Computer Vision 1*, 133–144.

Witkin, A. P. (1983). "Scale-space filtering". In *Proceedings of International Joint Conference on Artificial Intelligence*, Volume 2, pp. 1019–1022.

Won, C. S. and H. Derin (1992). "Unsupervised segmentation of noisy and textured images using Markov random fields". *CVGIP: Graphics Model and Image Processing 54*, 308–328.

Wong, A. K. C. and M. You (1985). "Entropy and distance of random graphs with application of structural pattern recognition". *IEEE Transactions on Pattern Analysis and Machine Intelligence 7*(5), 599–609.

Woods, J. W. (1972). "Two-dimensional discrete Markovian fields". *IEEE Transactions on Information Theory 18*, 232–240.

Wu, C. F. J. (1983). "On the convergence properties of the EM algorithm". *The Annals of Statistics 11*, 95–103.

Wu, Y. N., S. C. Zhu, and X. W. Liu (2000). "Equivalenceof julesz ensemble and frame models". *International Journal of Computer Vision 38*(3), 245–261.

Wu, Z. and R. Leahy (1993). "An approximation method of evaluating the joint likelihood for first-order GMRF's". *IEEE Transactions on Image Processing 2*(4), 520–523.

Yang, D. and J. Kittler (1994). "MFT based discrete relaxation for matching high order relational structures". In *Proceedings of International Conference on Pattern Recognition*, Volume 2, Jerusalem, Israel, pp. 219–223.

Yedidia, J. S., W. T. Freeman, and Y. Weiss (2000). "Generalized belief propagation". In *NIPS*, pp. 689–695.

Yedidia, J. S., W. T. Freeman, and Y. Weiss (2003). *"Understanding belief propagation and its generalizations"*, pp. 239–269. Morgan Kaufmann.

Yin, W. (2006). *The TV-L1 model: Theory, computation, and applications*. Ph.d, Columbia University.

Yuan, X. and S. Z. Li (2007). "Half Quadratic analysis for mean shift: with extension to a sequential data mode-seeking method". In *IEEE International Conference on Computer Vision*.

Yuille, A. L. (1987). "Energy function for early vision and analog networks". *A. I. Lab. Memo* No. 987, MIT, Cambridge, MA.

Yuille, A. L. (1990). "Generalized deformable models, statistical physics and matching problems". *Neural Computation 2*, 1–24.

Yuille, A. L., D. Cohen, and P. W. Hallinan (1989). "Feature extraction from faces using deformable templates". In *Proceedings of IEEE Computer Society Conference on Computer Vision and Pattern Recognition*, pp. 104–109.

Yuille, A. L. and J. J. Kosowsky (1994). "Statistical physics algorithms that converge". *Neural Computation 6*, 341–356.

Zabih, R. Graph cuts for energy minimization. http://www.cs.cornell.edu/∼rdz/graphcuts.html.

Zenzo, S. D. (1986). "A note on the gradient of a multi-image". *Computer Vision, Graphics, and Image Processing 33*(1), 116–125.

Zhang, J. (1988). *Two dimensional stochastic model-based image analysis*. Ph. D. thesis, Rensselaer Polytechnic Institute.

Zhang, J. (1992). "The mean field theory in EM procedures for Markov random fields". *IEEE Transactions on Image Processing 40*(10), 2570–2583.

Zhang, J. (1993). "The mean field theory in EM procedures for blind Markov random field image restoration". *IEEE Transactions on Image Processing 2*(1), 27–40.

Zhang, J. (1995). "Parameter reduction for the compound Gauss-Markov model". *IEEE Transactions on Image Processing 4*, –.

Zhang, J. and J. W. Modestino (1990). "A model-fitting approach to cluster validation with application to stochastic model-based image segmentation". *IEEE Transactions on Pattern Analysis and Machine Intelligence 12*, 1009–1017.

Zhu, S. C. and X. W. Liu (2000). "Learning in gibbsian feilds: How accurate and how fast can it be?". In *Proceedings of IEEE Computer Society Conference on Computer Vision and Pattern Recognition*, Hilton Head, NC, USA, pp. 104–109.

Zhu, S. C. and D. Mumford (1997). "Prior learning and gibbs reaction-diffusion". *IEEE Transactions on Pattern Analysis and Machine Intelligence 19*(11), –.

Zhu, S. C., Y. Wu, and D. Mumford (1997). "Minimax entropy principle and its applications to texture modeling". *Neural Computation 9*(8), –.

Zhu, S. C., Y. N. Wu, and D. Mumford (1998). "FRAME: Filters, random field and maximum entropy: – towards a unified theory for texture modeling". *International Journal of Computer Vision 27*(2), 1–20.

Zhuang, X., T. Wang, and P. Zhang (1992). "A highly robust estimator through partially likelihood function modeling and its application in computer vision". *IEEE Transactions on Pattern Analysis and Machine Intelligence 14*, 19–35.

Zucker, S. W. (1976). "Relaxation labeling and the reduction of local ambiguities". In *Proc. 3rd Int. Joint Conf. on Pattern Recognition*, pp. 852–861.

List of Notation

α, β	parameters for single- and pair- site clique potentials
c	clique
$\mathcal{C}$	set of cliques
d	observed data
$\mathcal{D}$	set of admissible values for data
$E(f)$	energy function
f	MRF configuration, set of assigned labels, or mapping from $\mathcal{S}$ to $\mathcal{L}$
$f_{\mathcal{N}_i}$	set of labels at sites in $\mathcal{N}_i$
$\mathbb{F}$	solution space
$g(\cdot)$	potential function
$G(f)$	gain function
$\mathcal{G}$	relational structure or graph
$h(\cdot)$	interaction function
i	index to site
i'	neighbor of i
I	index to labels
(i, j)	index to site in a lattice
$\mathcal{L}$	set of labels
$\mathcal{N}_i$	set of sites neighboring i
$N(\mu, \sigma^2)$	Gaussian distribution

$P(x)$, $p(x)$	probability, density, function of random variable x
P_i, p_i	vector of point
$\mathcal{S}$	set of sites
σ^2	noise variance
$\mathcal{T}$	pose transformation
θ	set of parameters in MRF model
θ_d	set of parameters in $P(d \mid f)$
θ_f	set of parameters in $P(f)$
$U(f)$	energy function for prior distribution
$U(f \mid d)$	energy function for posterior distribution
$U(d \mid f)$	energy function for likelihood distribution
$V_1(f)$	clique potential function
$V_1(f_i)$	single-site potential function
$V_2(f_1, f_{i'})$	pair-site potential function
Z	partition function

There are a few exceptions to the interpretation of the symbols. But they can be indentified easily from the context.

Index